AF389518

Artificial Neural Networks and Evolutionary Computation in Remote Sensing

Artificial Neural Networks and Evolutionary Computation in Remote Sensing

Editor

Taskin Kavzoglu

MDPI • Basel • Beijing • Wuhan • Barcelona • Belgrade • Manchester • Tokyo • Cluj • Tianjin

Editor
Taskin Kavzoglu
Department of Geomatics Engineering
Gebze Technical University
Turkey

Editorial Office
MDPI
St. Alban-Anlage 66
4052 Basel, Switzerland

This is a reprint of articles from the Special Issue published online in the open access journal *Remote Sensing* (ISSN 2072-4292) (available at: https://www.mdpi.com/journal/remotesensing/special_issues/ANN_RS).

For citation purposes, cite each article independently as indicated on the article page online and as indicated below:

LastName, A.A.; LastName, B.B.; LastName, C.C. Article Title. *Journal Name* **Year**, *Volume Number*, Page Range.

ISBN 978-3-03943-827-3 (Hbk)
ISBN 978-3-03943-828-0 (PDF)

Cover image courtesy of Taskin Kavzoglu.

Contents

About the Editor

Taskin Kavzoglu is a senior researcher in remote sensing with more than 20 years of research experience in Earth observation/remote sensing. Since his first publication in 1998 and he is now author of more than 90 papers in peer-reviewed journals and international conference proceedings. Prof. Dr. Kavzoglu received his B.Sc. in Geodesy and Photogrammetry as the first student in his class from Karadeniz Technical University (Turkey), M.Sc. in Geographical Information Systems (with Distinction) and Ph.D. in Remote Sensing from Nottingham University (UK). Best Ph.D. Thesis Award was given to his thesis entitled "An Investigation of the Design and Use of Feed-forward Artificial Neural Networks in the Classification of Remotely Sensed Images" by the Remote Sensing and Photogrammetry Society in the UK. Currently, he is a full-time professor at Gebze Technical University (Turkey) where he is the Acting Dean of Engineering Faculty. His research interests include: Earth observation, remote sensing of natural resources, deep learning, machine learning, data mining, change detection, object-based image analysis and landslide susceptibility mapping. He has chaired a national remote sensing symposium, acted as chairman in many national and international conferences and workshop sessions, has been an expert in projects evaluation for Turkish and Austrian Scientific Research Councils, is a reviewer for 110 journals including Remote Sensing, Remote Sensing of Environment, the International Journal of Remote Sensing, International Journal of Applied Earth Observation and Geoinformation, IEEE Transactions on Geoscience and Remote Sensing, Catena, Landslides and Geomorphology. He has been teaching courses related to remote sensing, geographical information systems and computer programming, supervising students at the undergraduate and graduate levels since 2001, acting as an external examiner for graduate thesis for Turkish and Malaysian universities, and has worked in many remote sensing and geographical information science projects. His current project titled "Design and Development of Spectral Library for Culturable Poplar Tree Plantation and Producing Thematic Maps Using Very High-Resolution Satellite Imagery" includes field campaigns in different regions of Turkey. He is a member of the Remote Sensing and Photogrammetry Society (UK), American Geophysical Union, Turkish Academy of Sciences, and Turkish Photogrammetry and Remote Sensing Society.

Preface to "Artificial Neural Networks and Evolutionary Computation in Remote Sensing"

Abstract: The Special Issue on "Artificial Neural Networks and Evolutionary Computation in Remote Sensing" presents a wide range of studies focusing on a variety of remote sensing models and techniques, demonstrating the benefits and potential of neural networks and evolutionary computation models. The Special Issue consists of eleven research articles, covering a broad range of research topics that employ UAV and satellite imagery, deep learning models to detect ships and airplanes, deploy a nanosatellite payload to select images eligible for transmission to ground, conduct fine image segmentation and extract improved land-use information, extract valley fill faces, and superstructural design for feedforward neural networks. This preface provides a brief summary of each article published in the issue.

1. Introduction

Biological evolution has long inspired scientists to seek solutions to difficult computational problems. In evolutionary computation, a subfield of artificial intelligence, the process of natural evolution is mimicked in some way to model the problem typically from samples, in which we search for optimal or near-optimal solutions. Evolutionary algorithms are usually applied to computationally intense optimization problems, such as the design and training of neural networks and combinatorial optimization. The combination of artificial neural networks and evolutionary search methods (e.g., genetic algorithm) called evolutionary artificial neural networks have also been employed in complex problems.

Since the introduction of perceptron by Rosenblatt in 1958, numerous studies in almost all scientific fields have been conducted to apply neural network models. The first studies using artificial neural networks (ANNs) for the analysis of remotely sensed data were reported in the late 80s. Over the past decade, there has been a considerable increase in both the quantity of remotely sensed data and the use of neural networks for remote sensing research problems. Initially called black-box methods, neural nets are now more popular with new network types and algorithms (particularly deep learning approaches). Until now, ANNs have been applied to many tasks, not only for statistical regressions or image classification, but also for image segmentation, feature extraction, data fusion, or dimensionality reduction.

Although significant progress has been made in the analysis of remotely sensed imagery using neural networks, seveal issues remain to be resolved. This Special Issue showcases the variety and relevance of the recent developments in the theory and application of artificial neural networks and evolutionary computation in remote sensing. Thus, the latest and most advanced ideas and findings related to the application of neural nets are presented to the remote sensing community.

The response to the Special Issue call was beyond expectations, and following an extensive peer-review process, eleven articles were accepted and included in the issue. The following section briefly summarizes each paper published in the Special Issue.

2. Overview of Contributions

In the study by Giuffrida et al. [1] funded by European Space Agency, the authors proposed a custom convolutional neural network (CNN) deployed for a nanosatellite payload to select images eligible for transmission to ground, called CloudScout. The latter is installed on the Hyperscout-2, in the frame of the Phisat-1 ESA mission, which exploits a hyperspectral camera to classify cloud-covered images and clear ones. The images transmitted to the ground are those that present less than 70% of cloudiness in a frame. The network was tested on a Sentinel-2 dataset, which was processed to emulate the Hyperscout-2 hyperspectral sensor. Results were promising with 92% accuracy. The study has the potential to enable a new era of edge applications and enhance remote sensing applications directly on-board the satellite.

Vasilakos et al. [2] conducted extensive research on the effectiveness of ensemble methods and voting methods using Sentinel-2 images for identifying land use/land cover types in the heterogeneous Mediterranean landscape of Lesvos Island, Greece. In the case study, six machine learning methods were used as base classifiers, which are decision tree, linear discriminant analysis, support vector machines, k-Nearest neighbor, random forest, and neural networks. An ensemble approach with nine voting methods was developed for increased accuracy over classification algorithms. Each base classifier was trained with its own dataset to create the accuracy metrics that were used within the voting methods. All the base classifiers and the ensemble methods were applied to an unseen testing dataset. The result shows that the combination of multiple classifiers based on the examined voting schemes does not always provide a better performance in land cover classification. The nine ensemble models were further statistically compared by applying McNemar's test. The SVM algorithm outperformed all the classifiers and was proven as the most accurate approach especially for this quite unbalanced dataset.

Kentsch et al. [3] applied deep learning (DL) to their UAV-acquired images for forestry applications in Japan and investigated the specific issue related to the effect of Transfer Learning (TL) and the deep learning architecture chosen or whether a simple patch-based framework may produce results in different practical problems. Two deep learning architectures, namely ResNet50 and UNet were applied to two in-house datasets called winter and coastal forest. The main objectives of the study are developing an algorithm using a ResNet50 to classify patches corresponding to tree species, evolving an efficient semantic segmentation algorithm for tree species, and evaluating the applicability of the MLP algorithm to another practical problem (detection of an invasive tree species in a coastal forest. The results revealed that transfer learning is necessary to obtain satisfactory results in the problem of MLP classification of deciduous vs evergreen trees in the winter orthomosaic dataset. The experiments also show that DL provided valuable information about complex classification problems when used as a "black box". The patch-based classifier provided reasonably good results to find patches containing black locust trees in the black pine coastal forest mosaic.

Xiong et al. [4] investigated the use of generative adversarial nets (GANs) for their generalization ability across locations and sensors with some modification to accomplish the idea "training once, apply to everywhere and different sensors" for remote sensing images. It is well-known that considering the trade-off between high spatial and high temporal resolution in remote sensing images, many learning-based models (e.g., convolutional neural network, sparse coding, Bayesian network) have been established to improve the spatial resolution of coarse images in both the computer vision and remote sensing fields. However, data for training and testing in these learning-based methods are usually limited to a certain location and specific sensor, resulting in the limited ability to generalize the model across locations and sensors. The study was based on super-resolution generative adversarial nets (SRGANs), the loss function, and the structure of a network that was a modified and improved SRGAN (called ISRGAN), making the model training more stable and enhancing the generalization ability across locations and sensors. Significant improvement was reported with the use of the proposed ISRGAN model.

Sildir et al. [5] tackled the problem of finding the optimal size for a neural network. The design of a neural network is not a simple task. The number of nodes in the hidden layer(s) should be large enough for the correct representation of the problem, but, at the same time, low enough to have adequate generalization capabilities [6]. The optimum number of hidden layer nodes depends on various factors. If the network is too small, it cannot learn from the data, resulting in a high training error, which is a characteristic of underfitting. if the network is too large, a well-known overfitting problem occurs. In other words, it becomes over-specific to the training data and likely to fail with the test data, producing lower classification accuracies. The authors proposed a novel superstructure-based mixed-integer nonlinear programming method for optimal structural design (pruning, and input selection) for multilayer perceptron (MLP). The proposed method applies statistical measures including the parameter covariance matrix to increase the test performance while permitting reduced training performance. Two public hyperspectral datasets were employed for land use/land cover classification problems using a fully-connected network and superstructure neural network. The test results revealed promising performances in that fully connected networks were pruned by over 60% in terms of the total number of connections resulting in an increase in classification accuracy

Zhang et al. [7] proposed a deep quadruplet network (DQN) for hyperspectral image classification in cases with a small number of samples. Conventional classifiers require improvements in hyperspectral image classification, especially under the condition of small-samples due to the high dimensionality of the datasets. By seeking a solution to this problem, the authors developed a quadruplet network, in particular, a new quadruplet loss function, and a deep 3D CNN with double branches consisting of dense convolution and dilated convolution. Three widely-known public hyperspectral datasets (Salinas, Indian Pines, University of Pavia) were employed in the study for testing the performance of the proposed CNN model. Results revealed that the proposed approach produced higher accuracies than existing methods, and the proposed quadruplet loss provided the highest contribution.

Maxwell et al. [8] applied mask regional-convolutional neural networks (Mask R-CNN), which provides context-based instance mapping, for extracting valley fill faces (VFFs) resulting from mountaintop removal (MTR) surface coal mining in the Appalachian region of the eastern United States. Mapping VFFs is of significant interest to environmental modelers. The objectives of the study were to assess the Mask R-CNN DL algorithm for mapping VFFs using LiDAR-derived digital elevation data and to investigate model performance and generalization by applying the model to LiDAR-derived data in new geographic regions and acquired with differing LiDAR sensors and acquisition parameters. The produced results suggested that models trained in one area could transfer well to other areas if similar data were available It was concluded that the combination of Mask R-CNN and LiDAR has great potential for mapping anthropogenic and natural landscape features.

Improving the accuracy of edge pixel classification is an important research topic in remote sensing. Li et al. [9] studied this problem using convolutional neural networks (CNNs) in the extraction of winter wheat spatial distribution from remote sensing imagery of Gaofen-2. To improve CNN classification results, the post-processing CNN (PP-CNN) model was proposed to consider prior knowledge obtained from statistical analysis. The improved RefineNet generated the initial segmentation and outputted a category probability vector for each pixel and then these initial segmentations were statistically analyzed using manual labels as a reference to determine the confidence threshold. Lastly, all pixels below the confidence threshold were post-processed to generate their final category label. It was reported that the proposed extraction strategy effectively improved the accuracy of crop extraction results.

Alganci et al. [10] conducted a comparative study on the performances of deep learning methods for airplane detection using very high-resolution satellite images. Three types of CNN models, namely Faster R-CNN, Single Shot Multi-box Detector (SSD), and You Look Only Once-v3 (YOLO-v3) were employed to perform the objectives of the study using a limited number of labeled data for airplane detection problem. A non-maximum suppression algorithm (NMS) was introduced at the end of the SSD and YOLO-v3 flows to remove the multiple detection occurrences near each detected object in the overlapping areas. The trained networks were applied to five independent VHR test images that cover airports and their surroundings to evaluate their performance objectively.

Ship detection using optical remote sensing images has become an active research topic with the increased spatial resolution of recent sensors. Current ship detection methods usually adopt the coarse-to-fine detection strategy, but these methods have some drawbacks including complex calculations, false detection on land, and difficulty in detecting the small size ship. Wu et al. [11] proposed a multi-scale detection strategy using the feature pyramid network (FPN) to achieve ship detection with different degrees of refinement. Thus, a new coarse-to-fine ship detection network (CF-SDN) that directly achieves an end-to-end mapping from image pixels to bounding boxes with confidences was developed. Experiments on optical images revealed promising results as the proposed method outperformed the well-known detection algorithms by identifying small-sized ships and dense ships near the port.

With the availability of large datasets with a high number of bands with higher resolution, pixel-based image classification lost its popularity and object-based image analysis using segmentation has been a new way of analyzing high dimensional data. The quality of segmentation considerably affects the resulting classification. In the segmentation process, obtaining fine edges for extracted objects is one f the key problems. To overcome this problem Zhang et al. [12] suggested the use of a new weight feature value convolutional neural network (WFCNN), consisting of an encoder, a decoder, and a classifier. The encoder is used to extract pixel-by-pixel features, while the decoder is used to fuse the coarse high-level semantic features and fine low-level semantic features. The classifier is used to complete the pixel-by-pixel classification. When training the model, the image patch and the corresponding label file are used as inputs. Once the model is successfully trained, the image to be segmented is used as the input, with the output being pixel-by-pixel label files. Results of WFCN were compared with SegNet, U-Net, and RefineNet.

Acknowledgments: As a Guest Editor, I would like to express my gratitude to all authors who submitted their manuscripts to the Special Issue. I also sincerely thank the reviewers for providing constructive and prompt feedback vital for a successful reviewing process. It was a rewarding experience working with the courteous editorial staff of the journal. Thank you for providing this opportunity to handle the Special Issue.

Conflicts of Interest: The author declares no conflict of interest.

Taskin Kavzoglu

Guest Editor

References

1. Giuffrida, G.; Diana, L.; de Gioia, F.; Benelli, G.; Meoni, G.; Donati, M.; Fanucci, L. CloudScout: A Deep Neural Network for On-Board Cloud Detection on Hyperspectral Images. *Remote Sens.* **2020**, *12*, 2205.

2. Vasilakos, C.; Kavroudakis, D.; Georganta, A. Machine Learning Classification Ensemble of Multitemporal Sentinel-2 Images: The Case of a Mixed Mediterranean Ecosystem. *Remote Sens.* **2020**, *12*, 2005.

3. Kentsch, S.; Lopez Caceres, M.L.; Serrano, D.; Roure, F.; Diez, Y. Computer Vision and Deep Learning Techniques for the Analysis of Drone-Acquired Forest Images, a Transfer Learning Study. *Remote Sens.* **2020**, *12*, 1287.

4. Xiong, Y.; Guo, S.; Chen, J.; Deng, X.; Sun, L.; Zheng, X.; Xu, W. Improved SRGAN for Remote Sensing Image Super-Resolution Across Locations and Sensors. *Remote Sens.* **2020**, *12*, 1263.

5. Sildir, H.; Aydin, E.; Kavzoglu, T. Design of Feedforward Neural Networks in the Classification of Hyperspectral Imagery Using Superstructural Optimization. *Remote Sens.* **2020**, *12*, 956.

6. Kavzoglu, T.; Mather, P.M. Pruning artificial neural networks: An example using land cover classification of multi-sensor images. *Int. J. Remote Sens.* **1999**, *20*, 2761–2785.

7. Zhang, C.; Yue, J.; Qin, Q. Deep Quadruplet Network for Hyperspectral Image Classification with a Small Number of Samples. *Remote Sens.* **2020**, *12*, 647.

8. Maxwell, A.E.; Pourmohammadi, P.; Poyner, J.D. Mapping the Topographic Features of Mining-Related Valley Fills Using Mask R-CNN Deep Learning and Digital Elevation Data. *Remote Sens.* **2020**, *12*, 547.

9. Li, F.; Zhang, C.; Zhang, W.; Xu, Z.; Wang, S.; Sun, G.; Wang, Z. Improved Winter Wheat Spatial Distribution Extraction from High-Resolution Remote Sensing Imagery Using Semantic Features and Statistical Analysis. *Remote Sens.* **2020**, *12*, 538.

10. Alganci, U.; Soydas, M.; Sertel, E. Comparative Research on Deep Learning Approaches for Airplane Detection from Very High-Resolution Satellite Images. *Remote Sens.* **2020**, *12*, 458.

11. Wu, Y.; Ma, W.; Gong, M.; Bai, Z.; Zhao, W.; Guo, Q.; Chen, X.; Miao, Q. A Coarse-to-Fine Network for Ship Detection in Optical Remote Sensing Images. *Remote Sens.* **2020**, *12*, 246.

12. Zhang, C.; Chen, Y.; Yang, X.; Gao, S.; Li, F.; Kong, A.; Zu, D.; Sun, L. Improved Remote Sensing Image Classification Based on Multi-Scale Feature Fusion. *Remote Sens.* **2020**, *12*, 213.

Article

CloudScout: A Deep Neural Network for On-Board Cloud Detection on Hyperspectral Images

Gianluca Giuffrida [1,*], Lorenzo Diana [1], Francesco de Gioia [1], Gionata Benelli [2], Gabriele Meoni [1], Massimiliano Donati [1] and Luca Fanucci [1]

[1]   Department of Information Engineering, University of Pisa, Via Girolamo Caruso 16, 56122 Pisa PI, Italy; lorenzo.diana@phd.unipi.it (L.D.); francesco.degioia@phd.unipi.it (F.d.G.); gabriele.meoni@ing.unipi.it (G.M.); massimiliano.donati@unipi.it (M.D.); luca.fanucci@unipi.it (L.F.)

[2]   IngeniArs S.r.l., Via Ponte a Piglieri 8, 56121 Pisa PI, Italy; gionata.benelli@ingeniars.com

*   Correspondence: gianluca.giuffrida@phd.unipi.it

Received: 31 May 2020; Accepted: 5 July 2020; Published: 10 July 2020

Abstract: The increasing demand for high-resolution hyperspectral images from nano and microsatellites conflicts with the strict bandwidth constraints for downlink transmission. A possible approach to mitigate this problem consists in reducing the amount of data to transmit to ground through on-board processing of hyperspectral images. In this paper, we propose a custom *Convolutional Neural Network (CNN)* deployed for a nanosatellite payload to select images eligible for transmission to ground, called *CloudScout*. The latter is installed on the Hyperscout-2, in the frame of the Phisat-1 ESA mission, which exploits a hyperspectral camera to classify cloud-covered images and clear ones. The images transmitted to ground are those that present less than 70% of cloudiness in a frame. We train and test the network against an extracted dataset from the Sentinel-2 mission, which was appropriately pre-processed to emulate the Hyperscout-2 hyperspectral sensor. On the test set we achieve 92% of *accuracy* with 1% of *False Positives (FP)*. The Phisat-1 mission will start in 2020 and will operate for about 6 months. It represents the first in-orbit demonstration of Deep Neural Network (DNN) for data processing on the edge. The innovation aspect of our work concerns not only cloud detection but in general low power, low latency, and embedded applications. Our work should enable a new era of edge applications and enhance remote sensing applications directly on-board satellite.

Keywords: earth observation; on-board; microsat; mission; nanosat; hyperspectral images; AI on the edge; CNN

---

1. Introduction

In the last years the number of micro and nanosatellites, respectively *microsat* and *nanosat*, has rapidly increased. These satellites allow testing, experimenting and proving several new ideas by reducing at the same time the overall costs of the missions [1,2]. The increase in the number of microsats and nanosats and the augmented resolution of modern sensors lead to an increase in bandwidth usage and therefore the need to exploit new techniques to efficiently manage the bandwidth resources. Generally, for many sensors, only a portion of the data has valuable information for the mission and it is exploitable for the purpose of the mission. In recent years, the advances in low-power computing platforms combined with new Artificial Intelligence (AI) techniques have paved the way to the "edge computing" paradigm [3]. In fact, through the use of new hardware accelerators, it is possible to bring efficient algorithms, such as Convolutional Neural Network (CNN), directly on board. One example is represented by cloud detection algorithms [4,5]. The latter allows to identify images whose content is shaded by the presence of clouds.

In this paper, we demonstrate the effectiveness of use CNN cloud detection algorithm directly on board satellites, which leads to several benefits including:

- On-board filtering of unuseful data, relaxing the strict bandwidth requirements typical of modern/future Earth Observation applications [6–8];
- Preliminary decision taken directly on board, without the need for a human operator;
- Mission reconfigurability, changing only the weights of the network[6];
- Continuous improvement of results, in terms of accuracy and precision, through new generated data.
- Reduction of operative costs and mission design cycles [6,7];
- Enabling the use of Commercial off-the-shelf (COTS) hardware accelerators for Deep Learning, featuring improved computation efficiency, costs, and mass compared to space-qualified components [6,7].

Moreover, recent radiation tests [6], performed on the COTS Eyes of Things (EoT) board [9] powered by the Intel Movidius Myriad 2, show it as the best candidate among the others.

Our CNN-based algorithm will be launched on board of the HyperScout-2 satellite, which is led by cosine Remote Sensing (NL) with the support of Sinergise (SL), Ubotica (IR) and University of Pisa (IT) in the framework of the European Space Agency (ESA) PhiSat-1 initiative. This represents the first in-orbit demonstrator of Deep Neural Network (DNN) applied to hyperspectral image [10–12]. Our network takes as input some bands of hyperspectral cubes produced by the HyperScout-2 sensor, identifying the presence of clouds through a binary response: *cloudy* or *not cloudy*. Since the EoT board has a specific low power hardware accelerator for Machine Learning (ML) on the edge, it is suitable to be integrated in microsat and nanosat.

The paper is structured as follows: in Section 2 we describe the goals of the PhiSat 1 mission, while in Section 3 we provide a description of the CNN model, the training procedure, and the dataset. In Section 4 results in terms of accuracy, number of False Positive (FP), and power consumption are shown both for the entire dataset and a critical dataset. In Section 5 a summary of the benefits brought by this technology is discussed and, finally in Section 6 overall conclusions are drawn.

2. Aim of the PhiSat-1 Mission

The aim of this mission is to demonstrate the feasibility and the usefulness in bringing AI on-board Hyperscout-2 satellite [12]. To this end, the mission involves the use of a CNN model suited on the Myriad 2 Vision Processing Unit (VPU) featured in the EoT board, which was chosen by the European Space Agency (ESA) as the best hardware to fly. The network is expected to classify hyperspectral satellite images, in two categories: *cloudy* and *not cloudy*. The main requirements for the network in this mission are:

- Maximum memory footprint of 5 MB: to update the network with respect to the uplink bandwidth limitation during the life of the mission;
- Minimum accuracy of 85%: to increase the quality of each prediction even in particular situations, e.g., clouds on ice, or clouds on salt-lake;
- Maximum FP of 1.2%: to avoid the loss of potentially good images.

This strategy allows downloading to ground only non-cloudy images, respecting the constraints imposed by the hardware accelerator and the budget of satellite resources i.e. power consumption, bandwidth, memory footprint, etc.

3. Methods

3.1. Machine Learning Accelerators Overview

In recent years, the interest in AI applications has grown very rapidly. These applications run both on the cloud, powered by Graphic Processing Unit (GPU)-farms that work as a global hardware accelerator, and on the edge through dedicated low-power hardware accelerators. A simple example of this mechanism is the "OK-Google" application. In fact, it is divided into two phases: the first part is requested by users on their personal smartphone using *keyword-spotting* [13] algorithm performed by the smartphone accelerator; then, during the second phase, the voice is sent to the cloud which uses its *"intelligence"* to complete the required tasks.

The cloud provides the greatest flexibility in all the cases where there are no bandwidth constraints or privacy issues; vice versa, in automotive, space, or real-time application, the cloud paradigm could not be the right choice [3].

Thus, several companies have developed their own AI hardware accelerators. The COTS accelerators are easily classifiable by their processors [14,15]: VPU, Tensor Processing Unit (TPU), the most known GPU and Field-Programmable Gate Array (FPGA). The first two processors have the best performance in terms of power per inference since they have been devised to speed up inferences. Instead, GPUs and FPGAs are more general purposes and they are the most powerful in term of computational capabilities.

A **TPU**: TPU is an innovative hardware accelerator dedicated to a particular data structure: *Tensors* [16]. Tensors are a base type of the TensorFlow framework [17] developed by Google. The standard structures and the dedicate libraries for GPU and VPU make tensors and consequently TensorFlow very powerful tools in the ML world. The Coral Edge TPU is an example of an edge hardware accelerator whose performances are very promising, especially in the static images processing acceleration e.g., CNN, Fully Convolutional Network (FCN).
The best performances of this hardware platform are reached exploiting TensorFlow Lite and 8 bits integer quantization, even if the latter could have a big impact on the model metrics.

B **GPU**: GPUs [18] are the most widely used to carry out both inference and training process of the typical ML models. Their computational power is entrusted to the parallel structure of the hardware that computes operations among matrices at a very high rate. Nvidia and AMD lead the market of the GPU for ML training, using respectively *CUDA Core* (Nvidia) and *Stream processor* (AMD), as shown in [14,15]. Moreover, several frameworks allow to use the potentiality offered by GPUs, including TensorFlow, TensorFlow Lite, and PyTorch. This hardware can quantize the model and run inferences supporting a wide range of computational accuracies e.g., 32 and 16 bits floating point, 16, 8, 4, and 2 bits integer. On the other hand, these solutions consume huge power, reaching a peak of 100 W and therefore cannot be used for on the edge applications.

C **FPGA**: FPGAs are extremely flexible hardware solutions, which could be completely customized. This customizability, however, represents the bottleneck for a fast deployment [19]. In fact, the use of an FPGA requires many additional design steps compared to COTS Application-Specific Integrate Circuit (ASIC), including the design of the architecture of the hardware accelerator and the quantization of the model, for approaches exploiting fixed-point representation. FPGAs are produced by numerous companies such as Xilinx, MicroSemi, Intel. Some FPGAs, like RTG4 or Brave, are also radiation-hard/tolerance, which means these boards can tolerate the radiations suffered during the life of the mission as explained in [20,21].

D **VPU**: VPUs represent a new class of processors able to increase the speed of visual processing as CNN, Scale-Invariant Feature Transform (SIFT) [22], Sobel and similar. The most promising accelerators in this category are the Intel Movidius Myriad VPUs. At the moment, there exist two versions of this accelerator, the Myriad 2 [23] and the Myriad-X. The *core* of both processors is the computational engine that uses groups of specialized vectors of Very Long Instruction Word

(VLIW) processors called Streaming Hybrid Architecture Vector Engine (SHAVE)s capable of consuming only a few watts (W) [24,25]. Myriad 2 has 12 SHAVEs instead the Myriad-X has 18 SHAVEs. The Myriad 2 and Myriad-X show better performance when they accelerate CNNs model or other supported layers than mobile CPUs or general-purpose low-power processors. To reduce the computational effort, all the registers and operation within the processor use 16 bits floating point arithmetic. Moreover, the Myriad 2 processor has already passed the *radiation tests* at CERN [6].

A more extensive comparison among the various COTS devices used for Deep Learning can be found in [26,27].

3.2. Eyes of Things

The EoT board was developed in the framework of H2020 European project called *Eyes of Things* by Spain's Universidad de Castilla-La Mancha [9] and it is powered by the Intel Myriad 2 VPU, which results to be a promising solution for Low Earth Orbit (LEO) missions [23,25]. Moreover, as described in [6], the EoT board with Myriad 2 VPU passed the preliminary *radiation tests*.

However, the core of the Myriad 2 VPU hardware accelerator, described in Section 3.1, natively supports:

- Convolutional layers
- Pooling layers
- Add and subtraction layers
- Dropout layers
- Fully connected layers

The Myriad 2 chip shows some features that match with the set of requirements described in Section 2. Notably, as briefly described in Section 3.1 and in [6,14], Myriad 2 shows one of the best compromises in terms of power per inference among the hardware accelerators available on the market. It also supports *in-hot reconfiguration* via uploading a new GRAPH file, which contains information about the model, the weights, and a set of hardware configurations (e.g., number of SHAVEs to use, batch size, layer fusion, etc.). Unluckily, one limitation of the EoT board is that it exploits only 8 SHAVEs, while the Myriad 2 processor has 12 SHAVEs. The reconfigurability is of great importance for space applications, as it enables a new generation of re-configurable AI satellites whose goals could be changed during the mission life.

In fact, we plan to improve the network accuracy during the mission life exploiting the data taken directly by Hyperscout-2 satellite. This factor is of fundamental importance when flying with new sensors for which there is no data set available.

3.3. Dataset and Training Preparation

3.3.1. Satellite Selection

The training of the CNN network is carried out through a supervised process in which the network calculates the difference between the set of images and the corresponding right decisions or labels. The dataset should represent the entire range of the images that will be presented to the network during the mission. In addition to this, the data could be augmented to represent all the possible disturbances introduced by the camera or some other acquisition errors that might happen during the life of the mission. Hence, training a network using a highly variegated set of images and their corresponding outputs labeled as accurate as possible, provides high-quality results able also to tolerate small errors, band misalignment, and light effects.

For all the missions that exploit new sensors, as HyperScout-2, there are not enough representative data to build a dataset; thus, a new dataset is simulated from the images captured by similar previous missions. The source satellite is selected by taking into account: the similarity of the data with respect to the new mission, the availability of data and labels, and the existence of analysis tools, dedicate to the source satellite, able to produce and enhance data for the new mission.

To this aim, the dataset was composed of 21546 Sentinel-2 hyperspectral cubes. These cubes consist of 13 bands; each of them represents a different wavelength. An example of a hyperspectral cube is shown in Figure 1. Each two-dimensional image in the hyperspectral cube represents the scene with respect to only one wavelength, allowing the division of the components through the light reflection on them such as fog, nitrogen, methane, etc.

Sentinel-2 satellites elaborate the images in three steps:

- Level-0 saves raw-data and appends them with annotation and meta-data;
- Level-1 is divided into three sub-steps: A, B, C. Step A decompresses the mission-relevant Level-0 data. Step B applies radiometric correction, then step C applies geometric corrections;
- Level-2 applies atmospheric correction and, if required, some other filters or scene classification.

The output of the Sentinel-2 Level-2A produces a Bottom-Of-Atmosphere (BOA) reflectance *hyperspectral cube* of data, while HyperScout-2 satellite has a sensor producing images in radiation bands [11,28]. Due to these differences in sensors, we need to transform the Sentinel-2 hyperspectral cubes from saturation to radiation data. This process exploited the additional information provided by the Sentinel-2 satellites during the Level-0 step, such as the relative position of the sun with respect to the satellite, the location of the satellite, and the maximum and minimum values of pixel saturation.

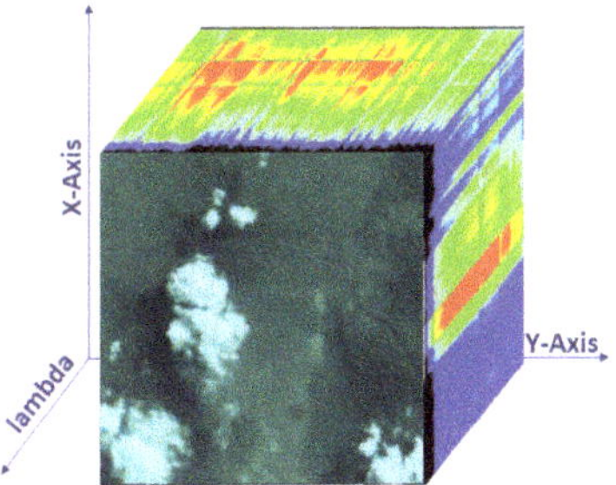

Figure 1. Example of hyperspectral Cube.

There are multiple reasons to use the Sentinel-2 dataset cubes even if they need some pre-processing activities:

- The data are provided by a member of the CloudScout mission consortium, Sinergise Ltd.;
- The data are accurately labeled, as shown in Figure 2, using a Sinergise's ML algorithm called *Sentinel Hub* [29];
- The Sentinel-2 spatial resolution (10 to 60 m) is compatible with HyperScout-2 (75 m). The downscale does not deteriorate the quality of the information;
- The swath of both satellites is about 300 km;
- 10 of the 13 bands of Sentinel-2 are compatible with HyperScout-2 camera both in wavelength and Signal-to-Noise Ratio (SNR);
- Sinergise provides also a tool to convert saturation images into radiance images.

Moreover, we introduced a random Gaussian noise to the Sentinel-2 images to obtain a dataset robust to small perturbation of SNR. Hence, the number of dataset images was doubled, noisy, and not noisy. The data were processed to obtain $512 \times 512 \times 3$ tiles, where each tile represents 30×30 km^2 with a resolution of 60×60 m^2 per pixels. These images represent the *original dataset*.

Fraction of classifications as clouds

	MAJA	Fmask	Sen2Cor	Sentinel Hub
Cloud	99.4%	95.8%	95.9%	100%
Cirrus	64.1%	89.8%	95.1%	73.7%
Land	0.7%	12.0%	9.6%	7.0%
Water	0.0%	0.0%	0.0%	0.0%
Snow	10.6%	71.5%	58.0%	18.3%
Shadow	5.8%	8.1%	8.4%	3.2%

Figure 2. Sentinel Hub results compared with the main cloud detector algorithms [29].

Unluckily, the original dataset contained some outliers. They represent a critical point for the training phase. Thus, they were removed using an outlier search. Some examples of these images are shown in Figures 3 and 4. The final dataset distribution is shown in Figure 5.

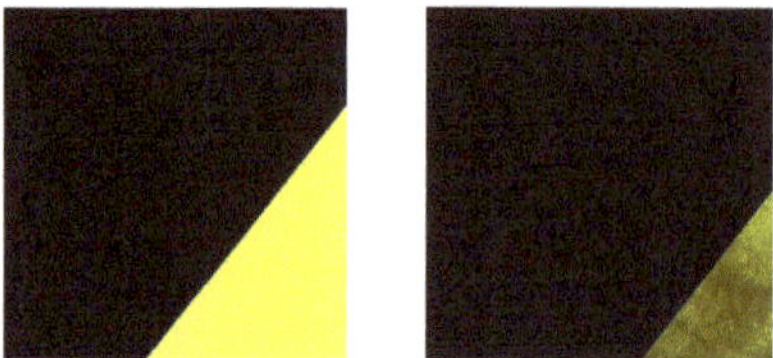

Figure 3. Example of incomplete images found in our dataset.

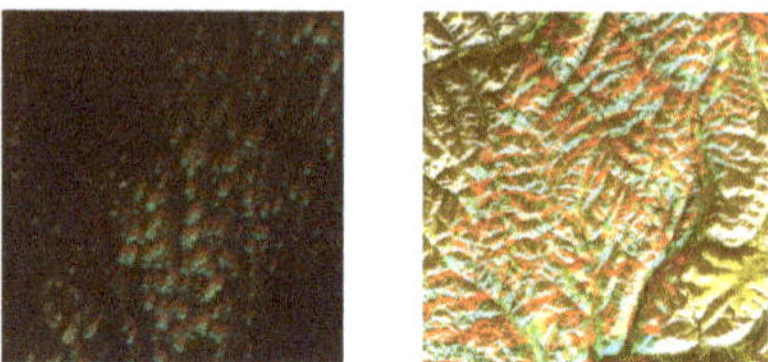

Figure 4. Example of images with not correctly aligned bands found in our dataset.

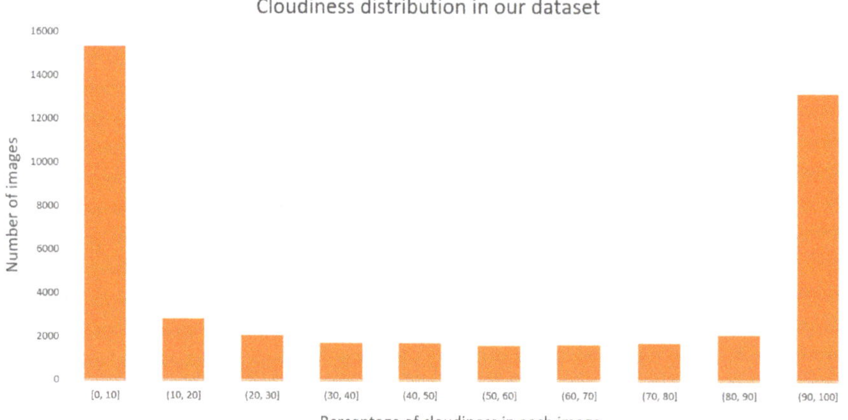

Figure 5. Histogram representing the cloudiness distribution in our dataset.

3.3.2. Preliminary Trainig Phase

The tile size was decided by taking into account the maximum image size supported by the *intra-layer* memory of the Myriad 2, which is a 121 MB memory dedicated to the intermediate results between two consecutive layers. The compiler for the EoT board was provided as part of the Neural Compute Software Development Kit (NCSDK) [30]. It supported only input images with at most three bands. This limitation derived from the datatype *ImageData* of *Caffe*, which was the framework used to develop DNN models [31] on EoT. For this reason, Sinergise Ltd. chosen the best three bands through a Principal Component Analysis (PCA), starting from the 13 ones available on Sentinel-2 [32]. The bands detected by PCA were the 1, 2, and 8 with respect to the Sentinel-2 bands ordering. bOur network gives as output the probability to belong to one of the two classes *cloudy* and *not cloudy*.

To obtain the two classes we labeled as *cloudy* those images containing a number of cloudy pixels higher or equal to a specific *threshold*. Vice versa we labeled as *not cloudy* the remaining images. The labeling process was executed twice to perform incremental training. The first time a threshold of 30% cloudiness (named TH30 dataset) was used, while the second time a 70% threshold (named TH70 dataset) was considered. The TH30 dataset is used to train the network to recognise the "cloud shape". The TH70 one, instead, it is used to change the decision layers (Fully Connected layers), by blocking the back-propagation [33] for the *feature extraction* layers and performing a fine-tuning on the *decision* layers. The two training steps were necessary due to the non-equal distribution of the images inside the original dataset as shown in Figure 5. In fact, the original dataset is composed mainly by fully cloudy covered images or cloudless images. This approach improves the generalization capabilities of the network.

The two datasets were divided as described in Table 1 and Table 2 respectively.

The difference in the number of images belonging to the training, validation, and test sets, of the two datasets derive again from the unbalance of the original dataset. We divided each of the two datasets to obtain a balanced training and validation sets. The following steps illustrate how the training, validation, and test sets were obtained for each dataset:

- Label the original dataset with the selected threshold;
- Compute the minimum between the number of *cloudy* and *not cloudy* images, call it N;
- Sample $0.7 \times N$ *cloudy* and *not cloudy* images, call it training set;
- Sample $0.15 \times N$ *cloudy* and *not cloudy* images, call it validation set;
- Exploit all the remaining images to populate the test set.

Table 1. TH30 dataset details.

Training	
Number of Images	31,926
	Mirror
Data augmentation used	Flip X axis
	Flip Y axis
	Noise injection
Validation	
Number of Images	5986
Data augmentation used	Mirror
Test	
Number of Images	5180
Data augmentation used	none

Table 2. TH70 dataset details.

Training	
Number of Images	26,834
	Mirror
Data augmentation used	Flip X axis
	Flip Y axis
	Noise injection
Validation	
Number of Images	5032
Data augmentation used	Mirror
Test	
Number of Images	11,226
Data augmentation used	none

3.4. CloudScout Network

3.4.1. EoT NCSDK

To develop our CNN model for the EoT board, we used the iterable method shown in Figure 6 where the model is considered acceptable if it meets the requirements about accuracy (greater than 85%) and FP rate (less than 1.2%), both in GPU and EoT.

GPU generally produces 32 bits floating point weights, while the Myriad 2 supports only 16-bits floating point weights. Thus, we converted the GPU weights to 16-bits floating point. This conversion may change the overall model behaviour in a non-predictable way due to *quantization* and *pruning* processes. Both of them are non-linear processes, which might affect the performance of the generated network in terms of accuracy and FP results.

Owing to that, it is necessary to design networks to minimize the impact of these processes on performance, or to design a network that already considers the quantization during the training [34,35].

In particular, the quantization process is required to port the developed model to the EoT board. For this purpose, we had to rely on the Movidius NCSDK [30] only. Hence, we decided to adopt the *quantization-aware* technique [35] which creates virtual constraint during the training to maintain the weights within the 16 bits. This process ends when the difference between the GPU weights and the weights generated by the NCSDK is under 10^{-2}.

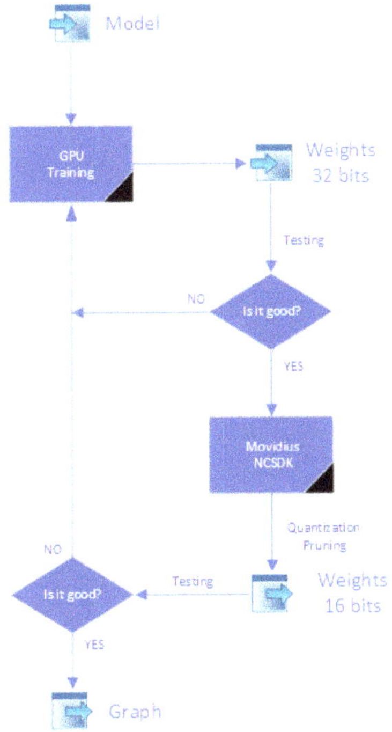

Figure 6. Iterative approach used to select the right network.

3.4.2. CloudScout Network Structure

The model is composed by two levels: *feature extraction* and *decision*, as shown in Figure 7.

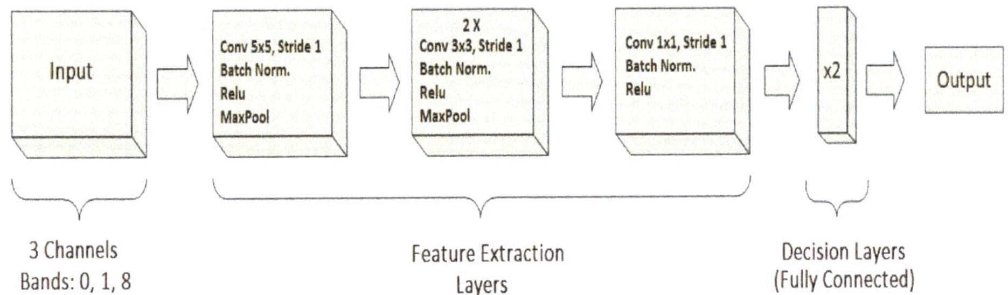

Figure 7. The CloudScout model block diagram.

The layers of the *feature extraction* level are used to discriminate different components of each image using convolutional layer with different kernel sizes and depth. This allows to separate clouds with respect to terrains, sea, sand, etc. *Feature extraction* level is composed by a convolutional size reduction layer followed by four group of three convolutional layers. Each group improves the generalization performance exploiting a sequence of kernels of $3 \times 3, 1 \times 1, 3 \times 3$, which provides the same performance of a 5×5 kernel, but they reduce the number of weights. Furthermore, the output given by the 1×1 convolutional layer with the bias set to true represents a local pixel level classification [36]. In order to allow future upload of the network during the mission and to meet the model size requirement (less than 5 MB) a *Global Max Pooling* layer was exploited. It extracts the maximum value from each filter produced by the last convolutional layer of the *feature extraction* level.

The *decision* layer collects and assigns a value to the data produced by the *Global Max Pooling* in the *feature extraction* level.

The last layer is a *softmax*, which computes the probability of belonging to the first or second class from the output of the *decision* level. The training was performed in two steps:

- train the CNN model against the dataset with labels generated by using TH30 dataset to improve the recognition of the shape of the clouds;
- train the CNN model against the dataset with labels generated by using TH70 dataset to improve the accuracy of the decision layers.

One of the given requirement was to obtain FP number of 1% with respect to the given dataset without affecting the accuracy. Since accuracy and Receiver Operating Characteristic (ROC) analysis assume FP and False Negative (FN) results to be equally important, we decided to exploit the F_2 metrics. This metric is a particular case of the F_β score, also known as the Sørensen–Dice coefficient or Dice Similarity Coefficient (DSC), defined in Equation (1).

$$F_\beta = (1 + \beta^2) \cdot \frac{precision \cdot recall}{(\beta^2 \cdot precision) + recall} \tag{1}$$

$$= \frac{(1 + \beta^2) \cdot TP}{(1 + \beta^2) \cdot TP + \beta^2 \cdot FN + FP}$$

This equation gives more relevance to *FP* errors than *FN* errors and allows to better estimate the influence of the *FP* within the network.

Moreover, to effectively reduce the number of *FP* during the training phase, we changed the standard *Binary-Cross Entropy* loss-function, shown in Equation (2), by doubling the penalty given in case of *FP* errors.

$$L(y, \hat{y}) = -\frac{1}{N} \cdot \sum_{i=0}^{N} (y_i \cdot \log(\hat{y}_i) + 2 \cdot (1 - y_i) \cdot \log(1 - \hat{y}_i)) \tag{2}$$

where y is the expected/labeled value, and $\hat{y}$ is the predicted value.

This produced a network more prone to classify images as cloudless, effectively reducing the number of *FP* results at the expense of producing more *FN* ones. However, the increased number of *FN* results, does not pose a threat to the overall performance of the system thanks to the high value of accuracy of 92% on the EoT as detailed in Section 4.

4. Results

In this Section, we show the results obtained from the CNN model described in Section 3.4 implemented on the EoT board featuring the Myriad 2 hardware accelerator.

As shown in Table 3 and in Table 4 we met all the requirements, described in Section 2, related to Accuracy and False Positive Rate. These results were obtained exploiting the entire 43,092 images contained in the dataset since the network quantization process could fully change the hyper-plane of the network. The network was trained by exploiting an initial learning rate of $lr = 0.01$ and using an exponential decay computed as shown in Equation (3).

$$lr = lr_{old} \cdot e^{-0.6 \cdot epoch} \tag{3}$$

where lr is the new learning rate computed at each epoch, lr_{old} is the learning rate used for the previous epoch, *epoch* is the actual number of epoch, and 0.6 is a constant chosen empirically. In Table 5 the model characterization run on the EoT board is reported, taking into account inference time, power consumption per inference, and memory footprint. Furthermore, as preliminary described in Section 2, the 2.1 MB memory footprint allows uploading a new version of weights during the mission operation through telecommand allowing on-the-fly flexibility and improvements.

Table 3. Performance of the CloudScout model on the EoT board (Myriad 2).

ROC analysis:	
TPR	0.83
FPR	0.02
False Positive	1.03 %
EoT board (Myriad 2) accuracy	92 %

Table 4. Confusion matrix of the CloudScout model run on the entire dataset.

	Cloudy	Not Cloudy
Cloudy	32.5 %	1 (FP)%
Not Cloudy	6.8 (FN) %	59.7 %

Table 5. Network characterization on EoT board (Myriad 2).

EoT board (Myriad 2) Inference Time	325 ms
Model memory footprint	2.1 MB
Power consumption per inference	1.8 W

The CloudScout network has been also tested using a dataset of 24 images representing some critical environments with or without clouds. The critical dataset is composed of images taken by Sentinel-2 and elaborated following the same criteria used in Section 3.3. In particular, the 24 images contain very difficult classification problems due to the presence of salt lakes or snow mountains mixed with clouds. On this dataset, the CloudScout network achieves 67% accuracy with only 8 misclassified images against the 24 of the entire critical dataset. The obtained results are described in the Confusion Matrix described in Table 6. It is noted that the number of FN classifications is much higher than the FP owing to the unbalance of the loss function, described in Section 2, used during the training.

Table 6. Confusion matrix of the model with respect to the critical test set.

	Cloudy	Not Cloudy
Cloudy	2	2 (FP)
Not Cloudy	6 (FN)	14

Some example of the images contained in the critical dataset are shown in Figures 8–12. Each figure shows the radiance image obtained after the elaboration to represent a CloudScout 2 image, its RGB visible image, and the corresponding binary cloud mask which shows in white the cloudy pixels while in black the not cloudy ones. The binary mask was used as ground truth only to compute the number of cloudy pixels inside the image. Figures 8–10, represent three of the most obvious wrong classifications while Figures 11 and 12 are two examples of good classifications. Here we briefly give a justification for the wrong classification of three images:

- In Figure 8 the network recognises ice as a continuum of the cloud. To avoid this behaviour, a possible solution could be to use a thermal band that provides high-quality cloud contours.
- The CloudScout network was not trained to recognise the fog, even if in some cases it could represent a big obstacle to visibility. This phenomenon is observable in Figure 9. Here, the image is fully covered by fog as further shown by the cloud mask, but our network does not interpret the fog as cloud obtaining, in fact, just a 1% probability of cloudiness in this picture.
- Another reason for wrong classifications is the use of a fixed threshold to define an image cloudy or not. Indeed, as shown in Figure 10, the cloudiness inside the image is about 65%. This value

is very close to the threshold selected for this project (70%). To achieve a better result for all the borderline cases should be useful to increase the granularity of the classification, developing a segmentation network.

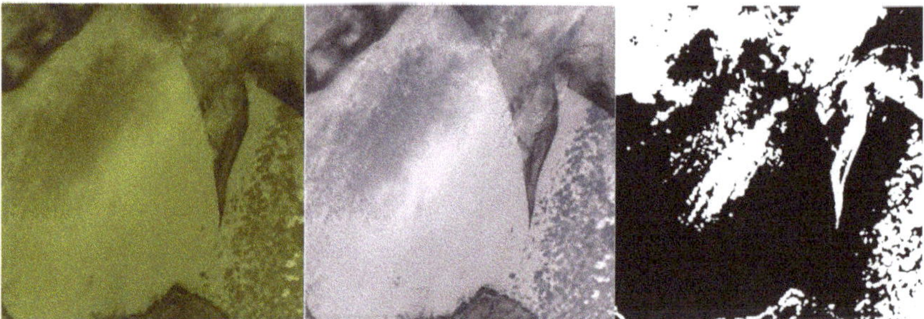

Figure 8. Aral sea clouds and ice; radiance image, RGB image, cloud mask. False Positive case.

Figure 9. Bonneville clouds; radiance image, RGB image, cloud mask. False Negative case.

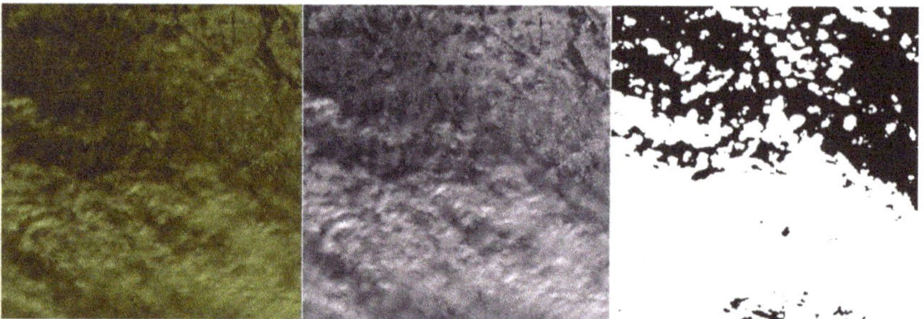

Figure 10. Erie ice and clouds; radiance image, RGB image, cloud mask. False Positive case.

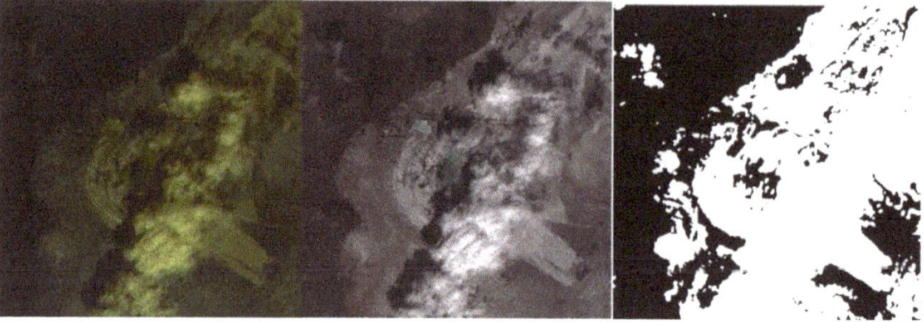

Figure 11. Bonneville clouds and shadows; radiance image, RGB image, cloud mask.

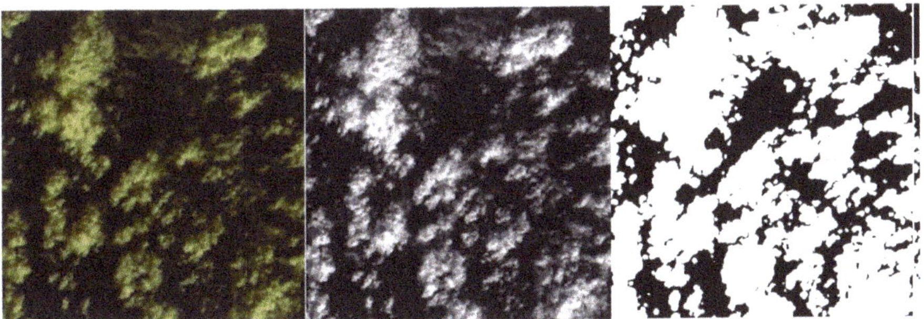

Figure 12. Pontchartrain clouds over water; radiance image, RGB image, cloud mask.

Finally, Table 7 shows the output details of the inferences performed on the critical dataset.

Table 7. The critical test set results.

Name	Clouds %	Prob. of Cloudiness % (Net Output)
Aral Sea clouds and ice Figure 8	35%	99 %
Bonneville clouds Figure 9	87%	1 %
Algeria mixed clouds and shadows	98%	0 %
Erie ice clouds Figure 10	65%	96 %
Pontchartrain clouds	72%	1 %
Amadeus January clouds	78%	1 %
Aral Sea clouds	87%	2 %
Aral Sea thin clouds	90%	0%
Aral Sea no clouds	0%	0%
Pontchartrain sun glint	3%	0%
Caspian sea cloud	64%	0%
Alps snow no clouds	9%	0%
Bonneville clouds shadows Figure 11	56%	1%
Greenland snow ice no-cloud	2%	18%
Alps snow cloud no-shadow	44%	12%
Amadeus no clouds	0%	0%
Alps snow cloud shadow	22%	0%
Greenland snow ice clouds	85%	99%
Amadeus no-clouds	3%	0%
Bonneville no clouds	29%	0%
Erie ice no-clouds	0%	9%
Amadeus clouds	99%	100%
Pontchartrain small clouds	14%	0%
Pontchartrain clouds over water Figure 12	66%	2%

5. Discussion

The advantages and limitations of the proposed approach concerning several aspects are discussed in this Section:

- **Relaxing Bandwidth limitations**: as described in Section 1, the introduction of DNNs onboard Earth Observation satellites allows filtering data produced by sensors according to the *Edge computing* paradigm. In particular, cloud-covered images can be discarded mitigating bandwidth limitations. More detailed information on these aspects is provided in [6–8].
  However, producing a binary response (cloudy/not cloudy) leads to a complete loss of data when clouds are detected, preventing the applications of *cloud removal techniques*, described in [37–39]. Such approaches allow reconstructing the original contents by exploiting low-rank decomposition using temporal information of different images, spatial information of different locations, or frequency contents of other bands. Such methods are generally performed on ground

because of their high complexity.

To enable the use of these approaches and to the reduction of downlink data through filtering data on the edge, DNN-based approaches performing image segmentation can be exploited [40]. In this way, pixel-level information on the presence of clouds can be exploited to improve compression performance through the substitution of cloudy areas through completely white areas. In this way, the application of cloud removal techniques can be performed after data downlink.

The implementation of a segmentation DNN for on-board cloud-detection represents a future work.

- **Power-consumption/inference time**: The use of DNNs allows leveraging modern COTS hardware accelerators featuring enhanced performance/throughput trade-offs compared to space-qualified components [6]. Table 5 shows the proposed model can perform an inference in 325 ms with an average power consumption of 1.8 W. Such reduced power consumption represents an important outcome for CubeSats, for which the limited power budget represents an additional limitation for downlink throughput [41].

- **Training procedure**: This work proposes to exploit a *synthetic dataset* for the training of the DNN model in view of the lack of data due by the novelty of the HyperScout 2 imager [11].
  Despite the functionality of the proposed approach has still to be demonstrated through the validation by means of HyperScout 2 data, the methodology described might be exploited in near future to realize a *preliminary* training of new applications used for novel technology for which a proper dataset is not available. Moreover, thanks to the possibility to reconfigure hardware accelerators for DNNs, the model can be improved after the launch thanks to a fine-tuning process performed through the actual satellite data [6].

- **Cloud detection performance**. There are different cloud detection techniques for satellite images, both at the pixel and image level. Yang at al. [42] divide the cloud detector techniques into three categories:

  - Threshold methods: some of the most known thresholding-based techniques are ISCPP [43], APOLLO [44], MODIS [45], ACCA [46], and some new methods which work well when ice and clouds coexist. However, these methods are very expensive for the CPU because of the application of several custom filters on the images. Hence, they are not good candidates to be used directly on-board.

  - Multiple image-based methods: Zhu and Woodcock [47] use multiple sequential images to overcome the limitations of thresholding-like algorithms. Again, processing this information requires an amount of power that is hardly available on board. In addition, this method requires the use of a large amount of memory.

  - Learning-base methods: these methods are the most modern. They exploit all the ML techniques such as Support Vector Machines (SVM) [48], Random Forest (RF) [49], and NN as our CloudScout or [42]. Contrary to SVM and RF methods, NNs have a standard structure that allows building ad-hoc accelerators able to speed up the inference process by reducing energy consumption.

All three categories provide an excellent solution for ground processing. However, the purpose of CloudScout is to provide reliable information directly on board, without requiring a huge amount of power. So, thanks to the technological advancement given by the EoT board, we developed a simple NN model that performs a hyperspectral image threshold directly on board. The needed to build a new custom model and do not exploit directly one taken from the literature is given by the limitation of the hardware itself. In fact, as described in Section 3.2, the accelerator has some hardware constraints that must be considered in order to obtain a valid result.

6. Conclusions

The Phi-sat mission and in particular Hyperscout 2 is a nano-sat system with the aim to demonstrate the possibility of using DNN in space, by exploiting COTS hardware accelerators directly on-board, such as the Myriad-2 VPU. Moreover, the reconfigurability provided by these devices enables the update of DNN models on the fly, allowing to improve the model performance exploiting new data acquired during the mission. Furthermore, thanks to the intrinsic modularity of such DNN models, it would be possible to change the mission goals during the mission life.

This may represent the beginning of a new era of re-configurable smart satellites equipped by programmable hardware accelerator (such as the Myriad 2) enabling the *on-demand* paradigm at payload level [6].

As described by [6], this fact would enable a significant cost reduction due to satellite design or even satellite platform reuse.

Other advantages are related to the smart use of transmission bandwidth that represents one of the main concern in nearly future in view of the growing number of satellites and the increasing resolution of on-board sensors[6,8]. The efficacy of AI and in particular of CNN for this goal will be demonstrated by the results of the HyperScout-2, which might represent a precursor for this new smart satellite era. Indeed, such results demonstrate that exploiting a CNN running on the Intel Movidius Myriad 2 we are able to detect with a 92% accuracy, 1.8W of power consumption, and 2.1 MB of memory footprint the cloudiness of hyperspectral images directly on board the satellite.

Author Contributions: Conceptualization, G.G., L.D. and G.B.; Data curation, G.M.; Formal analysis, G.G. and G.B.; Funding acquisition, L.F.; Investigation, G.M.; Methodology, G.G. and L.D.; Project administration, M.D.; Resources, M.D.; Software, G.G., L.D. and F.d.G.; Supervision, L.F.; Validation, L.D., F.d.G. and G.M.; Visualization, G.M.; Writing—original draft, G.G., L.D. and F.d.G.; Writing—review and editing, G.G. and G.M. All authors have read and agreed to the published version of the manuscript

Funding: This research was funded by Europen Space Agency, grant number 4000124907.

Acknowledgments: The CloudScout project on board of the HyperScout-2 satellite is funded by the European Space Agency (ESA) in the framework of the PhiSat-1 initiative and led by Cosine. We would like particularly to acknowledge the contribution of Gianluca Furano, Massimiliano Pastena and Pierre Philippe Mathieu from the European Space Agency; Marco Esposito and John Hefele from Cosine Remote Sensing(NL), who realized HyperScout-2 satellite; Matej Batic from Sinergise L.t.d.(SL) for providing Sentinel 2 dataset; Aubrey Dunne from Ubotica(IR) for the support provided for the use of EoT in the HyperScout-2 mission.

Conflicts of Interest: The authors declare no conflict of interest.

References

1. Sweeting, M.N. Modern small satellites-changing the economics of space. *Proc. IEEE* **2018**, *106*, 343–361. [CrossRef]
2. Madry, S.; Martinez, P.; Laufer, R. Conclusions and Top Ten Things to Know About Small Satellites. In *Innovative Design, Manufacturing and Testing of Small Satellites*; Springer: Berlin, Germany, 2018; pp. 105–111.
3. Shi, W.; Cao, J.; Zhang, Q.; Li, Y.; Xu, L. Edge computing: Vision and challenges. *IEEE Internet Things J.* **2016**, *3*, 637–646. [CrossRef]
4. Li, H.; Zheng, H.; Han, C.; Wang, H.; Miao, M. Onboard spectral and spatial cloud detection for hyperspectral remote sensing images. *Remote Sens.* **2018**, *10*, 152. [CrossRef]
5. Hagolle, O.; Huc, M.; Pascual, D.V.; Dedieu, G. A multi-temporal method for cloud detection, applied to FORMOSAT-2, VENμS, LANDSAT and SENTINEL-2 images. *Remote Sens. Environ.* **2010**, *114*, 1747–1755. [CrossRef]
6. Furano, G.; Meoni, G.; Dunne, A.; Moloney, D.; Ferlet-Cavrois, V.; Tavoularis, A.; Byrne, J.; Buckley L.; Psarakis, M.; Voss, K.O.; et al. Towards the use of Artificial Intelligence on the Edge in SpaceSystems: Challeng-es and Opportunities. *IEEE Aerosp. Electron. Syst.* **2020** *forthcoming.*
7. Kothari, V.; Liberis, E.; Lane, N.D. The Final Frontier: Deep Learning in Space. In Proceedings of the 21st International Workshop on Mobile Computing Systems and Applications, Austin, TX, USA, 3–4 March 2020; pp. 45–49.

8. Denby, B.; Lucia, B. Orbital Edge Computing: Nanosatellite Constellations as a New Class of Computer System. In Proceedings of the Twenty-Fifth International Conference on Architectural Support for Programming Languages and Operating Systems, Lausanne, Switzerland, 16–20 March 2020; pp. 939–954.

9. Deniz, O.; Vallez, N.; Espinosa-Aranda, J.; Rico-Saavedra, J.; Parra-Patino, J.; Bueno, G.; Moloney, D.; Dehghani, A.; Dunne, A.; Pagani, A.; et al. Eyes of Things. *Sensors* **2017**, *17*, 1173. [CrossRef]

10. Pastena, M.; Domínguez, B.C.; Mathieu, P.P.; Regan, A.; Esposito, M.; Conticello, S.; Dijk, C.V.; Vercruyssen, N. ESA Earth Observation Directorate NewSpace initiatives. Session V. 2019. Available online: https://digitalcommons.usu.edu/smallsat/2019/all2019/92/ (accessed on 7 July 2020).

11. Esposito, M.; Conticello, S.S.; Pastena, M.; Carnicero Domínguez, B. HyperScout-2: Highly Integration of Hyperspectral and Thermal Sensing for Breakthrough In-Space Applications. In Proceedings of the ESA Earth Observation ϕ-Week 2019, Frascati, Italy, 9–13 September 2019.

12. Esposito, M.; Conticello, S.; Pastena, M.; Domínguez, B.C. In-orbit demonstration of artificial intelligence applied to hyperspectral and thermal sensing from space. In *CubeSats and SmallSats for Remote Sensing III*; International Society for Optics and Photonics: San Diego, CA, USA, 2019; Volume 11131, p. 111310C.

13. Benelli, G.; Meoni, G.; Fanucci, L. A low power keyword spotting algorithm for memory constrained embedded systems. In Proceedings of the 2018 IFIP/IEEE International Conference on Very Large Scale Integration (VLSI-SoC), Verona, Italy, 8–10 October 2018; pp. 267–272.

14. Reuther, A.; Michaleas, P.; Jones, M.; Gadepally, V.; Samsi, S.; Kepner, J. Survey and benchmarking of machine learning accelerators. *arXiv* **2019**, arXiv:1908.11348.

15. Wang, Y.E.; Wei, G.Y.; Brooks, D. Benchmarking TPU, GPU, and CPU platforms for deep learning. *arXiv* **2019**, arXiv:1907.10701.

16. Jouppi, N.P.; Young, C.; Patil, N.; Patterson, D.; Agrawal, G.; Bajwa, R.; Bates, S.; Bhatia, S.; Boden, N.; Borchers, A.; et al. In-datacenter performance analysis of a tensor processing unit. In Proceedings of the 44th Annual International Symposium on Computer Architecture, Toronto, ON, Canada, 24–28 June 2017; pp. 1–12.

17. Abadi, M.; Barham, P.; Chen, J.; Chen, Z.; Davis, A.; Dean, J.; Devin, M.; Ghemawat, S.; Irving, G.; Isard, M.; et al. Tensorflow: A system for large-scale machine learning. In Proceedings of the 12th USENIX Symposium on Operating Systems Design and Implementation (OSDI 16), Savannah, GA, USA, 2–4 November 2016; pp. 265–283.

18. Nickolls, J.; Dally, W.J. The GPU Computing Era. *IEEE Micro* **2010**, *30*, 56–69. [CrossRef]

19. Dinelli, G.; Meoni, G.; Rapuano, E.; Benelli, G.; Fanucci, L. An FPGA-Based Hardware Accelerator for CNNs Using On-Chip Memories Only: Design and Benchmarking with Intel Movidius Neural Compute Stick. *Int. J. Reconfigurable Comput.* **2019**, *2019*, 7218758. [CrossRef]

20. Baze, M.P.; Buchner, S.P.; McMorrow, D. A digital CMOS design technique for SEU hardening. *IEEE Trans. Nucl. Sci.* **2000**, *47*, 2603–2608. [CrossRef]

21. Sterpone, L.; Azimi, S.; Du, B. A selective mapper for the mitigation of SETs on rad-hard RTG4 flash-based FPGAs. In Proceedings of the 2016 16th European Conference on Radiation and Its Effects on Components and Systems (RADECS), Bremen, Germany, 19–23 September 2016; pp. 1–4.

22. Lowe, D.G. Distinctive Image Features from Scale-Invariant Keypoints. *Int. J. Comput. Vis.* **2004**, *60*, 91–110. doi:10.1023/B:VISI.0000029664.99615.94. [CrossRef]

23. Barry, B.; Brick, C.; Connor, F.; Donohoe, D.; Moloney, D.; Richmond, R.; O'Riordan, M.; Toma, V. Always-on Vision Processing Unit for Mobile Applications. *IEEE Micro* **2015**, *35*, 56–66. [CrossRef]

24. Rivas-Gomez, S.; Pena, A.J.; Moloney, D.; Laure, E.; Markidis, S. Exploring the vision processing unit as co-processor for inference. In Proceedings of the 2018 IEEE International Parallel and Distributed Processing Symposium Workshops (IPDPSW), Vancouver, BC, Canada, 21–25 May 2018; pp. 589–598.

25. Myriad2VPU. Available online: https://www.movidius.com/myriad2 (accessed on 7 July 2020)

26. Antonini, M.; Vu, T.H.; Min, C.; Montanari, A.; Mathur, A.; Kawsar, F. Resource Characterisation of Personal-Scale Sensing Models on Edge Accelerators. In Proceedings of the First International Workshop on Challenges in Artificial Intelligence and Machine Learning for Internet of Things, New York, NY, USA, 16–19 November 2019; pp. 49–55.

27. Li, W.; Liewig, M. A Survey of AI Accelerators for Edge Environment. In *World Conference on Information Systems and Technologies*; Springer: Berlin, Germany, 2020; pp. 35–44.

28. Deschamps, P.; Herman, M.; Tanre, D. Definitions of atmospheric radiance and transmittances in remote sensing. *Remote Sens. Environ.* **1983**, *13*, 89–92. [CrossRef]

29. Sinergise Ltd. Available online: https://www.sinergise.com/en (accessed on 7 July 2020)

30. Intel Movidius SDK. Available online: https://movidius.github.io/ncsdk/ (accessed on 7 July 2020)

31. Jia, Y.; Shelhamer, E.; Donahue, J.; Karayev, S.; Long, J.; Girshick, R.; Guadarrama, S.; Darrell, T. Caffe: Convolutional Architecture for Fast Feature Embedding. *arXiv* **2014**, arXiv:1408.5093.

32. Drusch, M.; Del Bello, U.; Carlier, S.; Colin, O.; Fernandez, V.; Gascon, F.; Hoersch, B.; Isola, C.; Laberinti, P.; Martimort, P.; et al. Sentinel-2: ESA's optical high-resolution mission for GMES operational services. *Remote Sens. Environ.* **2012**, *120*, 25–36. [CrossRef]

33. Hecht-Nielsen, R. Theory of the backpropagation neural network. In *Neural Networks for Perception*; Elsevier: Amsterdam, The Netherlands, 1992; pp. 65–93.

34. Hubara, I.; Courbariaux, M.; Soudry, D.; El-Yaniv, R.; Bengio, Y. Quantized neural networks: Training neural networks with low precision weights and activations. *J. Mach. Learn. Res.* **2017**, *18*, 6869–6898.

35. Krishnamoorthi, R. Quantizing deep convolutional networks for efficient inference: A whitepaper. *arXiv* **2018**, arXiv:1806.08342.

36. Lin, M.; Chen, Q.; Yan, S. Network in network. *arXiv* **2013**, arXiv:1312.4400.

37. Ji, T.Y.; Yokoya, N.; Zhu, X.X.; Huang, T.Z. Nonlocal tensor completion for multitemporal remotely sensed images' inpainting. *IEEE Trans. Geosci. Remote Sens.* **2018**, *56*, 3047–3061. [CrossRef]

38. Kang, J.; Wang, Y.; Schmitt, M.; Zhu, X.X. Object-based multipass InSAR via robust low-rank tensor decomposition. *IEEE Trans. Geosci. Remote Sens.* **2018**, *56*, 3062–3077. [CrossRef]

39. Chen, Y.; He, W.; Yokoya, N.; Huang, T.Z. Blind cloud and cloud shadow removal of multitemporal images based on total variation regularized low-rank sparsity decomposition. *ISPRS J. Photogramm. Remote Sens.* **2019**, *157*, 93–107. [CrossRef]

40. Badrinarayanan, V.; Kendall, A.; Cipolla, R. Segnet: A deep convolutional encoder-decoder architecture for image segmentation. *IEEE Trans. Pattern Anal. Mach. Intell.* **2017**, *39*, 2481–2495. [CrossRef] [PubMed]

41. Selva, D.; Krejci, D. A survey and assessment of the capabilities of Cubesats for Earth observation. *Acta Astronaut.* **2012**, *74*, 50–68. [CrossRef]

42. Yang, J.; Guo, J.; Yue, H.; Liu, Z.; Hu, H.; Li, K. Cdnet: Cnn-based cloud detection for remote sensing imagery. *IEEE Trans. Geosci. Remote Sens.* **2019**, *57*, 6195–6211. [CrossRef]

43. Rossow, W.B.; Garder, L.C. Cloud detection using satellite measurements of infrared and visible radiances for ISCCP. *J. Clim.* **1993**, *6*, 2341–2369. [CrossRef]

44. Gesell, G. An algorithm for snow and ice detection using AVHRR data An extension to the APOLLO software package. *Int. J. Remote Sens.* **1989**, *10*, 897–905. [CrossRef]

45. Wei, J.; Sun, L.; Jia, C.; Yang, Y.; Zhou, X.; Gan, P.; Jia, S.; Liu, F.; Li, R. Dynamic threshold cloud detection algorithms for MODIS and Landsat 8 data. In Proceedings of the 2016 IEEE International Geoscience and Remote Sensing Symposium (IGARSS), Beijing, China, 10–15 July 2016; pp. 566–569.

46. Zhong, B.; Chen, W.; Wu, S.; Hu, L.; Luo, X.; Liu, Q. A cloud detection method based on relationship between objects of cloud and cloud-shadow for Chinese moderate to high resolution satellite imagery. *IEEE J. Sel. Top. Appl. Earth Obs. Remote Sens.* **2017**, *10*, 4898–4908. [CrossRef]

47. Zhu, Z.; Woodcock, C.E. Automated cloud, cloud shadow, and snow detection in multitemporal Landsat data: An algorithm designed specifically for monitoring land cover change. *Remote Sens. Environ.* **2014**, *152*, 217–234. [CrossRef]

48. Latry, C.; Panem, C.; Dejean, P. Cloud detection with SVM technique. In Proceedings of the 2007 IEEE International Geoscience and Remote Sensing Symposium, Barcelona, Spain, 23–28 July 2007; pp. 448–451.

49. Deng, J.; Wang, H.; Ma, J. An automatic cloud detection algorithm for Landsat remote sensing image. In Proceedings of the 2016 4th International Workshop on Earth Observation and Remote Sensing Applications (EORSA), Guangzhou, China, 4–6 July 2016; pp. 395–399.

Article

Machine Learning Classification Ensemble of Multitemporal Sentinel-2 Images: The Case of a Mixed Mediterranean Ecosystem

Christos Vasilakos *, Dimitris Kavroudakis and Aikaterini Georganta

Department of Geography, University of the Aegean, 81100 Mytilene, Greece; dimitrisk@aegean.gr (D.K.); geo15024@geo.aegean.gr (A.G.)
* Correspondence: chvas@aegean.gr; Tel.: +30-22510-36451

Received: 21 May 2020; Accepted: 20 June 2020; Published: 22 June 2020

Abstract: Land cover type classification still remains an active research topic while new sensors and methods become available. Applications such as environmental monitoring, natural resource management, and change detection require more accurate, detailed, and constantly updated land-cover type mapping. These needs are fulfilled by newer sensors with high spatial and spectral resolution along with modern data processing algorithms. Sentinel-2 sensor provides data with high spatial, spectral, and temporal resolution for the in classification of highly fragmented landscape. This study applies six traditional data classifiers and nine ensemble methods on multitemporal Sentinel-2 image datasets for identifying land cover types in the heterogeneous Mediterranean landscape of Lesvos Island, Greece. Support vector machine, random forest, artificial neural network, decision tree, linear discriminant analysis, and k-nearest neighbor classifiers are applied and compared with nine ensemble classifiers on the basis of different voting methods. kappa statistic, F1-score, and Matthews correlation coefficient metrics were used in the assembly of the voting methods. Support vector machine outperformed the base classifiers with kappa of 0.91. Support vector machine also outperformed the ensemble classifiers in an unseen dataset. Five voting methods performed better than the rest of the classifiers. A diversity study based on four different metrics revealed that an ensemble can be avoided if a base classifier shows an identifiable superiority. Therefore, ensemble approaches should include a careful selection of base-classifiers based on a diversity analysis.

Keywords: remote sensing; classification ensemble; machine learning; Sentinel-2; geographic information system (GIS)

---

1. Introduction

Remote sensing image classification is considered among the main topics of remote sensing that aims to extract land cover types on the basis of the spectral and spatial properties of targets in a study area [1]. The land cover/land use mapping is essential for many applications from a local to a global scale, i.e., environmental monitoring and management, detection of global change, desertification evaluation, support decision making, urban change detection, landscape fragmentation, and tropical deforestation [2–4]. A vast amount of remote sensing data is archived and can be accessed freely or with low cost while new data become available every day for the whole planet. The rapid growth of computational approaches, the evolution of sensors' characteristics and the availability of satellite data have fueled the development of novel methods in image classification. The most widely used methods are the supervised ones, including the traditional approaches of maximum likelihood and minimum distance, as well as more recently, the modern machine learning classifiers, especially as pixel-based classifiers [5–10].

According to the literature, non-parametric methods tend to perform better compared to the parametric methods [11,12]. The majority of articles compare machine learning algorithms on the basis of their accuracy performance and their advantages and disadvantages, aiming to identify the best algorithm for each classification case [11]. The most used algorithms with a wide range of application in classification are: Random Forest (RF); Support Vector Machines (SVM), Artificial Neural Networks (ANN) and Decision Trees (DT). In a recent study by Ghamisi et al. [13], hyperspectral data are classified with the usage of various classifiers including, amongst others, SVM, RF, ANN, and logistic regression. The comparison focuses on speed, the setup of various parameters, and the competence of automation. None of the classifiers have a clear advantage in terms of speed or accuracy. However, there is a significant number of SVM studies that have ascertained that the SVM algorithm presents a higher classification accuracy than the other algorithms [14–16]. The mathematical model of the SVM theory can distribute and separate the data more accurately than methods such as ANN, Maximum Likelihood Classifier (MLC), and DT [17]. In contrast, Lapini et al. [18] in a forest classification study based on Synthetic Aperture Radar (SAR) images at a Mediterranean ecosystem of central Italy, compared six classifiers, i.e., RF, AdaBoost with Decision Trees, k-Nearest Neighbor (KNN), ANN, SVM, and Quadratic Discriminant. According to their results, in almost all examined scenarios, RF performed better, while SVM was sensitive to unbalanced classes. In another recent study, authors compared KNN, RF, SVM, and ANN in a classification of Landsat-8 Operational Land Imager (OLI) image data of arid desert-oasis mosaic landscapes. ANN performed marginally better that other classifiers, while the RF had a stabile performance across several aspects, i.e., stability, ease of use and overall processing time [19].

Some authors suggest a hybrid approach on base classifiers. For example, in a recent work, Dang et al. [20] suggest that a combination of Random Forest and Support Vector Machine, namely Random Forest Machine, gives more accurate results than each algorithm separate and constitutes a promising tool. The study compares the results of Random Forest Machine with those of Random Forest and Support Vector Machine and observes higher efficiency on accuracy classification by this new hybrid approach. This new algorithm seems to be a promising tool for future applications.

In another work, RF, KNN, and SVM were compared to a land use/cover study based on Sentinel-2 Multi-Spectral Instrument (MSI) [16]. Various training datasets of different sizes were tested, representing the six classes of the study area in the Red River Delta of Vietnam. All classifiers showed an overall accuracy up to 95% with SVM presenting the highest of all, while remaining less sensitive to training data size. Comparison of machine learning methods have been recently further explored in the classification of boreal landscapes with Sentinel-2 data. In particular, the algorithms of SVM, RF Xgboost, and deep learning have been implemented at the multi temporal image with the higher accuracy corresponding to the SVM algorithm, with a total accuracy of 75.8% [21]. Sentinel-2 data was also used in an object based classification by comparing AdaBoost, RF, Rotation Forest, and Canonical Correlation Forest (CCF) classifiers [22]. Three different datasets were developed. The first dataset included only the 10 m bands, the second dataset included the bands with 20 m resolution, and the third dataset included the 10 m and pansharpened version of the 20 m bands. According to the results, the Rotation Forest and Canonical Correlation Forest outperformed for all datasets.

Except from the usage of unitemporal images, multitemporal classification has been extensively applied for more accurate results in land use/cover extraction [23–28]. The main reason is the seasonal variance of the vegetation's spectral reflectance, which changes according to the season and the growing stage for each vegetation type. The limited spectral information of a single image can be compensated by using multiple dates of the same type of images [29]. Kamusoko [30] compared five machine learning methods KNN, ANN, DT, SVM, and RF in a single date and multidate images of Landsat 5, concluding that multidate and RF method provided the best results among other combinations. In another study applied in a highly heterogeneous fragmented area and in a homogenous mountain area, the combination of maximum likelihood and multidate Sentinel-2 data performed better that SVM. The multidate input dataset was able to distinguish the classes of the highly fragmented area despite the spectral similarities between classes [31]. Thus, the multitemporal data are essential for discriminating the vegetation types, resulting in higher classification accuracy results [32].

Most remote sensing classification studies have relied on a single classifier or a comparison of a number of them [33–35]. Since all classifiers perform within an accuracy range, an ensemble approach may show improved accuracy levels and increased reliability in remote sensing image classification [36]. To this end, several methods are reported in the literature to address the issue of how to develop an ensemble classifier that combines the decisions from multiple base-classifiers [37–41] that can be used either on hard or soft classifications [42]. Three categories of methods can be identified in the literature [36,43]: (i) algorithms that are based on training data manipulation including the well-known "bagging" and "boosting" [44,45] applied on a single based classifiers, i.e., SVM and DT [46,47], (ii) algorithms that are based on a chain of classifiers that perform in a sequential mode, i.e., the output of a classifier is the input for the next one in the chain, and (iii) algorithms that are based on parallel processing of the base classifiers and the combination of their outputs. The main method to combine the decisions of the base classifiers is a weighted or unweighted voting [48,49]. The weights usually depend on the majority, the estimated probability and the accuracy metrics of the base classifiers. Shen et al. [1] compared the producer's accuracy and overall accuracy and they concluded that the overall accuracy had stability issues, while the producer's accuracy performed better in the classification of different land cover types.

This paper aims to apply a number of machine learning approaches, i.e., DT, Linear Discriminant Analysis (DIS), SVM, KNN, RF, and ANN to classify multitemporal Sentinel-2 images and add to whether an ensemble of these base classifiers can further enhance the output accuracy. The classification is applied to an insular environment at the Mediterranean coastal region. Even if various studies have been conducted for Mediterranean environments, an ensemble classification on multitemporal Sentinel-2 data, to the best of our knowledge, has not been examined for this type of ecosystem. Previous studies were focused either on specific types, i.e., on applying machine learning on forested areas [50] or wetlands [51]. Our implementation is somehow different. Each one of the base classifiers uses its own validation dataset rather than a common one, while the final evaluation of the ensemble is compared to base classifiers by using a common and unseen testing dataset.

2. Materials and Methods

This chapter presents a detailed description of our study area of Lesvos Island, Greece. A thorough description of the input data and the classification methods is followed by the accuracy metrics. Finally, the ensemble voting methods, the diversity measures, and the accuracy metrics are analytically presented.

2.1. Study Area and Data

The island of Lesvos is located at the northeastern Aegean Sea of Greece and covers an area of 1636 km^2 and the total length of shore 382 km. The island has a variety of geological formations, climatic conditions, and vegetation types (Figure 1). The climate conditions are categorized as "Mediterranean", with warm and dry summers and mild and moderately rainy winters. Annual precipitation average is 710 mm; the average annual air temperature is 17 °C with high oscillations between maximum and minimum daily temperatures. The terrain is rather hilly and rough, with a highest peak of 960 m a.s.l. Slopes greater than 20% are dominant, covering almost two-thirds of the island. The soils of Lesvos are widely cultivated, mainly with rain-fed crops such as cereals, vines, and olives.

Due to low productivity, many sites were abandoned 50–60 years ago; after abandonment, these areas were moderately grazed, and the shrub regeneration has been occasionally cleared by illegal burning to improve forage production [52]. The vegetation of these areas, defined on the basis of the dominant species, includes phrygana or garrigue-type shrubs in grasslands, evergreen-sclerophylous or maquis-type shrubs, pine forests, deciduous oaks, olive groves, and other agricultural lands.

Figure 1. The Lesvos Island at the north-east Aegean sea (source of location panels: Esri).

In order to perform the classification, three cloud-free satellite images were retrieved in JPEG 2000 format from Copernicus Open Access Hub [53], acquired by the Sentinel-2A (S2A) and Sentinel-2B (S2B) MSI satellites. The dataset consists of the dates 28/04/2018 (S2A), 12/07/2018 (S2B), and 04/11/2018 (S2A) and the product type is Level-2A. We selected three images of spring, summer, and autumn for our multitemporal approach. According to previous works, a combination of spring, summer, and autumn image provides the highest classification accuracy and high class separability [27,54]. Level-2A products are radiometrically, atmospherically and geometrically corrected, providing the bottom of atmosphere (BOA) reflectance in Universal Transverse Mercator (UTM)/WGS84 projection. We used 10 bands (Table 1) out of 13 available. The final image composition includes in total 30 bands.

Table 1. Spatial and spectral resolution of Sentinel-2.

Band	Central Wavelength (nm)	Bandwidth (nm)	Spatial Resolution (m)
Band 2—Blue	490	65	10
Band 3—Green	560	35	10
Band 4—Red	665	30	10
Band 5—Vegetation red edge	705	15	20
Band 6—Vegetation red edge	740	15	20
Band 7—Vegetation red edge	783	20	20
Band 8—Near infrared	842	115	10
Band 8A—Narrow near infrared	865	20	20
Band 11—Shortwave infrared	1610	90	20
Band 12—Shortwave infrared	2190	180	20

2.2. Methodology

The methodology consists of three stages: the ground truth data collection, the classification by applying six base classifiers including the estimation of accuracy and diversity metrics, and finally the application of nine ensemble voting methods.

2.2.1. Ground Truth Data Collection

As used by several other studies [55–57] the input dataset was created by visual interpretation of Google Earth's very high resolution images among with auxiliary data collected during field trips and a land cover map that was previously produced on the basis of a Worldview-2 image. A total of 1,119 homogenous polygons were identified and outlined with a total area of 127.4 km^2 across the island (Table 2). Within these polygons, random points were created to extract the values from the 30 layers of the image composition. The next step was to randomly split the dataset into training and testing partitions based on the 80% and the 20% of the initial cases. The training dataset was used to train all base classifiers and was further randomly split into a secondary training and validation dataset including the 75% and the 25% respectively.

Figure 2 presents the multitemporal spectral responses per class. The data of Figure 2 reveals significant differences in the spectral signatures within the date range especially in the near infrared (NIR) region except for pine forest. Furthermore, these data also address the phenology stages of deciduous, i.e., chestnut trees and agricultural lands. The variation of chlorophyll content in vegetation results in a significant variation of reflectance especially in infrared bands. These variations cause different phenological patterns for each vegetation cover type. According to previous studies, the different phenologies as described by these spectral responses is expected to improve the classification accuracy values compared to a single-date image especially in study areas where crops and vegetation are the dominant land cover types [58].

Table 2. Number of pixels per class and dataset.

Class	Number of Polygons	Number of Training Dataset	Number of Testing Dataset	Total Samples
Olive grove	342	3203	797	4000
Oak forest	72	1309	327	1636
Brushwood	73	4254	1088	5342
Built up	107	984	257	1241
Pinus brutia	113	4257	1090	5347
Chestnut forest	31	778	178	956
Pinus nigra	20	349	82	431
Maquis-type shrubland	107	936	226	1162
Barren land	44	212	61	273
Grassland	127	1327	312	1639
Other broadleaves	8	234	50	284
Agricultural land	54	729	188	917
Aquatic bodies	21	1070	254	1324
Total	1119	19642	4910	24552

2.2.2. Base Classifiers

On the basis of the literature, six base classifiers have been selected for the case study. A widely used non-parametric approach is the decision tree (DT) classifier characterized also by its intuitive simplicity [59]. Within DT the input data is recursively split, based on a set of rules, into smaller and more homogenous groups forming the branches of the tree until the end-nodes, which are the target values. In our case, these target nodes forming the leaves represent the classes. One of the major advantages of DT is that it does not have any prerequisites about input data distribution. Moreover, a good generalization can be achieved by pruning the DT that means to remove some branches or turning some branches into leaves. Therefore, pruning will increase the accuracy by avoiding overfitting [11,60].

Another widely used classification approach, which is based on Fisher's score optimization, is the DIS [61]. DIS approaches has been extensively used in the classification of hyperspectral data, either the initial or modified methods [62–64]. One of the major drawbacks of DIS in hyperspectral classification is that these data are ill-posed when the number of data are less than the number of bands. In our case, our data are more than sufficient to avoid this phenomenon during the classification of a 30-dimension space.

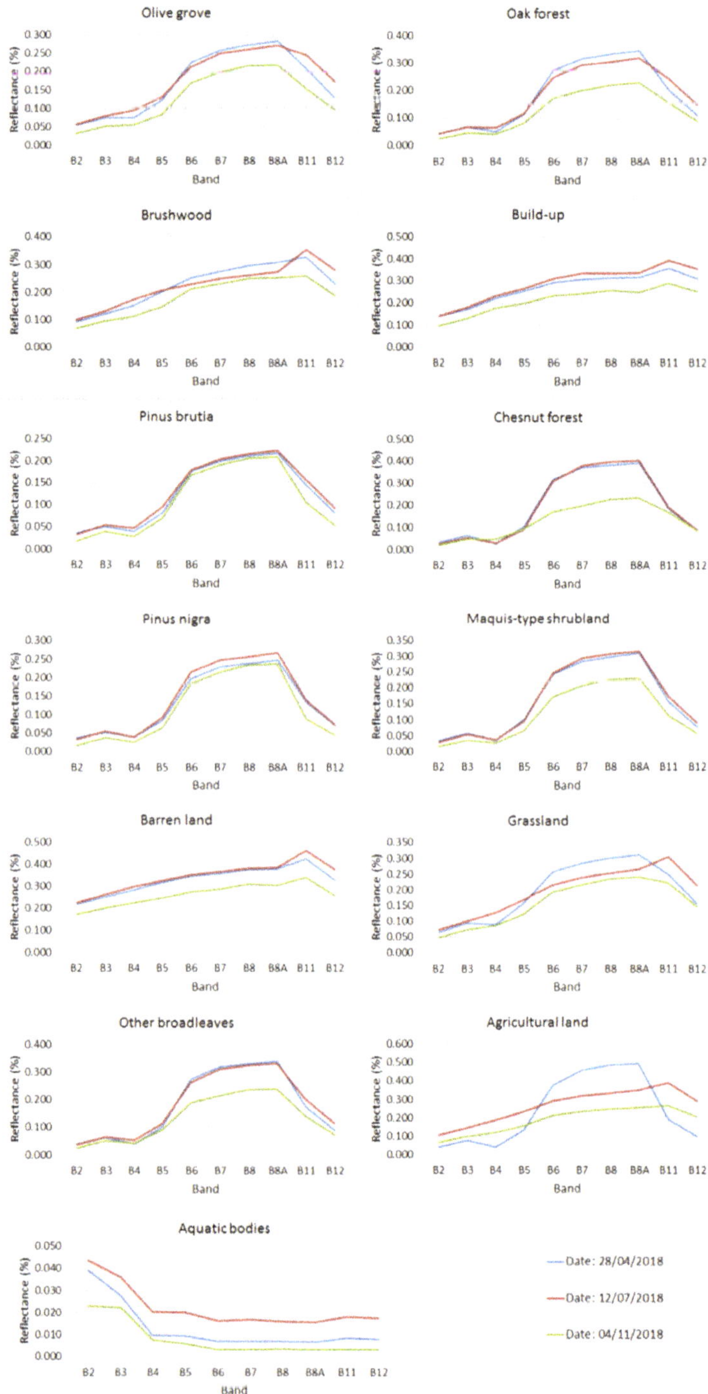

Figure 2. Spectral reflectance per class and per date.

A third method that we applied is the Support Vector Machine (SVM) which aims to find a hyperplane that separates categorical data in a high dimension space with the maximum possible margin between the hyperplane and the cases [65]. The cases that are closest to the hyperplane are called support vectors. However, in most cases, the classes are not linearly separable, hence a slack variable is included, and a kernel function is used to perform a non-linear mapping into the feature space. The most widely used kernels in remote sensing applications are the polynomials and radial basis functions (RBF) [11,66].

The forth approach was a k-Nearest Neighbor (KNN) classifier. This algorithm calculates the distances between an unclassified case and the nearest k training cases and classifies the unclassified case to the majority class of the nearest k training cases. The user can choose from a variety of distance metrics, however, the most widely used is the Euclidean Distance which can be applied either unweighted or with a weight [67].

The next applied classification method was the random forest (RF). The concept of DTs is expanded and enhanced through the RF algorithm. Multiple DTs are trained on the basis of a subset of the training data where each one is trained on the basis of its own random sample. A majority vote of all the DTs defines the final class of each case. One of the advantages is that RF does not make any assumption about the probability distribution of the input data [13]. A more detailed description of the RF algorithm in remote sensing applications can be found in [68–70].

Finally, we developed and applied an artificial neural network (ANN) classifier. ANNs have been very popular and have been extensively used in pattern recognition and in modeling complex problems. In the last 30 years, ANNs play a fundamental role in remote sensing land cover classification applications [71–74], while the new trend in the classification of very high resolution images are the convolutional neural networks (CNN) [75,76]. The CNNs have proven, in the last years, to be very powerful classifiers in image recognition, object detection, image segmentation, and instance segmentation [77]. However, CNNs are fundamentally based on spatial-contextual dependencies of the input data with the majority of them being trained on high resolution RGB images [78,79]. Opposite to patch-based CNNs, pixel-based CNNs have been developed. However, according to the literature, the common problems of pixel-based classification, i.e., the salt and pepper effect and the boundary fuzziness effect within the classification result are quite severe in CNN implementations [80]. Other disadvantages of CNNs are the higher processing time and resources. We believe that, for the pixel-based hard classification of the present multispectral and multitemporal approach, ANNs are more suitable given also the nature of the other base classifiers. ANNs are characterized by their architecture, their training algorithm, and their activation function. The most well-known and efficient type is the Multilayer Perceptron (MLP) with three layers: input, hidden, and output, while ANNs with one hidden layer are able to map any nonlinear function. Various gradient descent learning methods have been proposed.

The evaluation, comparison, and voting during the ensemble of the applied methods were based on the below metrics:

$$verall\ Accuracy\ (OA) = \frac{TP + TN}{TP + FP + TN + FN} \tag{1}$$

$$User's\ accuracy\ (UA) = \frac{TP}{TP + FN} \tag{2}$$

$$Producer's\ accuracy\ (PA) = \frac{TP}{TP + FP} \tag{3}$$

$$kappa = \frac{p_0 - p_e}{1 - p_e}\ with \tag{4}$$

$$p_0 = \frac{TP + TN}{TP + TN + FP + FN}\ and \tag{5}$$

$$p_e = \frac{(TP + FN) \times (TP + FP) + (FP + TN) \times (FN + TN)}{(TP + TN + FP + FN)^2} \tag{6}$$

$$F1 - score = \frac{2 \times TP}{2 \times TP + FP + FN} \tag{7}$$

$$MCC = \frac{TP \times TN - FP \times FN}{\sqrt{(TP + FP) \times (TP + FN) \times (TN + FP) \times (TN + FN)}} \tag{8}$$

where TP: true positive, TN: true negative, FP: false positive, and FN: false negative. The overall accuracy (OA) is a single and very basic summary measure of the probability of a case being correctly classified and is based on the sum of the diagonal elements of the confusion matrix. User's accuracy (UA) and producer's accuracy (PA) provide an accuracy performance for each class. UA is a performance measure of the credibility of the output map that is how well the map represents the actual cover types. On the other hand, PA measures the accuracy of how well the reference data is represented by the map. UA and PA are related with the commission and the omission errors respectively. The kappa coefficient on the other hand is a more advanced metric, which compares the observed accuracy against random chance. Opposite to OA, the kappa coefficient takes also into consideration the non-diagonal elements. Furthermore, the F1-score is a rather different measure of accuracy, defined as the weighted harmonic mean of both classification's precision and recall. It balances the use of precision and recall and provides a more realistic measure of performance. Finally, the Matthews correlation coefficient (MCC) is a more balanced metric that takes into account all parts of the confusion matrix and can handle under-represented classes. Each classifier produced a contingency matrix presenting the classification results of its validation dataset and the corresponding accuracy metrics. It should be noticed that for the calculation of the kappa, F1, and MCC for each class, each confusion matrix was converted to multiple binary matrices based on the 'one-vs-all' scheme.

Moreover, four different diversity statistics were calculated on the basis of the results of the base classifiers to the testing datasets, as depicted by the following 2×2 table of the relationship between a pair of classifiers C_i and C_k (Table 3) [81].

Table 3. Relational table between a pair of classifiers.

	C_i **Correct (1)**	C_i **Wrong (0)**
C_k **correct (1)**	N^{11}	N^{10}
C_k **wrong (0)**	N^{01}	N^{00}

Where N^{11} is the correctly classified cases by both classifiers, N^{10} is the correctly classified cases by C_k classifier, N^{01} is the correctly classified cases by Ci classifier, and N^{00} is the incorrectly classified cases by both classifiers. The diversity measures were the Q-statistic, the disagreement measure, the double-fault measure, and the inter-kappa statistic given by [36,81–83]:

$$Inter - kappa\ measure = \frac{2\left(N^{11}N^{00} - N^{01}N^{10}\right)}{(N^{11} + N^{10})(N^{01} + N^{00}) + (N^{11} + N^{01})(N^{10} + N^{00})} \tag{9}$$

$$Q_{i,k}\ statistic = \frac{N^{11}N^{00} - N^{01}N^{10}}{N^{11}N^{00} + N^{01}N^{10}} \tag{10}$$

$$disagreement\ measure = \frac{N^{01} + N^{10}}{N^{11} + N^{10} + N^{01} + N^{00}} \tag{11}$$

$$double - fault\ measure = \frac{N^{00}}{N^{11} + N^{10} + N^{01} + N^{00}} \tag{12}$$

2.2.3. Ensemble Voting Methods

After obtaining results from the six classifiers, an ensemble classifier should be constructed in order to classify the 4,910 cases of the testing dataset. For each testing case, all the base classifiers provided a prediction, and we tested nine different voting schemes to further evaluate:

- **'Mode'**: This voting method selects the suggestion with greater frequency in the six suggestions. In the cases with equal frequency, it selects the one with the higher sum of kappa.
- **'Max kappa'**: This voting method selects the suggestion with greater kappa.
- **'Greater Sum of Kappa'**: This voting method selects the suggestion with the greater sum of kappa aggregated on the suggestions. Identical suggestions are summed up and then compared with all other kappa values.
- **'Greater Mean Kappa'**: This method selects the suggestion with greater average kappa per suggestion. Identical suggestions are averaged and then compared with all other suggestions.
- **'Greater Weighted Sum Kappa'**: This method calculates the weighted sum of kappa which is the multiplication of the sum of kappa over the frequency of each suggestion group. Then, it selects the suggestion with the greater weighted sum of kappa.
- **'Greater mean F1'**: This voting method evaluates the average F1-score per suggestion and selects the one with the greater average F1. After grouping suggestions, we estimate the average F1 by group and compare the results. The result will be the one with the one with greater average F1-score.
- **'Greater sum F1'**: This voting method selects the suggestion with a greater aggregation of F1. After grouping the suggestions, we calculate the summation of F1 per group and compare the results. The result will be the suggestion group with the greater sum of F1-score.
- **'Greater mean MCC'**: This voting method evaluates the average MCC per suggestion group and then selects the one with the greater "mean MCC". After grouping the suggestions, we average their MCC value and compare the results. The result will be the suggestion group with the greater average MCC
- **'Greater sum MCC'**: This last voting method selects the suggestion with the greater average MCC. After grouping suggestions, we evaluate the summation of MCC per group before evaluating the result. The result will be the suggestion group with the greater sum of MCC.

For all the above voting methods, the metrics of kappa, F1, and MCC are the ones of the corresponding metrics calculated for each class of the validation dataset during the training phase. The final comparison and selection of the best voting method was based on the kappa value. It should be noticed that even if we have computed the OA, we did not use it during the ensemble phase. Due to the imbalanced input dataset, the overall accuracy does not have an adequate performance, thus we used the kappa, the F1-score, and the MCC.

The nine ensemble models were further statistically compared by applying McNemar's test [84]. McNemar's test has been widely used in comparison of classifiers performances [85]. All models were compared in pairs and the McNemar's value was given by:

$$\text{McNemar's value} = \frac{\left(|n_{01} - n_{10}| - 1\right)^2}{n_{01} + n_{10}} \tag{13}$$

where n_{01} is the number of samples misclassified only by algorithm A and n_{10} is the number of samples misclassified only by algorithm B. The null hypothesis is that both of the classification methods have the same error rate. McNemar's test is based on a x^2 test with one degree of freedom, where the critical x^2 value with a 95% confidence interval and a 5% level of significance is 3.841. If the computed McNemar's value for each pair is more than 3.841, then the null hypothesis is rejected, therefore, the two classification methods are significantly different.

The ArcGIS 10.2 [86] was used for the spatial processing and visualization of the data, the Matlab 2018a [87] for the base classifications, while the ensemble of the classifiers through the voting methods was carried out using R, including the packages caret, dplyr, and magrittr [88–91]. Figure 3 presents the overall workflow of the current research.

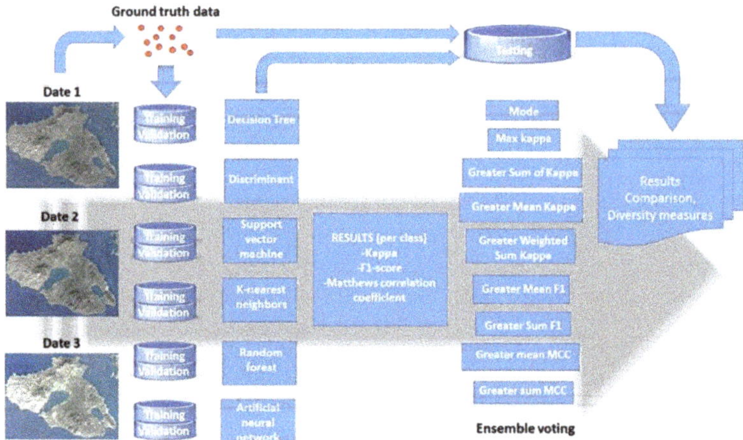

Figure 3. Workflow of the method followed for data classification and ensemble voting.

3. Results and Discussion

Each base classifier was carefully designed and trained under different settings. This section presents the training results of the base classifiers and the classification ensemble. A diversity analysis of the base classifiers and a significant test of the voting methods provide a more comprehensive view of the results.

3.1. Base Classifiers Training

For the training of the DT classifiers we tested 4, 10, and 100 as the maximum number of splits. The best results obtained with 100 maximum splits based on Gini's diversity index. The SVM classifier was trained with three different kernels; linear, RBF, quadratic and cubic. The cubic kernel provided the best accuracy and was further analyzed. We are aware that our approach is a multiclass imbalanced problem. It was decided that the best approach was to apply a "one-vs-one" instead of "one-vs-all" coding scheme in order to reduce the effect of the imbalance problem [92,93]. At the same time, the kappa, UA, PA, F1, and MCC provide a better interpretation of accuracy in an imbalanced dataset opposed to the overall accuracy. For the KNN we tested 1, 10, 100 nearest neighbors and the best results were provided with 10 neighbors. During the training, we applied the Euclidean distance unweighted and with a weight, but the overall accuracy increased when we used as a weight the inverse square of the distance for each case. The RF model tested with 30, 40, 50, 70, and 100 trees. The final model included 30 trees based on overall accuracy. Finally, during ANN training we applied different architectures with one hidden layer with 16 to 35 hidden nodes. We also tested two gradient descent learning methods: the Levenberg Marquardt [94] and the scaled conjugate gradient [95]. Each network was trained 10 times with different random initial weights. The model with the best performance had 16 hidden nodes and was trained with a scaled conjugate gradient for 272 epochs.

The classification accuracy of each classifier was evaluated based on the confusion matrix of the validation dataset presented in Tables A1–A6. Table 4 shows the user's (UA), producer's (PA) accuracy per class and the OA for each classifier while Figure 4 presents the heat map of UA and PA where the colors are normalized for each classifier. According to the results, the SVM outperformed all the classifiers according to the OA and the kappa (Figure 5). In most of the land cover classes, SVM presented the lower omission and commission errors while the DT and the DIS had the poorest performance with kappa 0.79 and 0.83, respectively. Aquatic bodies were almost perfectly classified by all classifiers, while brushwood, built up, *Pinus brutia*, and agricultural land classes also showed high accuracy. Figure 6 presents the diversity of UA and PA among all classifiers for each class. The base classifiers had significant different performances in omission error for other broadleaves, barren land,

and grassland classes and different performances in the commission error for oak forest, *Pinus nigra*, and other broadleaves. It should be noticed that the UA of SVM for other broadleaves is an outlier, i.e., its value is more than 1.5 times the interquartile range above the upper quartile.

Table 4. User accuracy (UA) and producer accuracy (PA) per class [1] for the validation dataset during the training phase of the base classifiers [2].

	DT		DIS		SVM		KNN		RF		ANN	
	UA	PA	UA	PA	UA	PA	UA	PA	UA	PA	UA	PA
OG	0.77	0.78	0.84	0.87	**0.93**	**0.92**	0.84	0.89	0.90	0.85	0.91	0.88
OF	0.54	0.61	0.52	0.72	**0.79**	**0.85**	0.77	0.71	0.72	0.81	0.71	0.72
BW	0.95	0.87	0.91	0.91	0.97	**0.96**	**0.98**	0.92	0.97	0.93	0.96	0.94
BU	0.91	0.92	0.86	**0.98**	**0.98**	0.97	0.90	0.96	**0.98**	0.92	0.93	0.93
PB	0.97	0.89	**0.99**	0.88	0.98	**0.98**	0.98	0.94	0.98	0.95	0.97	0.93
CF	0.85	0.72	0.85	0.75	**0.89**	**0.90**	0.83	0.82	**0.89**	0.83	0.85	0.80
PN	0.37	0.53	0.46	0.73	**0.83**	0.82	0.67	0.76	0.66	**0.88**	0.57	0.79
MS	0.51	0.66	0.65	0.67	**0.79**	**0.75**	0.67	0.70	0.72	0.73	0.62	0.70
BL	0.76	0.82	0.85	0.65	**0.87**	**0.96**	0.81	0.86	0.72	**0.97**	0.82	0.85
GL	0.53	0.63	0.79	0.66	**0.83**	**0.84**	0.75	0.78	0.71	0.82	0.75	0.80
OB	0.22	0.52	0.19	0.46	**0.64**	0.76	0.21	**0.92**	0.34	0.91	0.30	0.53
AL	0.92	0.91	0.90	0.97	**0.95**	**0.98**	0.93	0.97	0.93	0.94	0.93	0.93
AB	1	1	1	1	1	1	0.99	1	1	1	1	1
Overall Accuracy	0.82		0.85		0.93		0.89		0.90		0.89	

[1] With bold fonts the maximum UA and PA per class. OG: olive grove, OF: oak forest, BW: brushwood, BU: built up, PB: *Pinus brutia*, CF: chesnut forest, PN: *Pinus nigra*, MS: maquis-type shrubland, BL: barren land, GL: grassland, OB: other broadleaves, AL: agricultural land, AB: aquatic bodies. [2] DT: decision tree, DIS: discriminant, SVM: support vector machine, KNN: k-nearest neighbors, RF: random forest, ANN: artificial neural network.

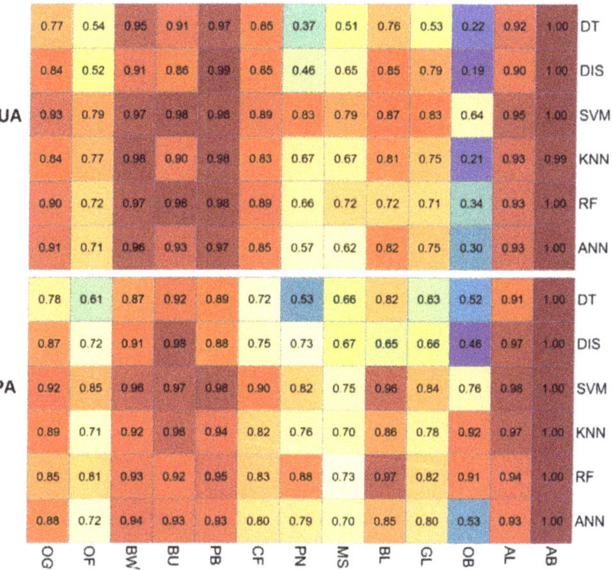

Figure 4. User accuracy (UA) and producer accuracy (PA) heatmap for the validation dataset during the training phase of the base classifiers.

Results are consistent with what has been found in previous studies. Shang et al. [96] applied SVM, RF and AdaBoost for the classification over an Australian eucalyptus forest. According to their results all three machine-learning algorithms outperformed the results produced by DIS. DT and DIS methods have also shown a poor performance in the comparison for the classification of Sentinel-2 data where RF outperformed followed by SVM and ANN [97]. However, the diversity of the results in the literature, reveals that the applied methods are data-driven and depended on the classification scheme, the number of training data and the type input data i.e. whether only the bands are taken into account or vegetation indices and other auxiliary data are used [19].

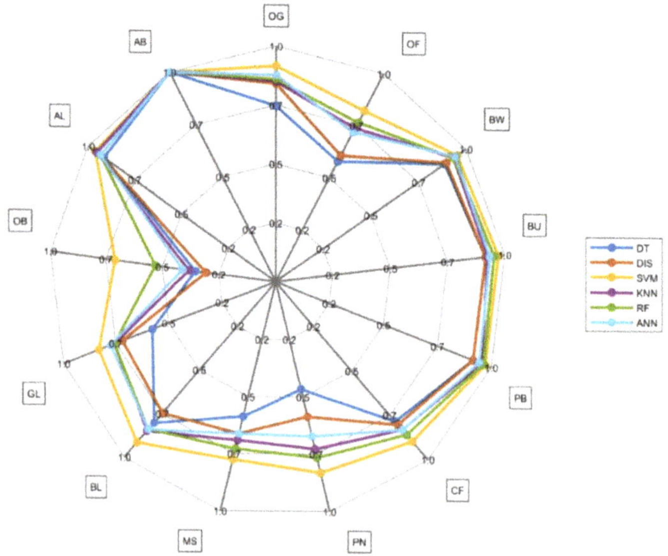

Figure 5. Kappa coefficient per class for the validation dataset during the training phase of the base classifiers.

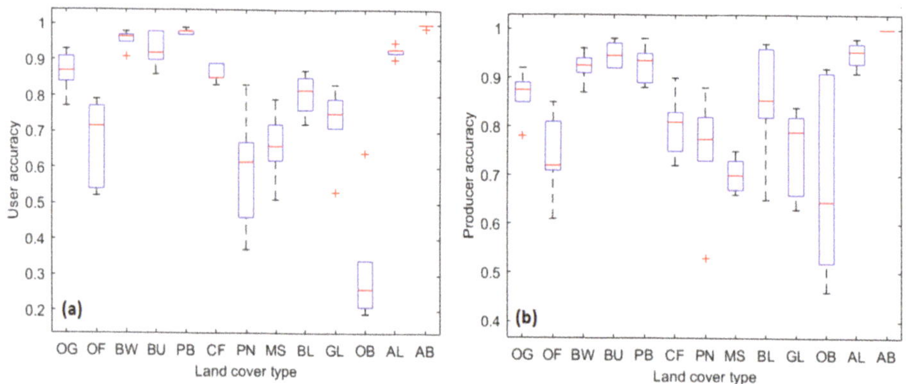

Figure 6. Distribution of (**a**) user's accuracy and (**b**) producer's accuracy per land cover class for the validation dataset.

3.2. Classification Ensemble

During the ensemble, we tested the nine voting methods and the base classifiers with the testing dataset. Figure 7 shows the k coefficient for the base classifiers and the voting methods applied in the testing dataset. According to the results, the SVM outperforms not only the base classifiers but also all the voting methods. However, all the voting methods based on sums as well as the method based on the majority of the votes performed better that all the rest of the base classifiers. It is worth mentioning that DT present the lower k among all classifiers, while DIS has the second to last performance.

Table 5 presents the kappa coefficient per class for each classifier for the testing dataset, while the Figure 8 presents the corresponding heatmap. It is observed that SVM shows a better performance in almost all classes. The voting methods based on sums and the majority of the votes performed slightly better in the built-up class. The confusion matrices of the testing dataset of the best classifier and the best ensemble methods are presented in Tables A7–A9. According to the results, it is evident that the combination of the classifiers does not provide always a better performance compared to the base classifiers. In a crop classification study in a fragmented arable landscape by Salas et al. [98], the authors concluded that when no classifier is clearly performing better than the others then an ensemble approach can be the best alternative. In our case SVM, RF, ANN, and KNN show a similar performance, however the SVM method performs better than all the applied voting methods. Therefore, a diversity study was applied in order to identify any potential dissimilar performance between SVM and the other base classifiers.

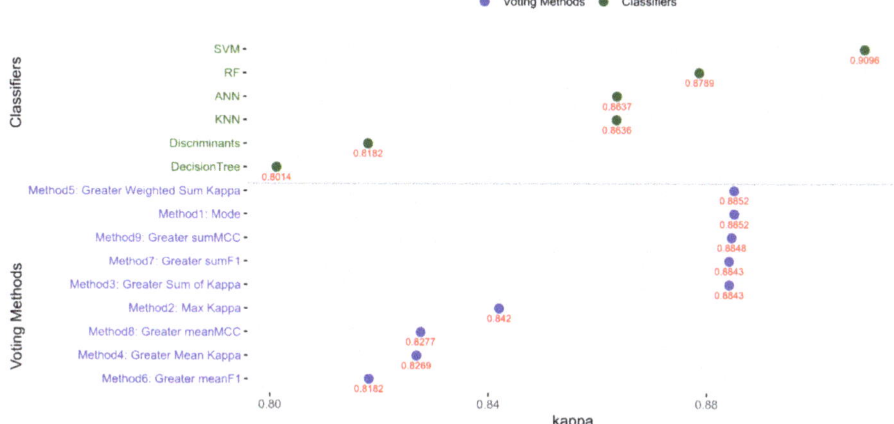

Figure 7. Kappa coefficients of the base classifiers and the voting methods for the testing datasets.

Table 5. Kappa coefficient per class [1] of all classifiers [2] for the testing dataset.

	OG	OF	BW	BU	PB	CF	PN	MS	BL	GL	OB	AL	AB
DT	0.72	0.61	0.88	0.90	0.93	0.81	0.40	0.59	0.77	0.54	0.32	0.91	**1.00**
DIS	0.84	0.59	0.88	0.91	0.91	0.75	0.51	0.64	0.75	0.62	0.05	0.93	**1.00**
SVM	**0.90**	**0.83**	**0.94**	0.95	**0.97**	**0.87**	**0.83**	**0.80**	**0.87**	**0.78**	**0.68**	0.99	**1.00**
KNN	0.84	0.74	0.91	0.89	0.95	0.82	0.72	0.70	0.78	0.71	0.35	0.98	**1.00**
RF	0.84	0.75	0.93	0.94	0.96	0.84	0.73	0.72	0.84	0.72	0.54	0.96	**1.00**
ANN	0.86	0.71	0.92	0.94	0.95	0.82	0.53	0.69	0.86	0.70	0.33	0.95	**1.00**
Mode	0.87	0.76	0.93	**0.96**	0.95	0.85	0.68	0.74	0.85	0.73	0.40	0.96	**1.00**
MaxK	0.82	0.74	0.91	0.94	0.90	0.80	0.42	0.69	0.75	0.58	0.00	0.93	**1.00**
GSK	0.87	0.76	0.93	**0.96**	0.95	0.84	0.68	0.74	0.85	0.73	0.30	0.96	**1.00**
GMK	0.79	0.64	0.90	0.93	0.90	0.75	0.42	0.68	0.72	0.58	0.04	0.93	**1.00**
GWSK	0.87	0.76	0.93	**0.96**	0.95	0.85	0.68	0.74	0.85	0.73	0.40	0.96	**1.00**
GMF1	0.78	0.63	0.89	0.89	0.90	0.75	0.40	0.68	0.72	0.54	0.04	0.93	**1.00**
GSF1	0.87	0.76	0.93	**0.96**	0.95	0.84	0.68	0.74	0.85	0.73	0.30	0.96	**1.00**
GMMCC	0.79	0.64	0.90	0.93	0.90	0.75	0.42	0.68	0.73	0.59	0.04	0.93	**1.00**
GSMCC	0.87	0.76	0.93	**0.96**	0.95	0.84	0.68	0.74	0.85	0.73	0.35	0.96	**1.00**

[1] With bold fonts the maximum kappa per class. [2] Mode: mode, maxK: max kappa, GSK: greater sum of kappa, GMK: greater mean kappa, GWSK: greater weighted sum kappa, GMF1: greater mean F1, GSF1: greater sum F1, GMMCC: greater mean MCC, GSMCC: greater sumMCC.

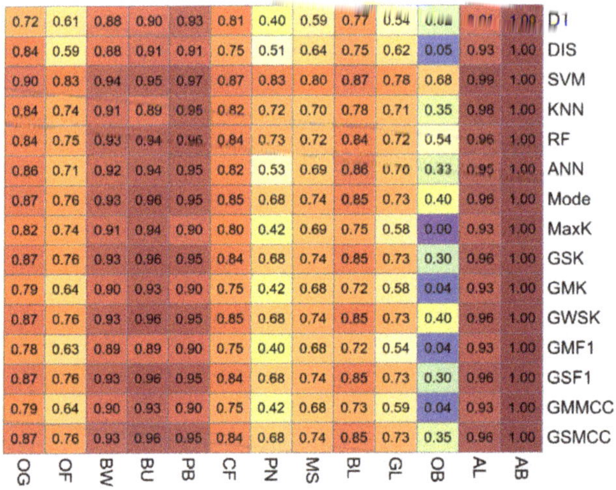

	OG	OF	BW	BU	PB	CF	PN	MS	BL	GL	OB	AL	AB
DT	0.72	0.61	0.88	0.90	0.93	0.81	0.40	0.59	0.77	0.54	0.08	1.00	1.00
DIS	0.84	0.59	0.88	0.91	0.91	0.75	0.51	0.64	0.75	0.62	0.05	0.93	1.00
SVM	0.90	0.83	0.94	0.95	0.97	0.87	0.83	0.80	0.87	0.78	0.68	0.99	1.00
KNN	0.84	0.74	0.91	0.89	0.95	0.82	0.72	0.70	0.78	0.71	0.35	0.98	1.00
RF	0.84	0.75	0.93	0.94	0.96	0.84	0.73	0.72	0.84	0.72	0.54	0.96	1.00
ANN	0.86	0.71	0.92	0.94	0.95	0.82	0.53	0.69	0.86	0.70	0.33	0.95	1.00
Mode	0.87	0.76	0.93	0.96	0.95	0.85	0.68	0.74	0.85	0.73	0.40	0.96	1.00
MaxK	0.82	0.74	0.91	0.94	0.90	0.80	0.42	0.69	0.75	0.58	0.00	0.93	1.00
GSK	0.87	0.76	0.93	0.96	0.95	0.84	0.68	0.74	0.85	0.73	0.30	0.96	1.00
GMK	0.79	0.64	0.90	0.93	0.90	0.75	0.42	0.68	0.72	0.58	0.04	0.93	1.00
GWSK	0.87	0.76	0.93	0.96	0.95	0.85	0.68	0.74	0.85	0.73	0.40	0.96	1.00
GMF1	0.78	0.63	0.89	0.89	0.90	0.75	0.40	0.68	0.72	0.54	0.04	0.93	1.00
GSF1	0.87	0.76	0.93	0.96	0.95	0.84	0.68	0.74	0.85	0.73	0.30	0.96	1.00
GMMCC	0.79	0.64	0.90	0.93	0.90	0.75	0.42	0.68	0.73	0.59	0.04	0.93	1.00
GSMCC	0.87	0.76	0.93	0.96	0.95	0.84	0.68	0.74	0.85	0.73	0.35	0.96	1.00

Figure 8. Kappa coefficients of the base classifiers and the voting methods for the testing datasets.

Table 6 presents the result of the four diversity statistics for all the possible pairs of the base classifiers for the testing dataset. According to the inter-kappa measure (Table 6a), all classifiers show a moderate agreement between them, except for SVM, which shows a fair agreement with DT and DIS and a moderate agreement with the rest of the classifiers. The high values of the Q-statistic (Table 6b) and the low values of the disagreement measure (Table 6c) suggest that there is not any significant diversity of the classifiers. The same conclusion results from the double-fault measure (Table 6d). However, from Figure 9, it is evident that the SVM presents a diverse performance especially based on the double-fault and the inter-kappa measures (Figure 9a,d). SVM's double-fault measures are tightly grouped while the values are quite low. Furthermore, the group of SVM's inter-kappa measures are lower, while the rest of the classifiers have a similar performance. Therefore, from the combination of the classification performance of the base classifiers with the diversity results is revealed that a voting method does not provide a better performance when a base classifier has a small but identifiable better performance than the rest of the classifiers.

Table 6. Diversity measures for all the possible pairs of base classifiers (**a**) inter-kappa measure, (**b**) Q-statistic, (**c**) disagreement measure, and (**d**) double-fault measure.

	DT	DIS	SVM	KNN	RF	DT	DIS	SVM	KNN	RF
DIS	0.523	-	-	-	-	0.895	-	-	-	-
SVM	0.356	0.367	-	-	-	0.870	0.871	-	-	-
KNN	0.540	0.483	0.477	-	-	0.930	0.896	0.924	-	-
RF	0.583	0.515	0.509	0.671	-	0.962	0.926	0.935	0.971	-
ANN	0.511	0.602	0.493	0.532	0.574	0.915	0.951	0.931	0.922	0.943

| (a) | | | | | | (b) | | | | |

	DT	DIS	SVM	KNN	RF	DT	DIS	SVM	KNN	RF
DIS	0.130	-	-	-	-	0.097	-	-	-	-
SVM	0.142	0.132	-	-	-	0.052	0.050	-	-	-
KNN	0.113	0.122	0.092	-	-	0.086	0.075	0.051	-	-
RF	0.099	0.110	0.081	0.064	-	0.087	0.074	0.050	0.078	-
ANN	0.120	0.094	0.089	0.096	0.083	0.083	0.089	0.052	0.068	0.068

| (c) | | | | | | (d) | | | | |

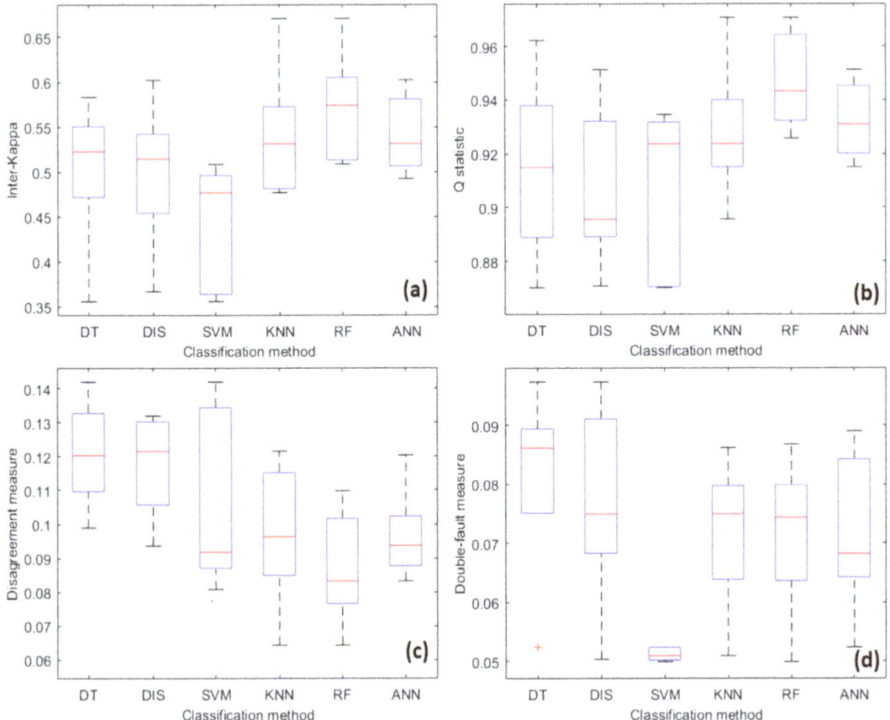

Figure 9. Distribution of diversity measures for all the possible pairs of base classifiers (**a**) inter-kappa measure, (**b**) Q-statistic, (**c**) disagreement measure, and (**d**) double-fault measure.

On the other hand, McNemar's test of the ensemble methods showed that the voting method based on greater kappa is significantly different from the rest of the voting methods (Table 7). More specifically, the x^2 exceeds the critical x^2 value of 3.84 and thus 'MaxK' is statistically significant at a 95% confidence interval for all the pair comparisons. Interestingly, the rest of the comparisons revealed that the null hypothesis cannot be rejected according to McNemar's test, hence the difference in accuracy between the ensemble methods is not statistically significant.

Table 7. McNemar's test x^2 values for pair comparisons of ensemble methods [1].

	MaxK	GSK	GMK	GWSK	GMF1	GSF1	GMMCC	GSMCC
Mode	13.483	0	1.028	0	1.125	0	1.936	0
MaxK	-	13.483	36.860	13.483	36.423	14.095	40.830	13.483
GSK	-	-	1.028	0	1.125	0	1.954	0
GMK	-	-	-	1.028	0	0.921	3.063	1.028
GWSK	-	-	-	-	1.125	0	1.954	0
GMF1	-	-	-	-	-	1.028	1.895	1.125
GSF1	-	-	-	-	-	-	1.787	0
GMMCC	-	-	-	-	-	-	-	1.936

[1] With bold fonts the McNemar's test x^2 values greater than value 3.84.

4. Conclusions

This work illustrates the potential use of a number of classifiers on identifying land cover types. The land cover type mapping is essential for the land management of the Mediterranean ecosystems. Long-term human activities along with geographic and climatic conditions have created a heterogeneous fragmented ecosystem that changes rapidly [99]. One of the main disturbances of Mediterranean ecosystems are wildfires. Land cover type mapping provides valuable information, i.e., vegetative fuel and socioeconomic inputs in wildfire risk assessment [100,101]. Furthermore, through remote sensing classification we can identify the change detection, possible land degradations and empower ecosystem monitoring [102,103].

An ensemble approach with nine voting methods has been developed for increased accuracy over classification algorithms using multi-temporal Sentinel-2 data from a mixed Mediterranean ecosystem. Each base classifier was trained with its own dataset in order to create the accuracy metrics that were used within the voting methods. All the base classifiers and the ensemble methods were applied to an unseen testing dataset. The result shows that the combination of multiple classifiers based on the examined voting schemes does not always provide a better performance in land cover classification. The SVM algorithm outperformed all the classifiers and was proven as the most accurate approach especially for this quite unbalanced dataset.

The diversity measures can explain the outperformance of SVM. The double-fault measure clearly shows that SVM significantly differs from the rest of the classifiers. Therefore, diversity measures should be thoroughly examined before building an ensemble method. The diversity metrics can be evidence in identifying possible overperformance within base classifiers hence an ensemble may not be always necessary. On the other hand, possible underperformances can be identified leading to the exclusion of some base classifiers. To sum up, our voting methods were influenced by the number of classifiers with a lower performance opposed to SVM. Hence, the accuracy of the ensembles are lower than the best base classifier, probably due to the 'curse of conflict' problem [104].

Potential further improvements of this methodology should include the incorporation of additional base algorithms and more ensemble methods. Moreover, opposed to pixel-based approaches, ensembles of segmentation approaches can be explored including traditional segmentation algorithms and CNNs. An interesting potential improvement of this work should be the comparison of multiple Mediterranean areas based on the very same ensemble of algorithms. Nevertheless, this work has proven that contemporary computational approaches along with advanced algorithmic measures show potential for land cover classification of unbalanced data.

Author Contributions: Conceptualization, C.V.; methodology, C.V. and D.K.; writing—Original draft preparation, C.V., D.K., and A.G. All authors have read and agreed to the published version of the manuscript.

Funding: This research received no external funding.

Acknowledgments: We thank the four anonymous reviewers for providing constructive comments for improving the overall manuscript.

Conflicts of Interest: The authors declare no conflict of interest.

Appendix A

Table A1. Confusion matrix for the validation dataset during the tr
for the decision tree (DT) model.

					Reference				
		OG	OF	BW	BU	PB	CF	PN	MS	BL
Predicted	OG	617	82	23	0	10	0	1	9	0
	OF	47	178	1	0	1	8	2	30	0
	BW	39	0	1012	10	0	0	0	0	4
	BU	4	2	2	225	0	0	0	1	9
	PB	33	2	0	0	1034	3	47	26	0
	CF	1	16	0	0	2	164	0	35	0
	PN	3	0	0	0	13	0	32	10	0
	MS	5	29	0	0	3	12	5	119	0
	BL	0	0	0	9	0	0	0	0	41
	GL	51	16	26	1	0	2	0	0	0
	OB	1	0	0	0	2	5	0	3	0
	AL	0	2	0	1	0	0	0	1	0
	AB	0	0	0	0	0	0	0	0	0
	Total	801	327	1064	246	1065	194	87	234	54
	PA	0.77	0.54	0.95	0.91	0.97	0.85	0.37	0.51	0.76
	kappa	0.73	0.55	0.88	0.91	0.91	0.77	0.43	0.56	0.79

Table A2. Confusion matrix for the validation dataset during the t
for the discriminant (DIS) model.

					Reference				
		OG	OF	BW	BU	PB	CF	PN	MS	BL
Predicted	OG	674	63	9	1	3	0	0	8	0
	OF	17	171	0	0	0	9	0	20	0
	BW	32	1	967	10	0	0	0	0	6
	BU	0	0	2	212	0	0	0	0	2
	PB	42	3	0	0	1053	0	44	31	0
	CF	1	27	0	0	0	164	0	16	0
	PN	3	3	0	0	5	0	40	4	0
	MS	5	36	0	0	4	20	3	152	0
	BL	0	1	1	23	0	0	0	0	46
	GL	26	14	85	0	0	0	0	0	0
	OB	1	8	0	0	0	1	0	3	0
	AL	0	0	0	0	0	0	0	0	0
	AB	0	0	0	0	0	0	0	0	0
	Total	801	327	1064	246	1065	194	87	234	54
	PA	0.84	0.52	0.91	0.86	0.99	0.85	0.46	0.65	0.85
	kappa	0.83	0.58	0.89	0.91	0.91	0.79	0.56	0.64	0.73

A3. Confusion matrix for the validation dataset during the training phase of the base classifiers
e support vector machine (SVM) model.

						Reference								
OG	**OF**	**BW**	**BU**	**PB**	**CF**	**PN**	**MS**	**BL**	**GL**	**OB**	**AL**	**AB**	**Total**	**UA**
743	35	8	0	4	0	0	7	0	13	1	1	0	812	0.92
23	259	0	0	0	4	0	15	0	2	1	1	0	305	0.85
5	1	1033	2	0	0	0	0	2	38	0	0	0	1081	0.96
3	0	0	242	0	0	0	0	5	0	0	0	0	250	0.97
6	0	0	0	1039	0	14	4	0	0	1	0	0	1064	0.98
0	2	0	0	0	172	0	13	0	0	5	0	0	192	0.9
1	0	0	0	12	0	72	2	0	0	1	0	0	88	0.82
3	24	0	0	7	16	1	186	0	0	10	1	0	248	0.75
0	0	0	2	0	0	0	0	47	0	0	0	0	49	0.96
17	6	23	0	0	0	0	0	0	275	1	6	0	328	0.84
0	0	0	0	3	2	0	7	0	0	37	0	0	49	0.76
0	0	0	0	0	0	0	0	0	3	1	173	0	177	0.98
0	0	0	0	0	0	0	0	0	0	0	0	267	267	1
801	327	1064	246	1065	194	87	234	54	331	58	182	267		
0.93	0.79	0.97	0.98	0.98	0.89	0.83	0.79	0.87	0.83	0.64	0.95	1	OA = 0.93	
0.91	0.81	0.95	0.97	0.97	0.89	0.82	0.76	0.91	0.82	0.69	0.96	1	kappa = 0.91	

A4. Confusion matrix for the validation dataset during the training phase of the base classifiers
e k-nearest neighbors (KNN) model.

						Reference								
OG	**OF**	**BW**	**BU**	**PB**	**CF**	**PN**	**MS**	**BL**	**GL**	**OB**	**AL**	**AB**	**Total**	**UA**
673	45	2	4	1	0	2	7	0	22	0	2	0	758	0.89
34	253	0	0	0	15	0	34	0	5	11	2	0	354	0.71
27	1	1038	15	0	0	0	0	3	50	0	0	0	1134	0.92
2	1	0	221	0	0	0	0	7	0	0	0	0	231	0.96
17	1	0	0	1047	0	24	13	0	2	9	0	2	1115	0.94
0	2	0	0	0	161	0	17	0	0	16	0	0	196	0.82
1	0	0	0	10	0	58	6	0	0	1	0	0	76	0.76
9	20	0	0	6	18	3	157	0	0	9	1	0	223	0.7
0	0	1	6	0	0	0	0	44	0	0	0	0	51	0.86
38	3	23	0	0	0	0	0	0	247	0	7	0	318	0.78
0	0	0	0	1	0	0	0	0	0	12	0	0	13	0.92
0	1	0	0	0	0	0	0	0	5	0	170	0	176	0.97
0	0	0	0	0	0	0	0	0	0	0	0	265	265	1
801	327	1064	246	1065	194	87	234	54	331	58	182	267		
0.84	0.77	0.98	0.9	0.98	0.83	0.67	0.67	0.81	0.75	0.21	0.93	0.99	OA = 0.89	
0.84	0.72	0.93	0.92	0.95	0.82	0.71	0.67	0.84	0.74	0.34	0.95	1	Kappa = 0.87	

Appendix A

Table A1. Confusion matrix for the validation dataset during the training phase of the base classifiers for the decision tree (DT) model.

		OG	OF	BW	BU	PB	CF	PN	MS	BL	GL	OB	AL	AB	Total	UA
															Reference	
Predicted	OG	617	82	23	0	10	0	1	9	0	46	0	1	0	789	0.78
	OF	47	178	1	0	1	8	2	30	0	5	14	4	0	290	0.61
	BW	39	0	1012	10	0	0	0	0	4	92	0	1	0	1158	0.87
	BU	4	2	2	225	0	0	0	1	9	1	0	1	0	245	0.92
	PB	33	2	0	0	1034	3	47	26	0	0	11	0	0	1156	0.89
	CF	1	16	0	0	2	164	0	35	0	0	10	0	0	228	0.72
	PN	3	0	0	0	13	0	32	10	0	0	2	0	0	60	0.53
	MS	5	29	0	0	3	12	5	119	0	0	6	0	0	179	0.66
	BL	0	0	0	9	0	0	0	0	41	0	0	0	0	50	0.82
	GL	51	16	26	1	0	2	0	0	0	177	0	7	0	280	0.63
	OB	1	0	0	0	2	5	0	3	0	0	13	1	0	25	0.52
	AL	0	2	0	1	0	0	0	1	0	10	2	167	0	183	0.91
	AB	0	0	0	0	0	0	0	0	0	0	0	0	267	267	1
	Total	801	327	1064	246	1065	194	87	234	54	331	58	182	267		
	PA	0.77	0.54	0.95	0.91	0.97	0.85	0.37	0.51	0.76	0.53	0.22	0.92	1	OA = 0.82	
	kappa	0.73	0.55	0.88	0.91	0.91	0.77	0.43	0.56	0.79	0.55	0.31	0.91	1	kappa = 0.79	

Table A2. Confusion matrix for the validation dataset during the training phase of the base classifiers for the discriminant (DIS) model.

		OG	OF	BW	BU	PB	CF	PN	MS	BL	GL	OB	AL	AB	Total	UA
															Reference	
Predicted	OG	674	63	9	1	3	0	0	8	0	17	0	0	0	775	0.87
	OF	17	171	0	0	0	9	0	20	0	3	13	5	0	238	0.72
	BW	32	1	967	10	0	0	0	0	6	42	0	0	0	1058	0.91
	BU	0	0	2	212	0	0	0	0	2	0	0	0	0	216	0.98
	PB	42	3	0	0	1053	0	44	31	0	1	18	0	0	1192	0.88
	CF	1	27	0	0	0	164	0	16	0	0	8	3	0	219	0.75
	PN	3	3	0	0	5	0	40	4	0	0	0	0	0	55	0.73
	MS	5	36	0	0	4	20	3	152	0	0	7	0	0	227	0.67
	BL	0	1	1	23	0	0	0	0	46	0	0	0	0	71	0.65
	GL	26	14	85	0	0	0	0	0	0	263	1	10	0	399	0.66
	OB	1	8	0	0	0	1	0	3	0	0	11	0	0	24	0.46
	AL	0	0	0	0	0	0	0	0	0	5	0	164	0	169	0.97
	AB	0	0	0	0	0	0	0	0	0	0	0	0	267	267	1
	Total	801	327	1064	246	1065	194	87	234	54	331	58	182	267		
	PA	0.84	0.52	0.91	0.86	0.99	0.85	0.46	0.65	0.85	0.79	0.19	0.9	1	OA = 0.85	
	kappa	0.83	0.58	0.89	0.91	0.91	0.79	0.56	0.64	0.73	0.7	0.26	0.93	1	kappa = 0.83	

Table A3. Confusion matrix for the validation dataset during the training phase of the base classifiers for the support vector machine (SVM) model.

		Reference													Total	UA
		OG	OF	BW	BU	PB	CF	PN	MS	BL	GL	OB	AL	AB		
Predicted	OG	743	35	8	0	4	0	0	7	0	13	1	1	0	812	0.92
	OF	23	259	0	0	0	4	0	15	0	2	1	1	0	305	0.85
	BW	5	1	1033	2	0	0	0	0	2	38	0	0	0	1081	0.96
	BU	3	0	0	242	0	0	0	0	5	0	0	0	0	250	0.97
	PB	6	0	0	0	1039	0	14	4	0	0	1	0	0	1064	0.98
	CF	0	2	0	0	0	172	0	13	0	0	5	0	0	192	0.9
	PN	1	0	0	0	12	0	72	2	0	0	1	0	0	88	0.82
	MS	3	24	0	0	7	16	1	186	0	0	10	1	0	248	0.75
	BL	0	0	0	2	0	0	0	0	47	0	0	0	0	49	0.96
	GL	17	6	23	0	0	0	0	0	0	275	1	6	0	328	0.84
	OB	0	0	0	0	3	2	0	7	0	0	37	0	0	49	0.76
	AL	0	0	0	0	0	0	0	0	0	3	1	173	0	177	0.98
	AB	0	0	0	0	0	0	0	0	0	0	0	0	267	267	1
	Total	801	327	1064	246	1065	194	87	234	54	331	58	182	267		
	PA	0.93	0.79	0.97	0.98	0.98	0.89	0.83	0.79	0.87	0.83	0.64	0.95	1	OA = 0.93	
	kappa	0.91	0.81	0.95	0.97	0.97	0.89	0.82	0.76	0.91	0.82	0.69	0.96	1	kappa = 0.91	

Table A4. Confusion matrix for the validation dataset during the training phase of the base classifiers for the k-nearest neighbors (KNN) model.

		Reference													Total	UA
		OG	OF	BW	BU	PB	CF	PN	MS	BL	GL	OB	AL	AB		
Predicted	OG	673	45	2	4	1	0	2	7	0	22	0	2	0	758	0.89
	OF	34	253	0	0	0	15	0	34	0	5	11	2	0	354	0.71
	BW	27	1	1038	15	0	0	0	0	3	50	0	0	0	1134	0.92
	BU	2	1	0	221	0	0	0	0	7	0	0	0	0	231	0.96
	PB	17	1	0	0	1047	0	24	13	0	2	9	0	2	1115	0.94
	CF	0	2	0	0	0	161	0	17	0	0	16	0	0	196	0.82
	PN	1	0	0	0	10	0	58	6	0	0	1	0	0	76	0.76
	MS	9	20	0	0	6	18	3	157	0	0	9	1	0	223	0.7
	BL	0	0	1	6	0	0	0	0	44	0	0	0	0	51	0.86
	GL	38	3	23	0	0	0	0	0	0	247	0	7	0	318	0.78
	OB	0	0	0	0	1	0	0	0	0	0	12	0	0	13	0.92
	AL	0	1	0	0	0	0	0	0	0	5	0	170	0	176	0.97
	AB	0	0	0	0	0	0	0	0	0	0	0	0	265	265	1
	Total	801	327	1064	246	1065	194	87	234	54	331	58	182	267		
	PA	0.84	0.77	0.98	0.9	0.98	0.83	0.67	0.67	0.81	0.75	0.21	0.93	0.99	OA = 0.89	
	kappa	0.84	0.72	0.93	0.92	0.95	0.82	0.71	0.67	0.84	0.74	0.34	0.95	1	Kappa = 0.87	

Table A9. Confusion matrix for the test dataset of the Mode ensemble model.

		Reference													
		OG	OF	BW	BU	PB	CF	PN	MS	BL	GL	OB	AL	AB	Total	UA
Predicted	OG	727	42	5	3	4	0	1	7	1	25	0	0	0	815	0.89
	OF	20	244	0	0	0	2	0	13	0	6	5	0	0	290	0.84
	BW	15	0	1037	3	0	0	0	0	3	35	0	1	0	1094	0.95
	BU	0	0	1	248	0	0	0	0	8	0	0	0	0	257	0.96
	PB	6	0	0	0	1074	0	26	9	0	2	7	0	0	1124	0.96
	CF	0	3	0	0	0	159	0	16	0	0	12	0	0	190	0.84
	PN	0	0	0	0	1	0	55	0	0	0	0	0	0	56	0.98
	MS	9	28	0	0	10	17	0	179	0	1	5	0	0	249	0.72
	BL	0	0	1	2	0	0	0	0	49	0	0	0	0	52	0.94
	GL	20	7	44	1	0	0	0	0	0	241	0	4	0	317	0.76
	OB	0	1	0	0	1	0	0	2	0	0	19	0	0	23	0.83
	AL	0	2	0	0	0	0	0	0	0	2	2	183	0	189	0.97
	AB	0	0	0	0	0	0	0	0	0	0	0	0	254	254	1
	Total	797	327	1088	257	1090	178	82	226	61	312	50	188	254	4910	
	PA	0.89	0.84	0.95	0.96	0.96	0.84	0.98	0.72	0.94	0.76	0.83	0.97	1	OA = 0.91	
	kappa	0.87	0.76	0.93	0.96	0.95	0.85	0.68	0.74	0.85	0.73	0.40	0.96	1	Kappa = 0.89	

References

1. Shen, H.; Lin, Y.; Tian, Q.; Xu, K.; Jiao, J. A comparison of multiple classifier combinations using different voting-weights for remote sensing image classification. *Int. J. Remote Sens.* **2018**, *39*, 3705–3722. [CrossRef]
2. Maulik, U.; Chakraborty, D. Remote Sensing Image Classification: A survey of support-vector-machine-based advanced techniques. *IEEE Geosci. Remote Sens. Mag.* **2017**, *5*, 33–52. [CrossRef]
3. Fathizad, H.; Hakimzadeh Ardakani, M.A.; Mehrjardi, R.T.; Sodaiezadeh, H. Evaluating desertification using remote sensing technique and object-oriented classification algorithm in the Iranian central desert. *J. Afr. Earth Sci.* **2018**, *145*, 115–130. [CrossRef]
4. Gómez, C.; White, J.C.; Wulder, M.A. Optical remotely sensed time series data for land cover classification: A review. *ISPRS J. Photogramm. Remote Sens.* **2016**, *116*, 55–72. [CrossRef]
5. Chen, K.S.; Tzeng, Y.C.; Chen, C.F.; Kao, W.L.; Ni, C.L. Classification of multispectral imagery using dynamic learning neural network. In Proceedings of the IGARSS '93—IEEE International Geoscience and Remote Sensing Symposium, Tokyo, Japan, 18–21 August 2013; IEEE: Piscataway, NJ, USA, 1994; pp. 896–898.
6. Lawrence, R. Classification of remotely sensed imagery using stochastic gradient boosting as a refinement of classification tree analysis. *Remote Sens. Environ.* **2004**, *90*, 331–336. [CrossRef]
7. Sohn, Y.; Rebello, N.S. Supervised and unsupervised spectral angle classifiers. *Photogramm. Eng. Remote Sens.* **2002**, *68*, 1271–1280.
8. Strahler, A.H. The use of prior probabilities in maximum likelihood classification of remotely sensed data. *Remote Sens. Environ.* **1980**, *10*, 135–163. [CrossRef]
9. Foody, G.M.; Mathur, A. A relative evaluation of multiclass image classification by support vector machines. *IEEE Trans. Geosci. Remote Sens.* **2004**, *42*, 1335–1343. [CrossRef]
10. Collins, M.J.; Dymond, C.; Johnson, E.A. Mapping subalpine forest types using networks of nearest neighbour classifiers. *Int. J. Remote Sens.* **2004**, *25*, 1701–1721. [CrossRef]
11. Maxwell, A.E.; Warner, T.A.; Fang, F. Implementation of machine-learning classification in remote sensing: An applied review. *Int. J. Remote Sens.* **2018**, *39*, 2784–2817. [CrossRef]
12. Qian, Y.; Zhou, W.; Yan, J.; Li, W.; Han, L. Comparing Machine Learning Classifiers for Object-Based Land Cover Classification Using Very High Resolution Imagery. *Remote Sens.* **2014**, *7*, 153–168. [CrossRef]
13. Ghamisi, P.; Plaza, J.; Chen, Y.; Li, J.; Plaza, A.J. Advanced Spectral Classifiers for Hyperspectral Images: A review. *IEEE Geosci. Remote Sens. Mag.* **2017**, *5*, 8–32. [CrossRef]
14. Ballanti, L.; Blesius, L.; Hines, E.; Kruse, B. Tree Species Classification Using Hyperspectral Imagery: A Comparison of Two Classifiers. *Remote Sens.* **2016**, *8*, 445. [CrossRef]
15. Seetha, M.; Muralikrishna, I.V.; Deekshatulu, B.L.; Malleswari, B.L.; Hegde, P. Artificial Neural Networks and Other Methods of Image Classification. *Theor. Appl. Inf. Technol.* **2008**, *4*, 1039–1053.

16. Thanh Noi, P.; Kappas, M. Comparison of Random Forest, k-Nearest Neighbor, and Support Vector Machine Classifiers for Land Cover Classification Using Sentinel-2 Imagery. *Sensors* **2017**, *18*, 18. [CrossRef] [PubMed]

17. Huang, C.; Davis, L.S.; Townshend, J.R.G. An assessment of support vector machines for land cover classification. *Int. J. Remote Sens.* **2002**, *23*, 725–749. [CrossRef]

18. Lapini, A.; Pettinato, S.; Santi, E.; Paloscia, S.; Fontanelli, G.; Garzelli, A. Comparison of Machine Learning Methods Applied to SAR Images for Forest Classification in Mediterranean Areas. *Remote Sens.* **2020**, *12*, 369. [CrossRef]

19. Ge, G.; Shi, Z.; Zhu, Y.; Yang, X.; Hao, Y. Land use/cover classification in an arid desert-oasis mosaic landscape of China using remote sensed imagery: Performance assessment of four machine learning algorithms. *Glob. Ecol. Conserv.* **2020**, *22*, e00971. [CrossRef]

20. Dang, V.-H.; Hoang, N.-D.; Nguyen, L.-M.-D.; Bui, D.T.; Samui, P. A Novel GIS-Based Random Forest Machine Algorithm for the Spatial Prediction of Shallow Landslide Susceptibility. *Forests* **2020**, *11*, 118. [CrossRef]

21. Abdi, A.M. Land cover and land use classification performance of machine learning algorithms in a boreal landscape using Sentinel-2 data. *GISci. Remote Sens.* **2020**, *57*, 1–20. [CrossRef]

22. Tonbul, H.; Colkesen, I.; Kavzoglu, T. Classification of poplar trees with object-based ensemble learning algorithms using Sentinel-2A imagery. *J. Geod. Sci.* **2020**, *10*, 14–22. [CrossRef]

23. Langley, S.K.; Cheshire, H.M.; Humes, K.S. A comparison of single date and multitemporal satellite image classifications in a semi-arid grassland. *J. Arid Environ.* **2001**, *49*, 401–411. [CrossRef]

24. Yuan, F.; Sawaya, K.E.; Loeffelholz, B.C.; Bauer, M.E. Land cover classification and change analysis of the Twin Cities (Minnesota) Metropolitan Area by multitemporal Landsat remote sensing. *Remote Sens. Environ.* **2005**, *98*, 317–328. [CrossRef]

25. Hütt, C.; Koppe, W.; Miao, Y.; Bareth, G. Best Accuracy Land Use/Land Cover (LULC) Classification to Derive Crop Types Using Multitemporal, Multisensor, and Multi-Polarization SAR Satellite Images. *Remote Sens.* **2016**, *8*, 684. [CrossRef]

26. Eisavi, V.; Homayouni, S.; Yazdi, A.M.; Alimohammadi, A. Land cover mapping based on random forest classification of multitemporal spectral and thermal images. *Environ. Monit. Assess.* **2015**, *187*, 291. [CrossRef] [PubMed]

27. Tigges, J.; Lakes, T.; Hostert, P. Urban vegetation classification: Benefits of multitemporal RapidEye satellite data. *Remote Sens. Environ.* **2013**, *136*, 66–75. [CrossRef]

28. Alcantara, C.; Kuemmerle, T.; Prishchepov, A.V.; Radeloff, V.C. Mapping abandoned agriculture with multi-temporal MODIS satellite data. *Remote Sens. Environ.* **2012**, *124*, 334–347. [CrossRef]

29. Key, T. A Comparison of Multispectral and Multitemporal Information in High Spatial Resolution Imagery for Classification of Individual Tree Species in a Temperate Hardwood Forest. *Remote Sens. Environ.* **2001**, *75*, 100–112. [CrossRef]

30. Kamusoko, C. Image Classification. In *Remote Sensing Image Classification in R*; Springer: Singapore, 2019; pp. 81–153.

31. Rujoiu-Mare, M.-R.; Olariu, B.; Mihai, B.-A.; Nistor, C.; Săvulescu, I. Land cover classification in Romanian Carpathians and Subcarpathians using multi-date Sentinel-2 remote sensing imagery. *Eur. J. Remote Sens.* **2017**, *50*, 496–508. [CrossRef]

32. Sharma, A.; Liu, X.; Yang, X. Land cover classification from multi-temporal, multi-spectral remotely sensed imagery using patch-based recurrent neural networks. *Neural Netw.* **2018**, *105*, 346–355. [CrossRef]

33. Pal, M.; Mather, P.M. A comparison of decision tree and backpropagation neural network classifiers for land use classification. In Proceedings of the International Geoscience and Remote Sensing Symposium (IGARSS), Toronto, ON, Canada, 24–28 June 2002.

34. Rogan, J.; Franklin, J.; Stow, D.; Miller, J.; Woodcock, C.; Roberts, D. Mapping land-cover modifications over large areas: A comparison of machine learning algorithms. *Remote Sens. Environ.* **2008**, *112*, 2272–2283. [CrossRef]

35. Pal, M.; Mather, P.M. Support vector machines for classification in remote sensing. *Int. J. Remote Sens.* **2005**, *26*, 1007–1011. [CrossRef]

36. Du, P.; Xia, J.; Zhang, W.; Tan, K.; Liu, Y.; Liu, S. Multiple Classifier System for Remote Sensing Image Classification: A Review. *Sensors* **2012**, *12*, 4764–4792. [CrossRef]

37. Briem, G.J.; Benediktsson, J.A.; Sveinsson, J.R. Multiple classifiers applied to multisource remote sensing data. *IEEE Trans. Geosci. Remote Sens.* **2002**, *40*, 2291–2299. [CrossRef]

38. Aguilar, R.; Zurita-Milla, R.; Izquierdo-Verdiguier, E.; de By, R.A. A Cloud-Based Multi-Temporal Ensemble Classifier to Map Smallholder Farming Systems. *Remote Sens.* **2018**, *10*, 729. [CrossRef]

39. Foody, G.M.; Boyd, D.S.; Sanchez-Hernandez, C. Mapping a specific class with an ensemble of classifiers. *Int. J. Remote Sens.* **2007**, *28*, 1733–1746. [CrossRef]

40. Lei, G.; Li, A.; Bian, J.; Yan, H.; Zhang, L.; Zhang, Z.; Nan, X. OIC-MCE: A Practical Land Cover Mapping Approach for Limited Samples Based on Multiple Classifier Ensemble and Iterative Classification. *Remote Sens.* **2020**, *12*, 987. [CrossRef]

41. Amani, M.; Salehi, B.; Mahdavi, S.; Brisco, B.; Shehata, M. A Multiple Classifier System to improve mapping complex land covers: A case study of wetland classification using SAR data in Newfoundland, Canada. *Int. J. Remote Sens.* **2018**, *39*, 7370–7383. [CrossRef]

42. Doan, H.T.X.; Foody, G.M. Increasing soft classification accuracy through the use of an ensemble of classifiers. *Int. J. Remote Sens.* **2007**, *28*, 4609–4623. [CrossRef]

43. Giacinto, G.; Roli, F. Ensembles of Neural Networks for Soft Classification of Remote Sensing Images. In Proceedings of the European Symposium on Intelligent Techniques, European Network for Fuzzy Logic and Uncertainty Modelling in Information Technology, Bari, Italy, 20–21 March 1997; pp. 166–170.

44. Drucker, H.; Cortes, C.; Jackel, L.D.; LeCun, Y.; Vapnik, V. Boosting and Other Ensemble Methods. *Neural Comput.* **1994**, *6*, 1289–1301. [CrossRef]

45. Breiman, L. Bagging predictors. *Mach. Learn.* **1996**, *24*, 123–140. [CrossRef]

46. Pal, M. Ensemble of support vector machines for land cover classification. *Int. J. Remote Sens.* **2008**, *29*, 3043–3049. [CrossRef]

47. Pal, M. Ensemble Learning with Decision Tree for Remote Sensing Classification. *World Acad. Sci. Eng. Technol.* **2007**, *36*, 258–260.

48. Battiti, R.; Colla, A.M. Democracy in neural nets: Voting schemes for classification. *Neural Netw.* **1994**, *7*, 691–707. [CrossRef]

49. Oza, N.C.; Tumer, K. Classifier ensembles: Select real-world applications. *Inf. Fusion* **2008**, *9*, 4–20. [CrossRef]

50. Puletti, N.; Chianucci, F.; Castaldi, C. Use of Sentinel-2 for forest classification in Mediterranean environments. *Ann. Silvic. Res.* **2018**, *42*, 32–38.

51. Chatziantoniou, A.; Psomiadis, E.; Petropoulos, G. Co-Orbital Sentinel 1 and 2 for LULC Mapping with Emphasis on Wetlands in a Mediterranean Setting Based on Machine Learning. *Remote Sens.* **2017**, *9*, 1259. [CrossRef]

52. Henderson, M.; Kalabokidis, K.; Marmaras, E.; Konstantinidis, P.; Marangudakis, M. Fire and society: A comparative analysis of wildfire in Greece and the United States. *Hum. Ecol. Rev.* **2005**, *12*, 169–182.

53. ESA Copernicus Open Access Hub. Available online: https://scihub.copernicus.eu/ (accessed on 10 February 2020).

54. Hill, R.A.; Wilson, A.K.; George, M.; Hinsley, S.A. Mapping tree species in temperate deciduous woodland using time-series multi-spectral data. *Appl. Veg. Sci.* **2010**, *13*, 86–99. [CrossRef]

55. Guirado, E.; Tabik, S.; Alcaraz-Segura, D.; Cabello, J.; Herrera, F. Deep-learning Versus OBIA for Scattered Shrub Detection with Google Earth Imagery: Ziziphus lotus as Case Study. *Remote Sens.* **2017**, *9*, 1220. [CrossRef]

56. Li, Q.; Qiu, C.; Ma, L.; Schmitt, M.; Zhu, X.X. Mapping the Land Cover of Africa at 10 m Resolution from Multi-Source Remote Sensing Data with Google Earth Engine. *Remote Sens.* **2020**, *12*, 602. [CrossRef]

57. Bwangoy, J.-R.B.; Hansen, M.C.; Roy, D.P.; De Grandi, G.; Justice, C.O. Wetland mapping in the Congo Basin using optical and radar remotely sensed data and derived topographical indices. *Remote Sens. Environ.* **2010**, *114*, 73–86. [CrossRef]

58. Lu, D.; Weng, Q. A survey of image classification methods and techniques for improving classification performance. *Int. J. Remote Sens.* **2007**, *28*, 823–870. [CrossRef]

59. Friedl, M.A.; Brodley, C.E. Decision tree classification of land cover from remotely sensed data. *Remote Sens. Environ.* **1997**, *61*, 399–409. [CrossRef]

60. Almuallim, H. An efficient algorithm for optimal pruning of decision trees. *Artif. Intell.* **1996**, *83*, 347–362. [CrossRef]

61. Aylmer Fisher FRS, R. Summary for Policymakers. In *Climate Change 2013—The Physical Science Basis*; Intergovernmental Panel on Climate Change, Ed.; Cambridge University Press: Cambridge, UK, 1936; pp. 1–30. ISBN 9788578110796.

62. Bandos, T.V.; Bruzzone, L.; Camps-Valls, G. Classification of Hyperspectral Images with Regularized Linear Discriminant Analysis. *IEEE Trans. Geosci. Remote Sens.* **2009**, *47*, 862–873. [CrossRef]

63. Feret, J.-B.; Asner, G.P. Tree Species Discrimination in Tropical Forests Using Airborne Imaging Spectroscopy. *IEEE Trans. Geosci. Remote Sens.* **2013**, *51*, 73–84. [CrossRef]

64. Clark, M.; Roberts, D.; Clark, D. Hyperspectral discrimination of tropical rain forest tree species at leaf to crown scales. *Remote Sens. Environ.* **2005**, *96*, 375–398. [CrossRef]

65. Cortes, C.; Vapnik, V. Support-vector networks. *Mach. Learn.* **1995**, *20*, 273–297. [CrossRef]

66. Kavzoglu, T.; Colkesen, I. A kernel functions analysis for support vector machines for land cover classification. *Int. J. Appl. Earth Obs. Geoinf.* **2009**, *11*, 352–359. [CrossRef]

67. Franco-Lopez, H.; Ek, A.R.; Bauer, M.E. Estimation and mapping of forest stand density, volume, and cover type using the k-nearest neighbors method. *Remote Sens. Environ.* **2001**, *77*, 251–274. [CrossRef]

68. Gislason, P.O.; Benediktsson, J.A.; Sveinsson, J.R. Random Forests for land cover classification. *Pattern Recognit. Lett.* **2006**, *27*, 294–300. [CrossRef]

69. Pal, M. Random forest classifier for remote sensing classification. *Int. J. Remote Sens.* **2005**, *26*, 217–222. [CrossRef]

70. Belgiu, M.; Drăguţ, L. Random forest in remote sensing: A review of applications and future directions. *ISPRS J. Photogramm. Remote Sens.* **2016**, *114*, 24–31. [CrossRef]

71. Mas, J.F.; Flores, J.J. The application of artificial neural networks to the analysis of remotely sensed data. *Int. J. Remote Sens.* **2008**, *29*, 617–663. [CrossRef]

72. Atkinson, P.M.; Tatnall, A.R.L. Introduction Neural networks in remote sensing. *Int. J. Remote Sens.* **1997**, *18*, 699–709. [CrossRef]

73. Yuan, H.; Van Der Wiele, C.; Khorram, S. An Automated Artificial Neural Network System for Land Use/Land Cover Classification from Landsat TM Imagery. *Remote Sens.* **2009**, *1*, 243–265. [CrossRef]

74. Kavzoglu, T.; Mather, P.M. The use of backpropagating artificial neural networks in land cover classification. *Int. J. Remote Sens.* **2003**, *24*, 4907–4938. [CrossRef]

75. Längkvist, M.; Kiselev, A.; Alirezaie, M.; Loutfi, A. Classification and Segmentation of Satellite Orthoimagery Using Convolutional Neural Networks. *Remote Sens.* **2016**, *8*, 329. [CrossRef]

76. Pires de Lima, R.; Marfurt, K. Convolutional Neural Network for Remote-Sensing Scene Classification: Transfer Learning Analysis. *Remote Sens.* **2019**, *12*, 86. [CrossRef]

77. Hoeser, T.; Kuenzer, C. Object Detection and Image Segmentation with Deep Learning on Earth Observation Data: A Review-Part I: Evolution and Recent Trends. *Remote Sens.* **2020**, *12*, 1667. [CrossRef]

78. Zuo, Z.; Shuai, B.; Wang, G.; Liu, X.; Wang, X.; Wang, B.; Chen, Y. Convolutional recurrent neural networks: Learning spatial dependencies for image representation. In Proceedings of the 2015 IEEE Conference on Computer Vision and Pattern Recognition Workshops (CVPRW), Boston, MA, USA, 7–12 June 2015; IEEE: Piscataway, NJ, USA, 2015; pp. 18–26.

79. Zhao, W.; Du, S. Spectral–Spatial Feature Extraction for Hyperspectral Image Classification: A Dimension Reduction and Deep Learning Approach. *IEEE Trans. Geosci. Remote Sens.* **2016**, *54*, 4544–4554. [CrossRef]

80. Lv, X.; Ming, D.; Chen, Y.; Wang, M. Very high resolution remote sensing image classification with SEEDS-CNN and scale effect analysis for superpixel CNN classification. *Int. J. Remote Sens.* **2019**, *40*, 506–531. [CrossRef]

81. Kuncheva, L.I.; Whitaker, C.J. Measures of diversity in classifier ensembles and their relationship with the ensemble accuracy. *Mach. Learn.* **2003**, *51*, 181–207. [CrossRef]

82. Petrakos, M.; Atli Benediktsson, J.; Kanellopoulos, I. The effect of classifier agreement on the accuracy of the combined classifier in decision level fusion. *IEEE Trans. Geosci. Remote Sens.* **2001**, *39*, 2539–2546. [CrossRef]

83. Thomas, G. Dietterich An Experimental Comparison of Three Methods for Constructing Ensembles of Decision Trees: Bagging, Boosting, and Randomization. *Mach. Learn.* **2000**, *40*, 139–157.

84. Edwards, A.L. Note on the "correction for continuity" in testing the significance of the difference between correlated proportions. *Psychometrika* **1948**, *13*, 185–187. [CrossRef]

85. Kavzoglu, T. Object-Oriented Random Forest for High Resolution Land Cover Mapping Using Quickbird-2 Imagery. In *Handbook of Neural Computation*; Academic Press: London, UK, 2017; ISBN 9780128113196.

86. Environmental Systems Research Institute. *ESRI ArcGIS Desktop: Release 10*; Environmental Systems Research Institute: Redlands, CA, USA, 2013.

87. The Mathworks Inc. The Mathworks Inc.: Massachusetts. 2018. Available online: https://www.Mathworks.com/Products/Matlab (accessed on 10 February 2020).

88. R Development Core Team. *R: A Language and Environment for Statistical Computing*; R Development Core Team: Vienna, Austria, 2017.

89. Wickham, H.; Francois, R. *The Dplyr Package*; R Core Team: Vienna, Austria, 2016.

90. Bache, S.M.; Wickham, H. Package 'magrittr'—A Forward-Pipe Operator for R. Available online: https://CRAN.R-project.org/package=magrittr (accessed on 10 February 2020).

91. Kuhn, M. Building Predictive Models in R Using the caret Package. *J. Stat. Softw.* **2008**, *28*, 1–26. [CrossRef]

92. Anthony, G.; Gregg, H.; Tshilidzi, M. Image classification using SVMs: One-Against-One vs One-against-All. In Proceedings of the 28th Asian Conference on Remote Sensing 2007, ACRS 2007, Kuala Lumpur, Malaysia, 12–16 November 2007.

93. Daengduang, S.; Vateekul, P. Enhancing accuracy of multi-label classification by applying one-vs-one support vector machine. In Proceedings of the 2016 13th International Joint Conference on Computer Science and Software Engineering (JCSSE), Khon Kaen, Thailand, 13–15 July 2016; IEEE: Piscataway, NJ, USA, 2016; pp. 1–6.

94. Marquardt, D.W. An Algorithm for Least-Squares Estimation of Nonlinear Parameters. *J. Soc. Ind. Appl. Math.* **1963**, *11*, 431–441. [CrossRef]

95. Møller, M.F. A scaled conjugate gradient algorithm for fast supervised learning. *Neural Netw.* **1993**, *6*, 525–533. [CrossRef]

96. Shang, X.; Chisholm, L.A. Classification of Australian Native Forest Species Using Hyperspectral Remote Sensing and Machine-Learning Classification Algorithms. *IEEE J. Sel. Top. Appl. Earth Obs. Remote Sens.* **2014**, *7*, 2481–2489. [CrossRef]

97. Pirotti, F.; Sunar, F.; Piragnolo, M. Benchmark of machine learning methods for classification of a sentinel-2 image. *ISPRS Int. Arch. Photogramm. Remote Sens. Spat. Inf. Sci.* **2016**, *XLI-B7*, 335–340. [CrossRef]

98. Salas, E.A.L.; Subburayalu, S.K.; Slater, B.; Zhao, K.; Bhattacharya, B.; Tripathy, R.; Das, A.; Nigam, R.; Dave, R.; Parekh, P. Mapping crop types in fragmented arable landscapes using AVIRIS-NG imagery and limited field data. *Int. J. Image Data Fusion* **2020**, *11*, 33–56. [CrossRef]

99. Gauquelin, T.; Michon, G.; Joffre, R.; Duponnois, R.; Génin, D.; Fady, B.; Bou Dagher-Kharrat, M.; Derridj, A.; Slimani, S.; Badri, W.; et al. Mediterranean forests, land use and climate change: A social-ecological perspective. *Reg. Environ. Chang.* **2018**, *18*, 623–636. [CrossRef]

100. Vasilakos, C.; Kalabokidis, K.; Hatzopoulos, J.; Kallos, G.; Matsinos, Y. Integrating new methods and tools in fire danger rating. *Int. J. Wildl. Fire* **2007**, *16*, 306. [CrossRef]

101. Vasilakos, C.; Kalabokidis, K.; Hatzopoulos, J.; Matsinos, I. Identifying wildland fire ignition factors through sensitivity analysis of a neural network. *Nat. Hazards* **2009**, *50*, 125–143. [CrossRef]

102. Bajocco, S.; De Angelis, A.; Perini, L.; Ferrara, A.; Salvati, L. The impact of Land Use/Land Cover Changes on land degradation dynamics: A Mediterranean case study. *Environ. Manag.* **2012**, *49*, 980–989. [CrossRef]

103. Otero, I.; Marull, J.; Tello, E.; Diana, G.L.; Pons, M.; Coll, F.; Boada, M. Land abandonment, landscape, and biodiversity: Questioning the restorative character of the forest transition in the Mediterranean. *Ecol. Soc.* **2015**. [CrossRef]

104. Song, C.; Pons, A.; Yen, K. Sieve: An Ensemble Algorithm Using Global Consensus for Binary Classification. *AI* **2020**, *1*, 16. [CrossRef]

 *remote sensing*

Article

Computer Vision and Deep Learning Techniques for the Analysis of Drone-Acquired Forest Images, a Transfer Learning Study

Sarah Kentsch [1,*], Maximo Larry Lopez Caceres [1], Daniel Serrano [2], Ferran Roure [2] and Yago Diez [3,*]

[1] Faculty of Agriculture, Yamagata University, Tsuruoka, Yamagata 997-8555, Japan; larry@tds1.tr.yamagata-u.ac.jp
[2] Eurecat, Centre Tecnològic de Catalunya, 08005 Barcelona, Spain; daniel.serrano@eurecat.org (D.S.); ferran.roure@eurecat.org (F.R.)
[3] Faculty of Science, Yamagata University, Yamagata 990-8560, Japan
* Correspondence: sarah@tds1.tr.yamagata-u.ac.jp (S.K.); yago@sci.kj.yamagata-u.ac.jp (Y.D.)

Received: 13 March 2020; Accepted: 16 April 2020 ; Published: 18 April 2020

Abstract: Unmanned Aerial Vehicles (UAV) are becoming an essential tool for evaluating the status and the changes in forest ecosystems. This is especially important in Japan due to the sheer magnitude and complexity of the forest area, made up mostly of natural mixed broadleaf deciduous forests. Additionally, Deep Learning (DL) is becoming more popular for forestry applications because it allows for the inclusion of expert human knowledge into the automatic image processing pipeline. In this paper we study and quantify issues related to the use of DL with our own UAV-acquired images in forestry applications such as: the effect of Transfer Learning (TL) and the Deep Learning architecture chosen or whether a simple patch-based framework may produce results in different practical problems. We use two different Deep Learning architectures (ResNet50 and UNet), two in-house datasets (winter and coastal forest) and focus on two separate problem formalizations (Multi-Label Patch or MLP classification and semantic segmentation). Our results show that Transfer Learning is necessary to obtain satisfactory outcome in the problem of MLP classification of deciduous vs evergreen trees in the winter orthomosaic dataset (with a 9.78% improvement from no transfer learning to transfer learning from a a general-purpose dataset). We also observe a further 2.7% improvement when Transfer Learning is performed from a dataset that is closer to our type of images. Finally, we demonstrate the applicability of the patch-based framework with the ResNet50 architecture in a different and complex example: Detection of the invasive broadleaf deciduous black locust (*Robinia pseudoacacia*) in an evergreen coniferous black pine (*Pinus thunbergii*) coastal forest typical of Japan. In this case we detect images containing the invasive species with a 75% of True Positives (TP) and 9% False Positives (FP) while the detection of native trees was 95% TP and 10% FP.

Keywords: deep learning; mixed forest; multi-label segmentation; semantic segmentation; unmanned aerial vehicles

---

1. Introduction

Forest ecosystems play an important role in water, carbon and nutrient cycling within the soil-vegetation-atmosphere continuum. In recent years climate change is exerting positive changes such as the early greening of forests in the northern hemisphere, shifting of forests to warmer environments northward [1] or negative ones, such as an increase of longer periods of drought [2], increase in the number of forest fires or extreme climate events [3]. Recent studies have focused on the ecological functions of mixed forests, since they show high resistance against insect outbreaks and a stronger capacity to recover from disturbances [4]. Detailed knowledge about mixed forest structure and composition [5] is needed in order to properly understand current status and future changes.

Forests in Japan occupy approximately 68% of the total territory, with most of them being natural deciduous broadleaf mixed forests [6]. Natural forests are affected by climate change while man-made forest ecosystems such as coastal forests are affected by invasive species that diminish their functions as windbreak [5,7]. The distribution of each species within a stand, or the interactions between the different tree species [4] makes them ecologically complex, especially in comparison to monoculture forests [8]. Until very recently, forest research has been carried out using labor and time-consuming land surveys [9]. They are costly and demand a high degree of organization training and expertise. Moreover, the characteristics of Japanese forests make them particularly challenging for land surveys as they are often located in steep mountain slopes that are difficult to access. Thus, new tools are needed in order to efficiently gain an overall understanding of species interaction and their response to climate change in order to design the proper response policy to ensure the sustainability of forests.

Unmanned Aerial Vehicles (UAVs) are rapidly becoming an essential tool in forestry applications [10–13] and they represent an easy-to-use, inexpensive tool for remote sensing of forests as they can fly close to tree canopies, which results in high image resolution (with one pixel representing a few centimeters). That can be processed by computer vision algorithms. An example of such an application is the building of orthomosaics by aligning the images for visualization and coherent processing of specific forest areas encompassing several hectares. From now on, and for the sake of brevity, we will refer to these *orthomosaics* simply as *mosaics* Additionally, emerging technologies such as Deep Learning (*DL*) allow the incorporation of expert knowledge to the automatic processing of images, which together with the availability of larger amount of data, has the potential to radically change the way land surveys are done. Namely, the time-consuming (and sometimes dangerous), intensive field surveys will likely become unnecessary, while those tasks requiring expert human knowledge are expected to be greatly increased.

1.1. State of the Art

Deep Learning is a part of Machine Learning that has gained much attention recently, resulting in a wide availability of data and software, making it the most easily accessible part of Artificial Intelligence. On the one hand, Deep Learning algorithms learn from examples: First, an *architecture* or set of nodes and connections among them is defined. The nodes in *Artificial Neural Networks* (ANN) are often grouped in *layers* and ANN with a large number of layers are called *Deep Neural Networks* (DNN). The type of each node, the number of nodes and the connections between them determine the behaviour of the network. Two main types of nodes exist: linear nodes, expressed as matrices and nodes that introduce non-linearity. The weights in linear nodes are typically initialised with random numbers. An important characteristic of ANN is that weights optimised for one problem can be used as the starting point to solve a new problem using less data. This is known as *Transfer Learning* [14]. After the network architecture has been defined and the weights have been initialised, the network is given example data. These data, known as *training data* contains instances of the problem being dealt with along with their solutions. The data are run through the network and the weights in all the linear nodes are changed following an optimization process.

Deep Learning was initially used to locate and classify different species of trees in mosaics built from UAV-acquired images in [10,11]. The authors used mosaics and Digital Elevation Models (DEM) to segment individual tree crowns [15], which were segmented and classified manually to build training data sets. These are two pioneering papers in the use of Deep Learning in forestry, however the insufficient amount of data appeared to hamper the sensitivity results reported in [10]. Since the number of images of individual trees used were about 900. This relatively small amount of data and the number of different tree species present in the forest resulted in a low number of images in each class, most notably in the test set (with some test classes having as few as 9 images). The data were also unbalanced, especially since the "other" class did not contain any trees. On the first set of results presented by the authors, there was low classification performance with 2 of the 6 tree classes not detected at all (0% reported sensitivity or recall, noted in the paper as "accuracy") and the others ranging from 20% to 85% reported sensitivity, using data augmentation. This resulted in reported sensitivity values for tree classes ranging from 68% to 95%. However, this has a significant risk of over-fitting that was not addressed in the paper. The relatively large number of epochs used for training (30) also supported this possibility. These problems are also an issue in [11], although some steps were taken to mitigate them. The data used comprised three different data acquisition campaigns of the same area. In this case each class of test set contained the same number of trees (30). Results using only one mosaic presented low accuracy (50% on average for the three studied tree classes), [11] while presented results using all their acquired data and obtained results with average accuracy of 80%. Therefore, they demonstrated that more data could improve the classification results but also exemplified the difficulty of this issue. Safonova et al. [12] first predicted potential regions of trees in UAV images before classifying them into four degrees of parasite infestation. In this case, RGB images were chosen manually to build a balanced training data set that was expanded using data augmentation. Non-tree containing parts of the mosaic were filtered out using non-DL computer vision processing. Reported results showed remarkable high accuracy values from 76 to 92 % with an extensive use of different DL architectures. The paper of [12] provided an example of how careful data annotation and balancing could lead to highly accurate results in terms of detection. Fromm et al. [13] used DNN on their data to detect conifer seedlings. They evaluated trainings-set size, the influence of seasonality and seedlings size by considering spatial resolutions. In this case, the authors worked on single images instead of mosaics. Those were produced after manually labelling 200 seedlings in different phenological stages. Three DL-models were trained with the best performance provided by ResNet101. The models were improved by Transfer Learning and suitable results were obtained from a data set of only 500 images.

Our Contribution

The main goal of this paper is to study the use of Deep Learning to gain information on forestry applications. By dividing mosaics built using UAV-acquired images into regular patches and using two well-known DL architectures (ResNet and UNet), we propose the following objectives (see Figure 1 for an overview):

1. Develop an algorithm (that uses a ResNet50) to classify patches corresponding to tree species (Multi-Label Patch (MLP) algorithm). Assess (a) Quality of the results obtained with the amount of data available (2800 images), and (b) Degree of improvement achieved by Transfer Learning.
2. Develop a semantic segmentation algorithm for tree species that is precise (DICE coefficient) and efficient (computation time), using three separate algorithmic approaches and two DL networks.
3. Evaluate the applicability of the MLP algorithm to another practical problem: Detection of an invasive tree species in a coastal forest.

To pursue these objectives we worked with two different in-house UAV-acquired datasets: Seven Winter mosaics of a mixed mountain forest and a mosaic of a pine tree plantation mixed with broad-leaf trees. We annotated the data and made it publicly available, at https://doi.org/10.5281/zenodo.3693326. The source code of our research is also available on demand. For our first objective, we used a MLP classification algorithm implemented using a ResNet50 architecture. We thoroughly studied the effects of learning rates (LR) on the output: (a) Not using Transfer Learning (b) Transfer Learning from ImageNet [16] and (c) Transfer Learning twice, first from ImageNet, then from the Planet satellite image dataset [17]. For the second objective we used three algorithms to obtain semantic (pixel-wise) segmentations of the deciduous vs. evergreen tree classes in winter mosaics: (1) The MLP algorithm as a standalone tool, (2) The MLP algorithm including a (non-DL) watershed segmentation refinement step and (3) A patch-based semantic segmentation algorithm using a UNet architecture.

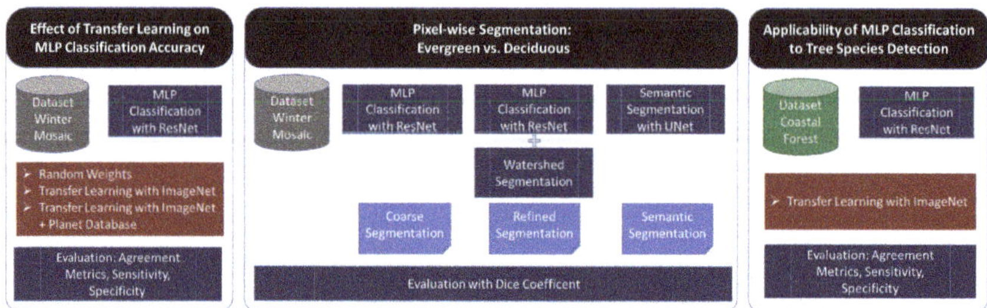

Figure 1. Overview of the contributions presented

The paper is organised as follows: Section 1.1 reviews the state of the art of DL applied to drone-acquired forestry images. Section 2 describes the data used in our study and provides details about the DL methodology used (problems solved, network architectures used). Section 3 contains experiments and a description of their results. In the first experiment, related to objective 1, we discuss the effects of Transfer Learning in tree species classification in winter mosaics. Then we discuss the effect of architecture and problem formalization by comparing our Resnet50 and UNet approaches (objective 2). Then we pursue objective 3 by using our MLP algorithm in a different application in the field of invasive tree species to show that the approach considered can be adapted to different practical problems. Section 4 provides interpretation of the importance of the results obtained and their implications for the research area. Finally, in section 5 we summarize the main results and outline the main conclusions obtained in our study.

2. Methodology

Most of the forests considered for this study are natural (i.e., unmanaged) and located in steep mountainous areas that make field surveys difficult. In addition, forests deal with invasive tree species which have a major impact on the structures, properties and function of forest ecosystems [18]. The abundance of precipitation in the climate of Japan makes data gathering with drone missions challenging which limits the amount of data. Furthermore, we faced the problem of unevenly distributed trees making image recognition and segmentation tasks more complicated (previous studies of forests were performed mostly in plantations or well-managed forests located in flat areas see, for example [19]). Consequently, acquiring and annotating large datasets made up of millions of images such as ImageNet [16] was not feasible for us. Our research was, thus, motivated by the reported capacity of Deep Learning networks to work with smaller datasets as well as it has benefits from Transfer Learning.

2.1. Data Acquisition

Having a sufficient amount of data is important for DL application. This study data were acquired using a drone (DJI Phantom 4), which is an easy-to-use, compact and lightweight drone equipped with a 12 megapixel stabilized camera that produces high-resolution geo-referenced images. The flights were performed with the autonomous mode, which standardized the acquisition protocol.

2.1.1. Location 1: Yamagata University Research Forest

Yamagata University Research Forest (*YURF*) is located on the main Island of Japan Honshu, covers an area of 750 ha and is crossed by the Wasada River (38°33′21.4″N 139°51′42.5″E). Sites close to the river are flat while the others are dominated by steep slopes reaching altitudes of 860 m. The forest is composed of mixed natural deciduous broad-leaf stands and man-made evergreen tree plantation [7]. The focus of this study is on mixed forest sites, which are located in lower altitude areas. For this location we decided to use the seasonality of mixed forests, by collecting images in the winter season aiming at gaining information on the locations of the stands of evergreen and deciduous trees.

Additionally, the seasonal data provided information such as the number of deciduous tree stems (Section 2.2). Therefore, images were captured at three different dates in late winter 2018 presenting differences in illumination and tree age. We performed seven flights in five separate study sites (Figure 2). For site 1, three flights were done at different times in winter season, covering areas of 3 to 8 ha. The duration of the flights ranged between 15 and 33 min. A flight altitude of 80 m was chosen for flat areas while altitudes up to 205 m were used in the areas with steeper slopes. In order to improve the Ground Sampling Distance (GSD), which is the distance between centre points of each image of the ground expressed in cm/px, two flight plans where performed in steeper slope areas. GSDs between 2.79 and 4.48 (cm/pix) were achieved. The overlap of the images was adjusted for each flight and varying between 90 and 96 % of front and side overlap. The number of raw images was between 233 and 556 per site. We used the following abbreviations for our sites: mosaics wm3, wm4 and wm5 belong to the site 1 while the rest are all from different sites (wM1 = site 3; wM2 = site 2; wM6 = site 6; wM7 = site 4).

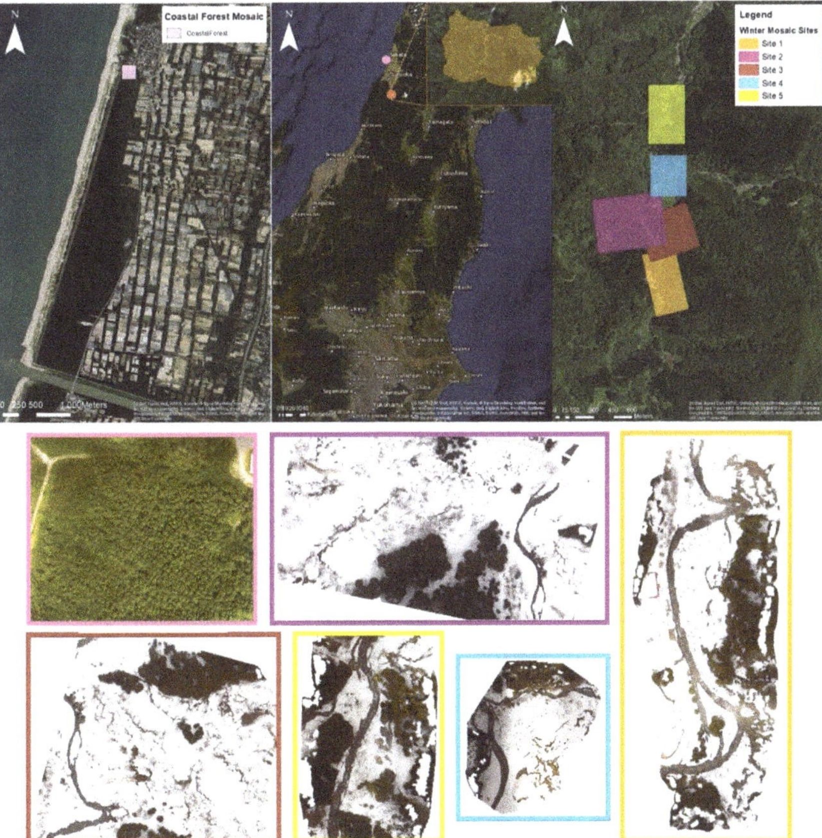

Figure 2. Location of the sites where data was acquired. The middle upper map shows the location of both study areas in Japan. YURF is marked as orange point and the coastal forest as pink point. The map also contains the (orange) area of YURF. The lower six images show the location of coastal forest site (left) and a map of the winter mosaic sites (right). At the bottom representative mosaics for each site are shown.

2.1.2. Location 2: Shonai Coastal Forest

The second study site was located in the coastal area of Shonai Region (Figure 2) (38°52′35.6″N 139°47′54.3″E). This is a 150-year-old black pine (*Pinus thunbergii*) coastal forest plantation [7]. Black pines are evergreen trees with a high tolerance to acidity, alkalinity and salty soils, as well as drought conditions [7,20]. This forest was planted to protect the surrounding area from strong winds and sand movements. From the early 1990s this forest has been invaded by black locust trees (*Robinia pseudoacacia*) that is expected to weaken their windbreak potential. This tree species is well-known for its rapid growth and the high biomass production with a high impact on the structure and function of tree communities because of its leguminous features [7]. The task with this forest is to precisely detect the location of invasive tree species in order to analyse their distribution and impact on the coastal forest. Therefore, we performed one flight in summer 2019 collecting, approximately, 1000 images at a flight altitude of 30 m covering an area of 2.8 ha.

2.2. Data Processing and Annotation

The raw images acquired by the drone were processed using Metashape software [21]. This software uses multi-view 3D reconstruction algorithms to align the pictures taken by the drone creating image mosaics and producing DEMs. We generated seven mosaics of the forest site in YURF and one mosaic for the coastal forest. The annotation process was done manually using GIMP [22] image manipulation software in the form of binary layers belonging to each class. The classes were decided by objects/characteristics that we were able to identify on the images. We used this annotation method, that needed 4 to 7 hours per mosaic as a simple, relatively fast way of annotating data for general applications. For the mixed forest winter mosaics five layers were annotated: River, Man-made, Uncovered, Deciduous and Evergreen (Figure 3). For the coastal mosaics four layers were annotated: "Black locust", "Soil", "Other trees" (containing black pine trees) and "Man-made" (Figure 4). The "other trees" classes corresponded mainly to pine trees but included also a small number of trees from other species. The manual annotations were used as ground truth data.

We then divided each of our mosaics into axis-aligned, square patches of the same side length (hereafter *size* of the patch) (Figure 3). We considered the annotations as well as the original data, obtaining a set of patches along with patch-wise binary masks for each annotated class.

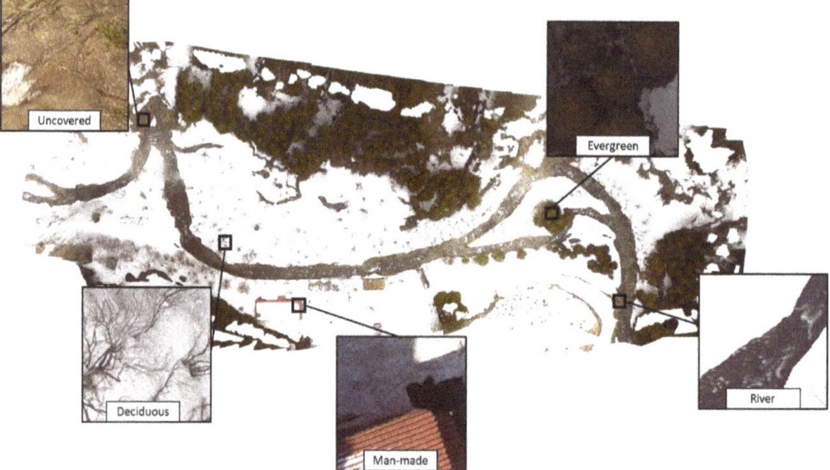

Figure 3. Winter mosaic. Examples of patches each containing a single different class.

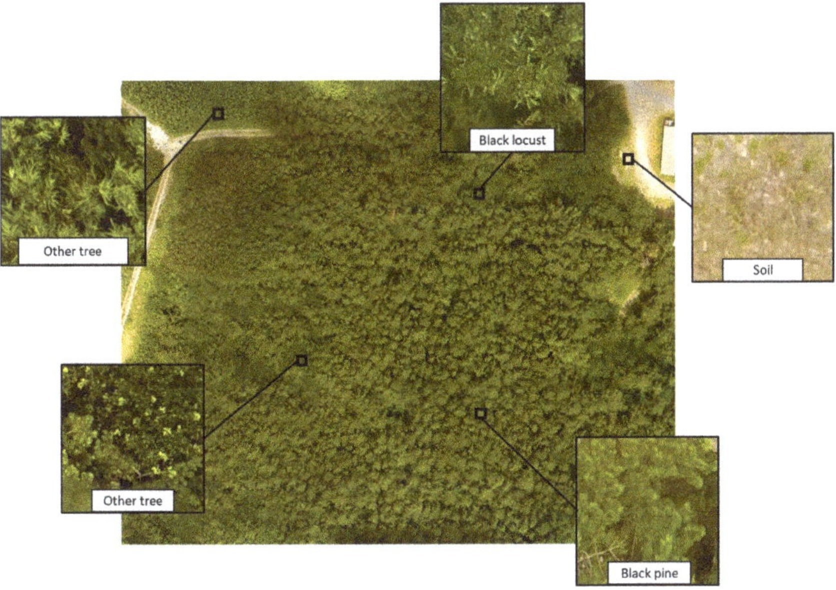

Figure 4. Coastal forest mosaic. Examples of patches each containing a single different class.

Any algorithm used in a practical setting is heavily influenced by the characteristics of the data. In our case, the disposition of the forest (see Section 2.1) prevented us from using Ground Control Points (GCP). GCPs are meant to be geo-localised easily distinguishable points that aid mosaicking software identify corresponding regions in different images during mosaic construction. Sites presenting dense and unmanaged forests (Figure 5B,F) prevent the placement of GCPs inside the forest making the use of GCPs impractical. Also derived from tree distribution, Figure 5B shows the difficulties to identify single trees. However, the use of winter mosaics (Figure 5A) shows an image of the same area where stems are visible and their numbers can be counted. Furthermore, the orography of the forest made data acquisition limited in terms of the amount of data that could be gathered. Figure 5C is an example of one of the steep slopes that are common in this area. This picture also shows deciduous trees in different conditions and frequently mixed with other classes. The understory vegetation present has the same colour of the deciduous tree and often appears mixed with them. A similar problem happens with the "River" class in Figure 5A. Another limitation likely resulting from not using GCPs is that Metashape produced some image registration artefacts (Figure 5C,D,G). This increased the difficulty of automatically processing the images. Images F to H present examples in the coastal forest. Image G shows black locust trees while image H shows examples of black pine and other trees of the area. The trees in this mosaic present only small differences in colour and structure. Image H shows that black locust trees are often smaller than black pine trees and therefore often covered by them.

2.3. Problem Formalization

The automatic division of the data and mosaic-level annotations into regular patches allowed us to formalize our problems in two different ways. First, we focused on identifying what classes where present in each mosaic patch, which is known as multi-label classification (Section 2.4). As a second formalization we focused on classifying each pixel in each mosaic patch. This problem is known as *semantic segmentation* [23]. The aforementioned patch multi-label classification algorithm already gave us an initial "coarse" approximation to this semantic segmentation. In order to refine it we used the classical watershed [24] image segmentation method (Section 2.4.1). We also considered the UNet DL architecture [25] (Section 2.5). A full pipeline of our work is provided in Figure 6, while Figure 7(left) shows a patch of one of the winter mosaics. The central part of the Figure 7 shows the manual annotation for this patch. Blue pixels are classified as the "Evergreen" class, yellow pixels to the "Uncovered" class and green pixels as the "River" class.

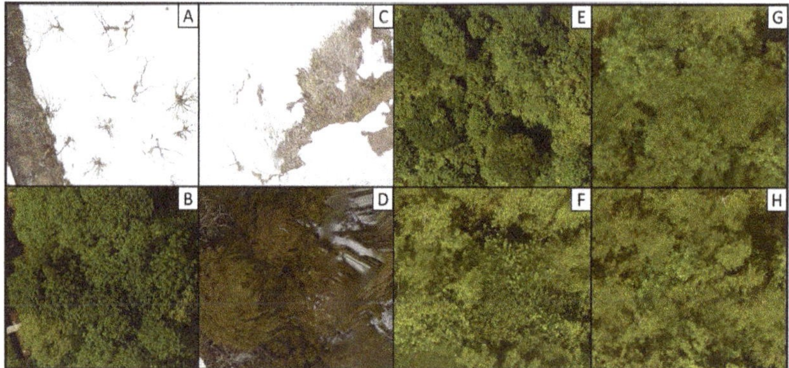

Figure 5. Example images of difficulties faced with the Data. Images (**A–E**) are examples of the mixed forest; (**F,H**) are examples from the coastal forest.

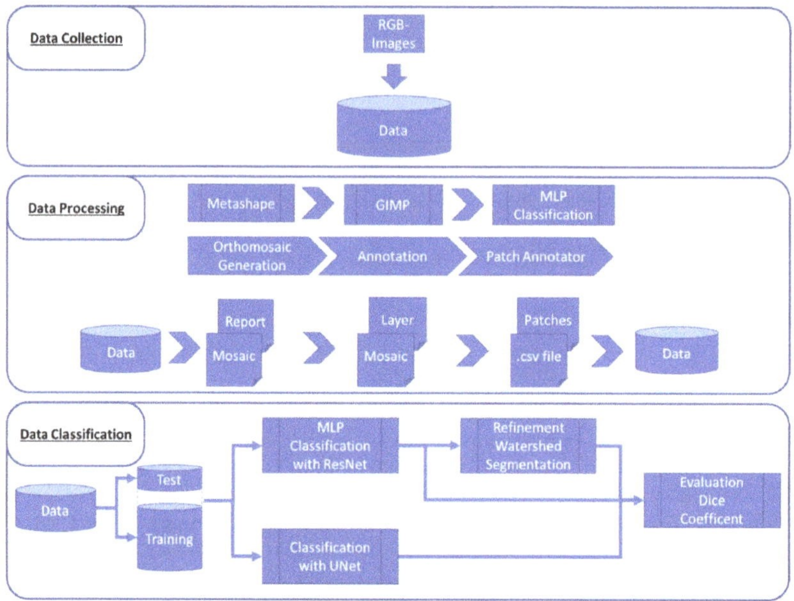

Figure 6. Algorithm Pipeline.

2.4. Multi-Label Patch Classification Using ResNet

Previous studies have shown the efficiency of DL networks to classify forestry images, specifically, [10,11] because they rely on the ResNet architecture [26]. We used the ResNet variant with 50 layers, known as ResNet50. Although the aforementioned approaches used a tree crown segmentation step that used the DEM produced by the drone, we decided to work only with RGB data to emphasize our focus on image data. We considered our data and annotations divided into patches and each patch was assigned a list containing the classes in it. For example, the list in the case of the patch visible in Figure 7 would be river, uncovered and evergreen. Thus, patches may belong to more than one class each with a probability value for each patch. Given high enough probabilities in more than one class, the patch would be "labelled" repeatedly. Thus, this algorithm will be referred to as the MLP based classifier. The ResNet50 network was then trained to classify these patches. A subset of the data was used for training and the remaining data was used to validate the quality of the trained model at predicting the correct classes. Eighty percent of the dataset was randomly chosen for training and the remaining 20% was used for testing (Section 3).

Figure 7. A small section (patch) of a winter mosaic. Patch of an RGB Image as captured by the drone (**left**). Mask with the annotations showing the class of each pixel (**middle**), in this case blue pixels belong to the "Evergreen" class, yellow pixels to the "Uncovered" class and green pixels to the "River" class. White pixels are assigned to the "void" class. This mask serves as the ground truth for the semantic segmentation formalization. Superposition of the two previous images (**right**). Consequently, the ground truth for this patch was the list [river, uncovered, evergreen] for the multi-label classification formalization.

DL architectures need less data than previous approaches, however, having sufficient data to produce results of satisfactory quality is still a problem in research areas such as forestry, where data acquisition is often problematic. Transfer Learning represents a way to improve the quality by initializing the weights of the matrices conforming the DL network to those obtained in the solution of a similar problem. In our study we used Transfer Learning from: (1) a general-purpose object classification problem codified in the ImageNet database [16] and (2) the closer problem of multi-label classification in satellite images of the amazon forests codified in the Planet dataset [17] (see Section 3.1).

2.4.1. Segmentation Refinement Using watersheds

The multi-label patch classification algorithm described generated an initial "coarse" segmentation. All pixels in any patch containing a class would be considered to belong to that class. This patch-wise masks for all classes constitute a segmentation of the mosaic. The "coarse" nature of this segmentation produces two problems. On the one hand, over-segmentation: by assigning all the pixels in the patch to all the classes present, would surely assign pixels to classes they did not belong to. These extra pixels make the masks of classes larger. On the other hand, the coarse class masks therefore generated intersect. This is undesirable in the semantic segmentation problem as each pixel should be classified into one single class. In order to improve the initial segmentation we implemented a refinement step based on the watershed image segmentation algorithm [24]. This algorithm uses binary images representing initial masks consisting of doubtless labelled pixels. Parts of the image that we could confidently assigned to the background were also determined. Any pixel not falling into any of these two cases were labelled as "unknown". Labelled regions were visualised as ridges and unknown areas as basins. Then, water was pictured as expanding from the ridges into the basins until two of the growing ridges meet and watershed lines were determined. These lines defined the segmentation.

In our case, we worked at mosaic level starting with the coarse segmentation of the "River" class. A binary image was generated from it by painting all pixels black that belonged to all patches where the river class had been predicted. All pixels out of these patches were marked as background. Then, this coarse mask was shrunk using a distance transform and pixels that had been deleted by this process were labelled as unknown. Further, all the connected components of the mask were computed and stored in a dictionary with their position and an identifier, along with the information that they belonged to the "River" class. Then, we considered the second class, the "Deciduous" class. The previous process was repeated but storing the results from the previous step. Regions that previously formed the background were overwritten, regions that were unknown before or where two labels were assigned were labelled as unknown. The process was repeated for all classes and the images were partitioned into (1) Background (2) unknown and (3) initial regions. Each initial region had the information attached as to what class it belonged to. The watershed algorithm was subsequently run to create a finer segmentation without intersection between the different classes.

2.5. Forest Mosaic Segmentation Using UNet

The UNet Deep Learning architecture [25] was originally developed for medical image segmentation ([27]) but has since then been used in a variety of applications. The UNet architecture is composed of two parts known as "paths": the encoder and the decoder path. The encoder path extracts features using convolutional layers and reduces the size of the images using max pooling layers. At the end of the encoder path the images are greatly reduced in size and the transformations they were subjected to are stored in the weights of the matrices along the path. The decoder path, moves back to full size ones by replacing pooling operators with upsampling operators. High resolution features from the encoder path were combined with the upscaled output in order to localize them. Successive convolution layers learn to re-assemble the output more precisely based on this information. An algorithm was implemented using the UNet architecture to perform semantic segmentation of the winter mosaics. We considered the data and pixel-wise-annotation patches described in Section 3 and used them to train a UNet network. Whenever a new mosaic needed processing, (1) we divided it in patches, (2) predicted the semantic segmentation of each patch using the trained UNet model and (3) joined all patch segmentations together to obtain a semantic segmentation for the whole mosaic.

2.6. Evaluation Criteria

We used several metrics to evaluate different aspects of our algorithms performance. First, the capacity of the MLP algorithm to correctly predict the labels in every patch was evaluated: **Full Agreement (FA)**: Stands for the percentage of patches where the predicted labels matched exactly those in the ground truth. **Full Agreement with False Positives (FAFP)**: In this case all ground truth classes were correctly predicted but some intrusive classes were added. **Partial Agreements (PA)**: Patches where some but not all of the correct classes were predicted and where some classes might have been added. Finally, the complementary of this measure (when the sets of ground truth and predicted labels had no intersection) was considered **No Agreement (NA)** .

In order to target the predictive capacities of our algorithms we considered patch labels for the MLP algorithm and pixel labels for the semantic segmentation algorithm. For all of them we considered the relation between predicted values and real values as stated in the ground truth and broke them into the usual classifications of **True Positives: TP, False Positives: FP, True Negatives: TN, False Negatives: FN**. Furthermore, and in order to focus on the classes of most practical interest (evergreen and deciduous for winter mosaics and black locust and other trees for the coastal forest mosaic) classification measures were computed on them: Sensitivity, Specificity and Accuracy. In order to evaluate semantic segmentation results, the DICE coefficient that considers the ground truth mask and the predicted mask to define TP, FP, TN and FN pixels was used. Subsequently, it compares the number of correctly predicted positives (TP) with the number of errors (FP + FN).

$$SENS = \frac{TP}{TP+FN} \quad SPEC = \frac{TN}{TN+FP} \quad ACC = \frac{TP+TN}{TP+TN+FP+FN} \quad DICE = \frac{2TP}{2TP+FP+FN}$$

3. Results

In this section, we present experiments using real data corresponding to seven winter mosaics and one summer mosaic from the coastal forest covering a total area of 38.5 ha. Of the five winter mosaics from YURF, three corresponded to the same site on different days and under different lighting conditions. All the algorithms described throughout the paper were implemented using the python programming language. The ResNet and UNet DL architectures were implemented using the Fastai [28] library. The watershed algorithm was implemented using the opencv computer vision library [29]. All experiments where run in workstation using a Linux Ubuntu operating system with 10 dual-core 3GHz processors and an NVIDIA GTX 1080 graphics board. For experiments 1 and 3 the data were randomly divided into (80%, 20%) training and validation/testing. For experiment 2 a leave-one-out approach was done with each of the seven winter mosaics using one for validation/testing and the other six for training.

3.1. Experiment 1: Transfer Learning and Multi-Label Patch Classification

The data were randomly divided into (80%, 20%) training and validation/testing. The patch size was chosen to be 150, see Section 3.2 for the effect of this parameter. With this parameter, the annotated patches contained the deciduous label in 52.26% of the cases, the evergreen class in 39.33%, uncovered in 28.29%, river in 12.84% and man-made in 1.26%. Notice that, as patches often belong to more than one class, these percentages add to more than 100%.

In order to assess the impact of Transfer Learning several "starting models" were built and trained using our images. Each model was considered in two forms, frozen and unfrozen. When a frozen model was re-trained only the final layers of the model were changed. When an unfrozen model was re-trained, all of the layers were modified. The starting models that we considered were:

1. **Random**: In order to test whether Transfer Learning was necessary, we included a model initialised with random weights. Only results of the unfrozen random model were presented as the frozen random model had poor results.
2. **RN50F, RN50UNF**: We also considered a ResNet50 with preloaded ImageNet [16] weights. The inclusion of this model allowed us to study whether or not a general-purpose classification model could be fit to solve our problem using a relatively low number of images. This model was re-trained frozen (RN50F) and unfrozen (RN50UNF).
3. **RN50 + PLANET-UNFF, RN50 + PLANET-UNFUNF, RN50 + PLANET-FF, RN50 + PLANET-FUNF**: We also considered the ResNet model again and re-trained it using the PLANET dataset of satellite images of the Amazon rainforest [17]. In order to assess whether better results could be obtained when training from a problem (classification of satellite images of tropical vegetation) being more similar to our data. The ResNet model was considered frozen and unfrozen as before, RN50 + PLANET-F, RN50 + PLANET-UNF. These two models were subsequently retrained with our images each considered frozen and unfrozen producing 4 models: RN50 + PLANET-UNFF, RN50 + PLANET-UNFUNF, RN50 + PLANET-FF, RN50 + PLANET-FUNF.

The learning rate of a DL model is a parameter that controls the step size of the optimizer that changes the weights in each iteration of the training phase. We tested various LR values in all the different Transfer Learning approaches. In order to present a comprehensive picture, among all values tested we present (Figure 8) those from 1×10^{-5} to 0.9 with 10 sampling points at each exponent value $(1 \times 10^{-5}, 2 \times 10^{-5}...9 \times 10^{-5}, 1 \times 10^{-4}, 2 \times 10^{-4}...)$.

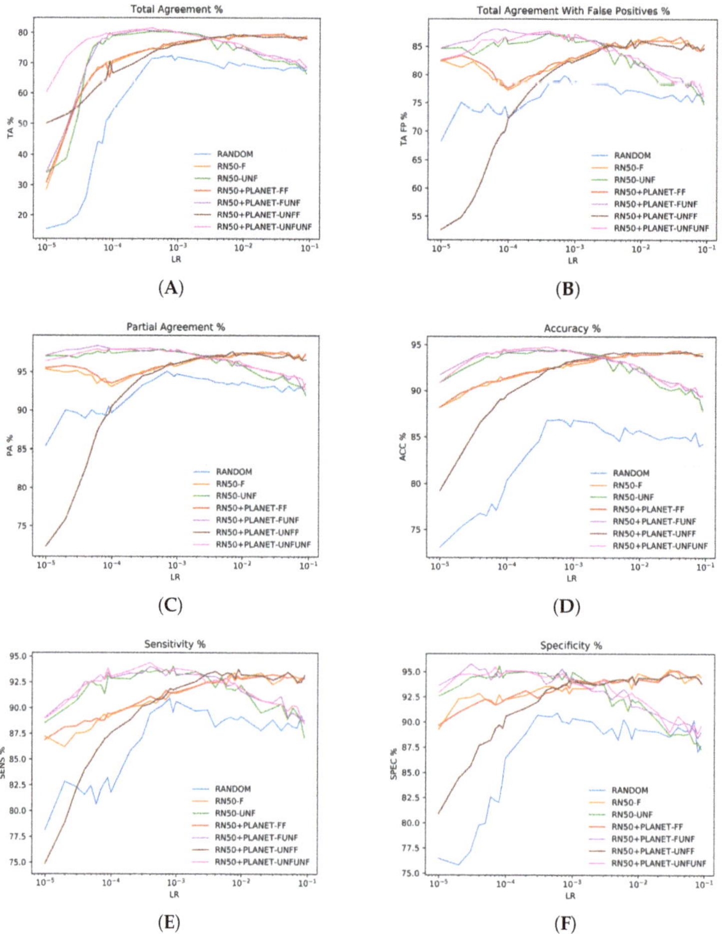

Figure 8. (**A**) Total Agreement (TA), (**B**) Accumulated value when we add to the previous the Total Agreement with False Positives(TAFP), (**C**) Accumulated value when we add to the previous the Partial Agreement (PA), (**D**) Accuracy (ACC), (**E**) Sensitivity (SENS), and (**F**) Specificity (SPEC) for winter mosaic data. Patch Size is 150. Frozen models end with an F, unfrozen models with UNF.

Agreement: Agreement results (TA, TAFP, PA) provided us with a general picture of the capacity to all the trained models to classify our patches with all the possible labels (River, Deciduous, Uncovered, Evergreen and Man-made). The best TA results were obtained by the RN50+PLANET-UNFUNF model with a value of 81.58 at a learning rate of 4×10^{-3}. The first trend that could be observed in plots A,B,C in Figure 8 is that the learning rate that provides better result for a model is determined by whether the model was frozen or not. Specifically, the three unfrozen models obtained best results with smaller learning rates of around 1×10^{-4} while frozen models achieve their best results with learning rates of 0.04. This is consistent with previous results reported in [10] where the best learning rate for a frozen model trained using ImageNet weights was 0.01.

The importance of Transfer Learning was indicated in the TA peaks of the different approaches. The model initialised with random weights peaked at 72.53 TA. a lower value than the other models. Frozen models achieved good results with models RN50F, RN50+PLANET-FF and RN50+PLANET-UNFF peaking at 79.51, 79.42 and 79.62 TA, respectively. This represented an improvement of 9.78% over the model with random weights. Similar trends could be observed for the accumulated TAFP and PA plots with peaks of 87.76 TAFP and 98.12 PA for the RN50+PLANET-UNFF model. Likewise, similar results could be observed with unfrozen models. Best results of RN50UNF, RN50+PLANET-FUNF and RN50+PLANET-UNFUNF peaked at 80.39, 80.98 and 81.58 TA. This was an improvement of 10.83%, 11.66% and 12.48% over the random weights model. Best overall results were obtained from the model that was built using first an unfrozen version of the ImageNet model to train the Planet dataset and then leaving the resulting model unfrozen to train with our images. The general tendencies observed here were confirmed using difference of means hypothesis tests (*t*-tests as data size was > 25). Even models presenting smaller differences in the TA peak (like the difference between the performance of the RN50UNF and RN50+PLANET-UNFF) were found to present statistically significantly different means with significance level 0.05.

Classification of deciduous and evergreen classes: The best TA was obtained with the RN50+PLANET-UNFUNF and a value of 81.58%. Patches without Total Agreement presented errors when classifying the three classes that were not of direct practical interest (river, uncovered, man-made). The number of patches containing them was low compared to patches presenting the other two classes (deciduous and evergreen. However, the results of our two classes of interest had high sensitivity, specificity and accuracy values (plots D, E, F in Figure 8, average values for the two classes) so we did not perform data balancing. Best average accuracy results between the two classes (94.80%) were obtained by the RN50+PLANET-UNFUNF model (that also obtained the top TA value). The clearly defined boundaries of evergreen trees resulted in an accuracy of 97.24%, while the deciduous tree class presented less defined edges and obtained 92.36%. Sensitivity and specificity values of the same model were also high with an average of 94.38% (94.75% Ev, 94.01% Dec) and 94.50% (98.73% Ev, 90.27% Dec). The larger difference appeared in specificity, where we observed an 8% difference. This can be explained by misclassifications of the deciduous class into mainly the uncovered class. Still, the classification results were high with values of up to 97%. This indicated that our patch-dividing and classification approach using a multi-label ResNet DL classifier was successful with our amount of data.

3.2. Experiment 2: Semantic Segmentation

We present results from three approaches, the patch based coarse segmentation produced by the algorithm in Section 2.4, the refinement of that segmentation using watershed segmentation and the patch-based UNet algorithm presented in Section 2.5. The first goal of this experiment was to test the performance of the algorithms in real-life conditions. We trained our models with data from six mosaics and validated (tested) with the one that had been left out. This approach relates to the use case where an already trained system receives a new mosaic for automatic classification. Of our seven mosaics three belonged to the same site. Consequently, in most cases (4/7), test images were of trees that the trained system had never seen before. In some other (3/7) images of trees previously seen under different conditions where used for testing. These two sets of results allowed us to discuss about the generalization power of our algorithm and the possibility of over-fitting. As a second major goal, we studied what the effect of choosing one problem formalization or DL architecture over another had in the insights gained from the data.

Patch size and learning rate: The proximity of the multi-label patch classifier (seen as a coarse segmentation) output to the manual annotations depended on two factors: First, the accuracy of the classification model, where wrongly classified patches would result either in False Positive or False Negative regions in our coarse mask and second the size of the patches, which produced an approximation error that grew with the size of the patches. At the same time, smaller patches took longer to compute. We considered patch sizes from 500 to 25 for both ResNet Patch based classification (known as "coarse" segmentation) and the version with Watershed refinement (noted "Refined"). The learning rate presented was the highest among all learning rates studied in Section 3.1 for each patch size. The training time needed for the Resnet varied from under 10 minutes for patch size 500 to two hours fifty minutes for patch size 100 up to over 40 hours for patch size 25. Concerning the UNet, we tried several patch sizes and learning rates where the patch size of 500 led to better results. In this study we only present five illustrative examples of learning rates. Tables 1 and 2 present the DICE coefficient for all the algorithm variants. The average training time of the UNet was of over 11 hours.

DICE coefficient: Concerning semantic segmentation for all sites, the best results for the UNet were (0.709,0.893) DICE for (deciduous,evergreen) for LR=0.0005. The ResNet50 coarse segmentation obtained (0.790,0.883) for patch size 25 with the refined version (watershed post-processing) reaching (0.733,0.855). On the other hand the UNet semantic segmentation achieved and average value of 0.893 for the "Deciduous" class, showing how its pixel wise approach adapts better to this class that presents less well defined borders. The learning rates presented for the UNet show some representative examples of all the learning rates considered.

Table 1. Comparison of semantic segmentation approaches for the "Deciduous" class in winter mosaics (wM*). The first half of the table contains the results for the UNet model, the second half contains results from the patch based multi-label ResNet classifier. The first column contains the size of the patches used. Rows marked "Coarse" use only the ResNet classifier while rows marked "Refined" also use watershed refinement. Last two columns are the average values (AVG) between all mosaics and between mosaics of site 1.

Deciduous		wM1	wM2	wM3	wM4	B	wM6	wM7	AVG	AVG Site1
UNet										
LR 0.3		0.253	0.222	0.000	0.402	0.254	0.329	0.397	0.265	0.219
LR 0.04		0.607	0.171	0.372	0.537	0.494	0.275	0.308	0.395	0.468
LR 0.003		0.650	0.631	0.571	0.448	0.429	0.254	0.326	0.473	0.483
LR 0.0005		0.824	0.772	0.797	0.609	0.596	0.664	0.699	0.709	0.667
LR 6×10^{-5}		0.792	0.740	0.784	0.591	0.637	0.639	0.620	0.686	0.671
Resnet										
Patches 500	Coarse	0.665	0.572	0.659	0.569	0.666	0.492	0.521	0.592	0.631
	Refined	0.704	0.597	0.695	0.501	0.583	0.510	0.568	0.594	0.593
Patches 300	Coarse	0.581	0.515	0.604	0.555	0.594	0.441	0.398	0.527	0.584
	Refined	0.585	0.505	0.501	0.552	0.666	0.450	0.449	0.530	0.573
Patches 200	Coarse	0.684	0.576	0.687	0.609	0.670	0.559	0.514	0.614	0.656
	Refined	0.720	0.613	0.708	0.510	0.597	0.555	0.616	0.617	0.605
Patches 100	Coarse	0.795	0.702	0.777	0.683	0.767	0.702	0.698	0.732	0.742
	Refined	0.796	0.703	0.780	0.681	0.764	0.707	0.704	0.733	0.741
Patches 50	Coarse	0.860	0.779	0.782	0.711	0.790	0.761	0.753	0.777	0.761
	Refined	0.667	0.614	0.514	0.570	0.670	0.432	0.509	0.568	0.585
Patches 25	Coarse	0.896	0.839	0.752	0.711	0.795	0.78	0.757	0.790	0.753
	Refined	0.706	0.6	0.448	0.562	0.611	0.494	0.488	0.558	0.540

The Refined ResNet algorithms, watershed post-processing helped to improve the coarse segmentation up to a certain patch size. For these models, the training time was the fastest among the three models and the results obtained were not far from the best obtained. This represents an example of specialized computer vision algorithms that can complement the knowledge gained by using DL. These algorithms, however, cannot easily be used by non-experts and require careful fine-tuning as exemplified by the failure of the watershed refinement to produce satisfactory results for small patch size. This was most likely due to small misclassified regions growing into larger regions due to poor parameter choice than actual limitations of the proposed approach. However, running the watershed refinement took less than a minute for any of the tested mosaics. Training time of the UNet network or any of the networks with smaller patch sizes was much larger than that.

The difference of the average DICE coefficient among sites were small. Mosaics 3, 4 and 5, representing site 1, show differences of 6% for the with UNet network defined "Deciduous" class. An improvement of 1% for the same mosaics and classes with the refined ResNet algorithm was attained.

Table 2. Comparison of semantic segmentation approaches for the "Evergreen" class in winter mosaics (wM*). The first half of the table contains the results for the UNet model, the second half contains results from the patch based multi-label ResNet classifier. The first column contains the size of the patches used. Rows marked "Coarse" use only the ResNet classifier while rows marked "Refined" also use watershed refinement. Last two columns are the average values (AVG) between all mosaics and between mosaics of site 1.

Evergreen		wM1	wM2	wM3	wM4	wM5	wM6	wM7	AVG	AVG Site1
UNet										
LR 0.3		0.856	0.822	0.364	0.576	0.708	0.785	0.620	0.676	0.549
LR 0.04		0.894	0.806	0.798	0.767	0.828	0.769	0.612	0.782	0.797
LR 0.003		0.855	0.802	0.803	0.626	0.761	0.763	0.648	0.751	0.730
LR 0.0005		0.931	0.896	0.906	0.815	0.897	0.928	0.876	0.893	0.873
LR 6×10^{-5}		0.914	0.857	0.859	0.794	0.866	0.896	0.818	0.858	0.840
Resnet										
Patches 500	Coarse	0.607	0.650	0.511	0.528	0.593	0.644	0.644	0.597	0.544
	Refined	0.676	0.710	0.548	0.375	0.604	0.700	0.729	0.620	0.510
Patches 300	Coarse	0.697	0.729	0.651	0.634	0.661	0.723	0.691	0.684	0.648
	Refined	0.804	0.345	0.744	0.708	0.676	0.800	0.809	0.698	0.709
Patches 200	Coarse	0.722	0.767	0.638	0.619	0.697	0.736	0.710	0.698	0.651
	Refined	0.845	0.850	0.759	0.722	0.769	0.807	0.721	0.782	0.750
Patches 100	Coarse	0.865	0.844	0.796	0.767	0.806	0.847	0.803	0.818	0.789
	Refined	0.887	0.869	0.884	0.689	0.873	0.907	0.877	0.855	0.815
Patches 50	Coarse	0.914	0.888	0.867	0.820	0.865	0.904	0.852	0.873	0.851
	Refined	0.844	0.744	0.880	0.514	0.523	0.865	0.735	0.729	0.639
Patches 25	Coarse	0.923	0.875	0.881	0.84	0.89	0.926	0.843	0.883	0.870
	Refined	0.424	0.529	0.621	0.514	0.55	0.656	0.675	0.567	0.562

3.3. Experiment 3: Detection of Invasive Tree Species in the Coastal Forest

We tested whether the general approach regarding data handling, annotation and problem formalization considered in this paper could be used to solve other practical problems. The winter mosaic application is an example of a classification problem of trees with no leaves (deciduous) against trees with full leaf cover (evergreen). This is an interesting problem that has multiple applications as for instance the classification of trees attacked by insects outbreaks in relation to healthy trees [12]. We collected data related to a problem of class differentiation in a dense green forest. In this section the MLP algorithm was trained for pine trees, with green-yellowish leaves from black locust, presenting a light green broadleaf. The shape of the leaves was visible on the mosaic but not easy to tell apart even by experts (Figure 9). As the main interest in practice is to detect the occurrences of black locust, we grouped pine trees and a small number of trees from other tree species into a class called "other trees". This is a complex problem in image classification terms as the differences between these "other trees" and the black locust were minor. Moreover, the distribution of the invasive species within the forest was irregular and often made up of very small patches sometimes including single trees. Furthermore, the coastal forest dataset showed a ratio of 90:10 between the "other trees" and "black locust" class.

As in experiment 2, we ran the experiment only with the ResNet50 with the ImageNet weights considered frozen and unfrozen. The models were retrained using the coastal forest data and the results with the highest accuracy were obtained with the unfrozen model. Figure 10 presents representative examples of the results obtained.

A 75% rate of True Positives for black locust and less than 10% False Positives (90.826 True Negative rate) was achieved (Figure 10 right side). At the same time, the other trees were detected with over 95% True Positive rate and about 10% of False Positives. These results provided valuable insight into the problem of detecting black locust trees in mixed forests. Furthermore, the remainder of the results indicated that, by using DL tools in a more sophisticated approach or complementing it with computer vision algorithms, these insights could likely be improved. For example, learning rate 2×10^{-3}, obtained a high TA (90.8) with only 62% sensitivity for the "black locust" class. This version of the classifier presents low False Positive recognizing values for both classes and, thus, high accuracy by largely ignoring the black locust class, which is less frequent. This was already apparent in Figure 9 but we computed that the "other trees" class appeared 10 times more frequently than the "black locust" class. The TA value in this case (42.78) was low but the sensitivity value of the "black locust" class was the highest in the whole experiment reaching almost 85%. This resulted in a high confusion between the "black locust" and "other trees" classes, with the specificity of the "black locust" class dropping to about 65%.

Figure 9. Example of the coastal forest. Mosaic with marked black locust areas (**left**); Example of Japanese black pine trees (**middle**); Example of black locust trees (**right**).

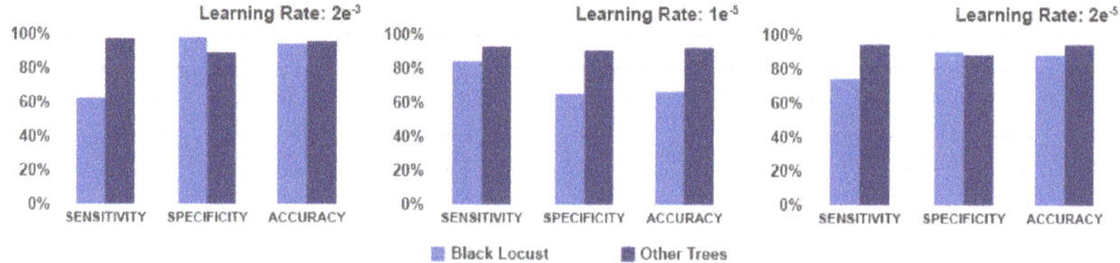

Figure 10. Results for the MLP classification of the Coastal Forest Mosaic for three learning rates (LR). The bar charts show classification measure for the "black locust" and "other trees" classes: Sensitivity or True Positive rate (SENS), Specificity or True Negative Rate (SPEC) and Accuracy (ACC)

4. Discussion

In our first experiment, the effects of Deep Learning were quantified concerning the problem of multi-label patch classification of winter mosaics. The available seven mosaics were an insufficient amount of data to reach high accuracies for tree species classification when training a ResNet network from scratch. Precisely, starting from random weights led to a TA value of 72.53. Our experiments showed that Transfer Learning from ImageNet (a general-purpose dataset) was essential to obtain high quality results with a 9.78% improvement in Total Agreement (up to 79.63 TA). Ref. [13] reported a similar increase in conifer seedling detection metrics ranging between 3% and 10%. Additionally, a further 2.7% improvement in our Total Agreement (reaching 81.58 TA) was observed when Transfer Learning from the Planet dataset, which is more closely related to our images (Section 3.1, Figure 8). Thus, Transfer Learning is necessary to obtain reliable results for our winter mosaics problem. Furthermore, the smaller improvement when Transfer Learning from the Planet dataset was performed (2.7% over 9.78%) suggests that dedicating too many resources to find closely related problems may not be cost-effective. Furthermore, these results indicated that making data and annotations from forestry-related Deep Learning research publicly available could speed up the development of this research area by decreasing the amount of data needed by future contributions.

In the last step of experiment 1 we focused on the results of only evergreen and deciduous trees since tree species classification was the aim of our study. High accuracy, sensitivity and specificity values were achieved. Specifically, regarding sensitivity, evergreen trees reached values of 94.75% and deciduous trees reached 94.01%. Comparing these results to previous work is difficult as previous studies have used single-label (rather than) multi-label classification. Also, they have used different DL networks, applied in problems with different levels of complexity [10–12]. Nevertheless, and for the sake of context, we present our obtained sensitivity values with these previous works in Table 3. Specifically, [10,11] reached average sensitivities of 89% and 81%, which were lower than the ones achieved in our experiment. Ref. [12] reached an average value of 91.84%, which was close to our results but the single classes showed a high variability from 81.25% to 100%.

Table 3. Comparison with State of the Art methods.

Onishi and Ise 2018		Safonova et al. 2019		Natesan et al. 2019		Our results	
Classes	Sens.%	Classes	Sens.%	Classes	Sens.%	Classes	Sens.%
Deciduous broadleaf	95.83	Healthy	100.00	Red Pine	67.00	Deciduous	94.75
Deciduous coniferous	84.62	Colonized	86.11	White Pine	77.00	Evergreen	94.01
Evergreen broadleaf	68.00	Recently died	81.25	None Pine	97.00		
Chamaecyparis obtusa	91.95	Deadwood	100.00				
Pinus elliottii/taeda	94.12						
Pinus strobus	88.89						
Others	88.20						
Average	89.00		91.84		81.00		94.38

Our second experiment focused on segmentation approaches of the "Deciduous" and "Evergreen" classes. Best DICE values for these two classes with UNet were (0.709,0.893) and with ResNet (0.790,0.883). These numbers showed that the best results were obtained by UNet for the "Evergreen" class while the "Deciduous" class was better detected by the MLP ResNet approach. The less-defined borders and the overlap between colours of the deciduous trees and the "Uncovered" class makes it difficult to properly segment the trees with the pixel-oriented UNet. In this sense, formalising the problem as a patch labelling problem allowed us to gain some flexibility in the definition of the classes and identify the pixels belonging to this class more precisely. On the other hand, formalising the problem as a semantic segmentation problem allowed us to use the internal coherence of the "Evergreen" class. In addition, the comparison between the DICE coefficients of all mosaics suggests that our algorithms were able to segment totally new mosaics as well as those that they had already seen under other lighting conditions. Overfitting, then, seems not to be present in our results and we could confirm that our algorithms can segment deciduous and evergreen trees in Japanese mixed forest with reasonably good results. Furthermore, the watershed-based refining step provided a fast compromise between the two pure-DL formalizations. Consequently, using larger patch size with watershed refinement might be a solution in situations where training time is a limiting factor. This problem-specific computer vision algorithm illustrates the effect that a more sophisticated use of DL and computer vision techniques can have on forestry research.

In our last experiment we assessed the performance of our simple patch-based approach for solving practical problems such as tree class differentiation. We obtained a 75% rate of True Positives for the "black locust" class with under 10% of False Positives while simultaneously obtaining (95%, 10%) values for the "other trees" class. This showed that DL can provide valuable information about complex classification problems even when used as a "black box". Nevertheless, a deeper analysis exposed an important imbalance in the number of instances of the two classes (black locust and other trees), suggesting that by using techniques such as data re-balancing (oversampling/down-sampling) or computer vision post-processing steps could improve the results even further by reducing the rate of False Positives. Onishi and Ise [10] already confirmed that statement showing an increase of their sensitivity results from an average of 83.1% to 89%. Safonova et al. [12] showed an increase of 12.1% in their average sensitivity data since their data was less unbalanced than [10]. Based on these studies, data re-balancing has the potential to improve the classification results not only for experiment 3, but also for experiment 1. These issues will be considered in our future research.

5. Conclusions

In this work, we analyzed the current role and development possibilities of Deep Learning to solve practical problems in forestry research. Our study provided a simple pipeline based on drone-acquired images for classifying tree species. Our experiments showed that Transfer Learning was essential to obtain good results for patch classification. Better results were obtained when a more closely related dataset was used.

We were also able to obtain semantic segmentations for winter mosaics that reached high DICE values when compared to the ground truth. The effect of the DL model was made apparent by the fact that best results for the "Evergreen" class were obtained by UNet and for the "Deciduous" class by ResNet. Watershed post-processing could be used to reduce the computation time of the most cost-intensive algorithms.

Finally, our experiments also showed that DL provided valuable information about complex classification problems when used as a "black box". Our patch-based classifier provided reasonably good results to find patches containing black locust trees in the black pine coastal forest mosaic. The methodology studied in this paper can, thus, be used to gain insight in other forestry applications.

Author Contributions: S.K., Y.D., M.L.L.C., F.R. and D.S. conceived the conceptualization and methodology, supported the writing—review and editing. S.K., Y.D. and F.R. developed the software, performed the validation, investigation and writing—original draft preparation. S.K. and Y.D. carried out formal analysis. S.K. and M.L.L.C. administrated the data. S.K. and F.R. were in charge of the visualisations. M.L.L.C. and Y.D. supervised the project and provided resources. M.L.L.C. directed the project administration. All authors have read and agreed to the published version of the manuscript.

Funding: This research received no external funding.

Conflicts of Interest: The authors declare no conflict of interest.

References

1. Chen, I.C.; Hill, J.K.; Ohlemüller, R.; Roy, D.B.; Thomas, C.D. Rapid Range Shifts of Species Associated with High Levels of Climate Warming. *Science* **2011**, *333*, 1024–1026. [CrossRef] [PubMed]
2. Esquivel-Muelbert, A.; Baker, T.R.; Dexter, K.G.E.A. Compositional response of Amazon forests to climate change. *Glob. Chang. Biol.* **2019**, *25*, 39–56. [CrossRef] [PubMed]
3. Kherchouche, D.; Slimani, S.; Touchan, R.; Touati, D.; Malki, H.; Baisan, C.H. Fire human-climate interaction in Atlas cedar forests of Aurès, Northern Algeria. *Dendrochronologia* **2019**, *55*, 125–134. [CrossRef]
4. Coll, L.; Ameztegui, A.; Collet, C.; Löf, M.; Mason, B.; Pach, M.; Verheyen, K.; Abrudan, I.; Barbati, A.; Barreiro, S.; et al. Knowledge gaps about mixed forests: What do European forest managers want to know and what answers can science provide? *For. Ecol. Manag.* **2018**, *407*, 106–115. [CrossRef]
5. Anderegg, W.R.L.; Anderegg, L.D.L.; Kerr, K.L.; Trugman, A.T. Widespread drought-induced tree mortality at dry range edges indicates that climate stress exceeds species' compensating mechanisms. *Glob. Chang. Biol.* **2019**, *25*, 3793–3802. [CrossRef]
6. Shimada, T. *State of Japan's Forests and Forest Management—2nd Country Report of Japan to the Montreal Process*; Forestry Agency: Tokyo, Japan, 2009.
7. Lopez C, M.L.; Mizota, C.; Nobori, Y.; Sasaki, T.; Yamanaka, T. Temporal changes in nitrogen acquisition of Japanese black pine (*Pinus thunbergii*) associated with black locust (*Robinia pseudoacacia*). *J. For. Res.* **2014**, *25*, 585–589. [CrossRef]
8. Grotti, M.; Chianucci, F.; Puletti, N.; Fardusi, M.J.; Castaldi, C.; Corona, P. Spatio-temporal variability in structure and diversity in a semi-natural mixed oak-hornbeam floodplain forest. *Ecol. Indic.* **2019**, *104*, 576–587. [CrossRef]
9. Frayer, W.E.; Furnival, G.M. Forest Survey Sampling Designs: A History. *J. For.* **1999**, *97*, 4–10. [CrossRef]
10. Onishi, M.; Ise, T. Automatic classification of trees using a UAV onboard camera and deep learning. *arXiv* **2018**, arXiv:abs/1804.10390.
11. Natesan, S.; Armenakis, C.; Vepakomma, U. Resnet-Based Tree Species Classification Using UAV Images. *ISPRS Int. Arch. Photogramm. Remote. Sens. Spat. Inf. Sci.* **2019**, *XLII-2/W13*, 475–481. [CrossRef]

12. Safonova, A.; Tabik, S.; Alcaraz-Segura, D.; Rubtsov, A.; Maglinets, Y.; Herrera, F. Detection of Fir Trees (*Abies sibirica*) Damaged by the Bark Beetle in Unmanned Aerial Vehicle Images with Deep Learning. *Remote. Sens.* **2019**, *11*, 643. [CrossRef]

13. Fromm, M.; Schubert, M.; Castilla, G.; Linke, J.; McDermid, G. Automated Detection of Conifer Seedlings in Drone Imagery Using Convolutional Neural Networks. *Remote. Sens.* **2019**, *11*, 2585. [CrossRef]

14. Pan, S.J.; Yang, Q. A Survey on Transfer Learning. *IEEE Trans. Knowl. Data Eng.* **2010**, *22*, 1345–1359. [CrossRef]

15. Diez, Y.; Kentsch, S.; Caceres, M.L.L.; Nguyen, H.T.; Serrano, D.; Roure, F. Comparison of Algorithms for Tree-top Detection in Drone Image Mosaics of Japanese Mixed Forests. In Proceedings of the 9th International Conference on Pattern Recognition Applications and Methods, Valletta, Malta, 22–24 February 2020; Volume 1: ICPRAM, INSTICC; SciTePress: Setubal, Portugal, 2020; pp. 75–87. [CrossRef]

16. Krizhevsky, A.; Sutskever, I.; Hinton, G.E. ImageNet Classification with Deep Convolutional Neural Networks. In *Advances in Neural Information Processing Systems 25*; Pereira, F., Burges, C.J.C., Bottou, L., Weinberger, K.Q., Eds.; Curran Associates, Inc.: Dutchess County, NY, USA, 2012; pp. 1097–1105.

17. Planet: Understanding the Amazon from Space. Avaliable online: https://www.kaggle.com/c/planet-understanding-the-amazon-from-space (accessed on 19 August 2019)

18. Richardson, D.; Binggeli, P.; Schroth, G. *Invasive Agroforestry Trees—Problems and Solutions*; Island Press: Washington, DC, USA, 2004; pp. 371–396.

19. Mubin, N.A.; Nadarajoo, E.; Shafri, H.Z.M.; Hamedianfar, A. Young and mature oil palm tree detection and counting using convolutional neural network deep learning method. *Int. J. Remote. Sens.* **2019**, *40*, 7500–7515. [CrossRef]

20. Lopez, M.L.C.; Nakano, S.; Ferrio, J.P.; Hayashi, M.; Nakatsuka, T.; Sano, M.; Yamanaka, T.; Nobori, Y. Evaluation of the effect of the 2011 Tsunami on coastal forests by means of multiple isotopic analyses of tree-rings. *Isot. Environ. Health Stud.* **2018**, *54*, 494–507. [CrossRef] [PubMed]

21. Agisoft. Agisoft Metashape 1.5.5, Professional Edition. Available online: http://www.agisoft.com/downloads/installer/ (accessed on 19 March 2020)

22. Team, T.G. GNU Image Manipulation Program. Available online: http://gimp.org (accessed on 19 March 2020).

23. Guo, Y.; Liu, Y.; Georgiou, T.; Lew, M.S. A review of semantic segmentation using deep neural networks. *Int. J. Multimed. Inf. Retr.* **2018**, *7*, 87–93. [CrossRef]

24. Beucher, S.; Meyer, F. *The Morphological Approach to Segmentation: The Watershed Transformation*; Marcel Dekker Inc.: New York, NY, USA, 1993; Volume 34, pp. 433–481. [CrossRef]

25. Ronneberger, O.; Fischer, P.; Brox, T. U-Net: Convolutional Networks for Biomedical Image Segmentation. *arXiv* **2015**, arXiv:abs/1505.04597.

26. He, K.; Zhang, X.; Ren, S.; Sun, J. Deep Residual Learning for Image Recognition. In Proceedings of the 2016 IEEE Conference on Computer Vision and Pattern Recognition (CVPR), Las Vegas, NV, USA, 27–30 June 2016; IEEE Computer Society: Los Alamitos, CA, USA, 2016; pp. 770–778. [CrossRef]

27. Funke, J.; Tschopp, F.; Grisaitis, W.; Sheridan, A.; Singh, C.; Saalfeld, S.; Turaga, S.C. Large Scale Image Segmentation with Structured Loss Based Deep Learning for Connectome Reconstruction. *IEEE Trans. Pattern Anal. Mach. Intell.* **2019**, *41*, 1669–1680. [CrossRef] [PubMed]

28. Howard, J.; Thomas, R.; Gugger, S. Fastai. Available online: https://github.com/fastai/fastai (accessed on 18 April 2020).

29. Bradski, G. The OpenCV Library. *Dr. Dobb's J. Softw. Tools* **2000**, *25*, 120–125.

Improved SRGAN for Remote Sensing Image Super-Resolution Across Locations and Sensors

Yingfei Xiong [1,2], **Shanxin Guo** [1,3], **Jinsong Chen** [1,3,*], **Xinping Deng** [1,3], **Luyi Sun** [1,3], **Xiaorou Zheng** [1,2] **and Wenna Xu** [1,2]

[1]   Center for Geo-Spatial Information, Shenzhen Institutes of Advanced Technology,
Chinese Academy of Sciences, Shenzhen 518055, China; yf.xiong@siat.ac.cn (Y.X.); sx.guo@siat.ac.cn (S.G.);
xp.deng1@siat.ac.cn (X.D.); ly.sun@siat.ac.cn (L.S.); xiaorou.zheng@siat.ac.cn (X.Z.); wn.xu@siat.ac.cn (W.X.)

[2]   University of Chinese Academy of Sciences, Beijing 100049, China

[3]   Shenzhen Engineering Laboratory of Ocean Environmental Big Data Analysis and Application,
Shenzhen 518055, China

*   Correspondence: js.chen@siat.ac.cn; Tel.: +86-755-86392331

Received: 6 March 2020; Accepted: 13 April 2020; Published: 16 April 2020

Abstract: Detailed and accurate information on the spatial variation of land cover and land use is a critical component of local ecology and environmental research. For these tasks, high spatial resolution images are required. Considering the trade-off between high spatial and high temporal resolution in remote sensing images, many learning-based models (e.g., Convolutional neural network, sparse coding, Bayesian network) have been established to improve the spatial resolution of coarse images in both the computer vision and remote sensing fields. However, data for training and testing in these learning-based methods are usually limited to a certain location and specific sensor, resulting in the limited ability to generalize the model across locations and sensors. Recently, generative adversarial nets (GANs), a new learning model from the deep learning field, show many advantages for capturing high-dimensional nonlinear features over large samples. In this study, we test whether the GAN method can improve the generalization ability across locations and sensors with some modification to accomplish the idea "training once, apply to everywhere and different sensors" for remote sensing images. This work is based on super-resolution generative adversarial nets (SRGANs), where we modify the loss function and the structure of the network of SRGANs and propose the improved SRGAN (ISRGAN), which makes model training more stable and enhances the generalization ability across locations and sensors. In the experiment, the training and testing data were collected from two sensors (Landsat 8 OLI and Chinese GF 1) from different locations (Guangdong and Xinjiang in China). For the cross-location test, the model was trained in Guangdong with the Chinese GF 1 (8 m) data to be tested with the GF 1 data in Xinjiang. For the cross-sensor test, the same model training in Guangdong with GF 1 was tested in Landsat 8 OLI images in Xinjiang. The proposed method was compared with the neighbor-embedding (NE) method, the sparse representation method (SCSR), and the SRGAN. The peak signal-to-noise ratio (PSNR) and structural similarity (SSIM) were chosen for the quantitive assessment. The results showed that the ISRGAN is superior to the NE (PSNR: 30.999, SSIM: 0.944) and SCSR (PSNR: 29.423, SSIM: 0.876) methods, and the SRGAN (PSNR: 31.378, SSIM: 0.952), with the PSNR = 35.816 and SSIM = 0.988 in the cross-location test. A similar result was seen in the cross-sensor test. The ISRGAN had the best result (PSNR: 38.092, SSIM: 0.988) compared to the NE (PSNR: 35.000, SSIM: 0.982) and SCSR (PSNR: 33.639, SSIM: 0.965) methods, and the SRGAN (PSNR: 32.820, SSIM: 0.949). Meanwhile, we also tested the accuracy improvement for land cover classification before and after super-resolution by the ISRGAN. The results show that the accuracy of land cover classification after super-resolution was significantly improved, in particular, the impervious surface class (the road and buildings with high-resolution texture) improved by 15%.

Keywords: super-resolution; SRGAN; model generalization; image downscaling

1. Introduction

Detailed and accurate information on the spatial variation of land cover and land use is a critical component of local ecology and environmental research. For these tasks, high spatial resolution images are required to capture the temporal and spatial dynamics of the earth's surface processes [1]. Considering the trade-off between high spatial and high temporal resolution in remote sensing images, at present, there are two main methods for obtaining high spatio-temporal resolution images: 1) the multi-source image fusion models [2,3] and 2) the image super-resolution models [4,5].

Compared to the image fusion models, the methods in the super-resolution field do not require auxiliary high spatial resolution data with the same location and a similar observation date to predict the detailed spatial texture, which makes these methods more approachable for different scenarios in both the computer vision and remote sensing fields.

The basic assumption of the image super-resolution model is that missing details in a low spatial resolution image can be either reconstructed or learned from other high spatial resolution images if these images follow the same resampling process as was used to create the low spatial resolution image. Based on this assumption, in the last decade, efforts have been made to focus on accurately predicting the point spread function (PSF), which represents the mixture process that forms the low-resolution pixels. There are mainly three groups of methods: 1) interpolation-based methods, 2) refactoring-based methods, and 3) learning-based methods (Table 1).

Firstly, the interpolation-based methods [6–8] are based on a certain mathematical strategy to calculate the pixel value of the target point to be restored from the relevant points, which is of low complexity and high efficiency. However, the edge effect in the resulting image is obvious, and the image details cannot be recovered since there is no new information produced in the interpolation process.

Secondly, the refactoring-based methods [9–11] model the imaging process and integrate different information from the same scene to obtain high-quality reconstruction results. Usually, these methods trade time differences for spatial resolution improvement, which usually requires pre-registration and a large amount of calculation.

Thirdly, the learning-based methods [12–20] overcome the limitation of the difficulty by determining the resolution improvement multiple of the reconstruction method and can be oriented towards a single image, which is the main development direction of the current super-resolution reconstruction. In this category, commonly used methods include the neighbor-embedding method (NE) [21], the sparse representation method (SCSR) [22] and the deep learning method.

Table 1. The comparison of super-resolution methods.

Types of Method	Basic Hypothesis	Representative Models	Advantages	Disadvantage
Interpolation-based methods	The value of the current pixel can be represented by the nearby pixels	The nearest neighbor interpolation [6] The bilinear interpolation [7] The bicubic interpolation [8]	low complexity and high efficiency	No image texture detail can be predicted and usually makes the image smoother looking
Refactoring-based methods	The physical properties and features can be recovered by image reconstruction (RE) technology, and these rules of the point spread function (PSF) can be further applying for the detail recovering	Joint MAP registration [9] Sparse regression and natural image prior [11] Kernel regression PSF deconvolution [23]	The different information on the same scene is fused to obtain high-quality reconstruction results	Requires pre-registration and a large amount of calculation
Learning-based methods	The point spread function can be created by learning from a large number of image samples [24]	Neighbor-embedding (NE) [21] Convolutional neural network (SRCNN) [25] Bayesian networks [26] Kernel-based methods [27] SVM-based methods [28] Sparse representation (SCSR) [22] Super-resolution GAN(SRGAN) [29]	Getting better performance when training samples are more like the target image, and can achieve a higher PSNR when a large number of samples are involved	Highly time consuming, requiring big training datasets and usually limited model generalization ability across datasets

The learning-based method usually requires the high representative of the training samples to cover the data variation of the whole population. In practice, a large number of training samples from different sources are usually collected to achieve this goal. However, in the remote sensing field, it is almost impossible to prepare such a training sample set because the variation of remote sensing data not only depends on the variation of the object but also on the different locations and different satellite sensors. Due to this limitation, many learning-based methods are limited to a certain location and specific sensor, resulting in the limited generalization ability of the model across locations and sensors. This limitation remains a challenge for producing one super-resolution model for different locations and different satellite sensors.

In recent years, with the rapid development of artificial intelligence, especially neural network-based deep learning methods, deep learning has been widely applied in the field of computer vision, due to its obvious advantages for nonlinear process fittings of large samples. One benefit of these models is their ability to handle large sample sets while retaining a good generalization ability. In the field of image super-resolution, the super-resolution CNN model (SRCNN) was first presented by Dong [25] in 2014. Compared with traditional image super-resolution, this method achieves a higher peak signal-to-noise ratio, but when the image on the sampling ratio is high, the reconstructed image will be too smooth, and details will be lost. To overcome this shortage, a super-resolution generative adversarial network model (SRGAN) was presented by replacing the original CNN structure with the generative adversarial network (GAN) [29]. As the newest learning model from the deep learning field, the SRGAN shows many advantages for capturing high-dimensional nonlinear features over large samples. However, the generalization ability of the SRGAN model in remote sensing images across different locations and sensors remains unknown.

In this study, we test whether the GAN-based method can improve the generalization ability across locations and sensors by making some modifications so that we can train once and apply the results everywhere and with different sensors. Our work is based on the SRGAN model by modifying the loss function and the structure of the network of the SRGAN. The major contributions of this study are as follows:

- We propose the improved SRGAN (ISRGAN), which stabilizes the model training and enhances the generalization ability across locations and sensors;
- We test the performance of the ISRGAN from two sensors (Landsat 8 OLI and Chinese GF 1) at different locations (Guangdong and Xinjiang in China). Specifically, for the cross-location test, the model training in Guangdong with the Chinese GF 1 data was tested with the GF 1 data in Xinjiang. For the cross-sensor test, the same model training in Guangdong with GF 1 was tested with the Landsat 8 OLI data in Xinjiang;
- The proposed method is compared to the neighbor-embedding (NE), sparse representation (SCSR), and the original SRGAN methods, and the results show that the ISRGAN achieved the best performance;
- The model provided in this study shows a good generalization ability across locations and sensors. This model can be further applied to other satellite sensors and different locations for image super-resolution to achieve the goal of "training once, apply to everywhere and different sensors."

The structure of this paper is as follows: The overall workflow (Section 2.1), the review of the original SRGAN (Section 2.2), the problem of the original SRGAN and the corresponding improvement (Section 2.3) are described in the Methods section (Section 2). The study area, dataset and assessment method in this study are explained in the Experiments section (Section 3). Section 4 describes the main results and findings in our study. In Section 5, we discuss the advantages and possible further works followed by the Conclusion section (Section 6).

2. Methods

2.1. Workflow

In this paper, the experiment on whether the super-resolution model has generalization capability across locations and sensors is mainly divided into three parts, as presented in Figure 1. The blue part represents the model training part, which used the GF 1 data in Guangdong. The green part indicates that the model was tested for generalization ability across regions, using the GF 1 data in Xinjiang. The orange part indicates that the model was tested for whether there was generalization ability across sensors, mainly using the Landsat 8 data in Xinjiang.

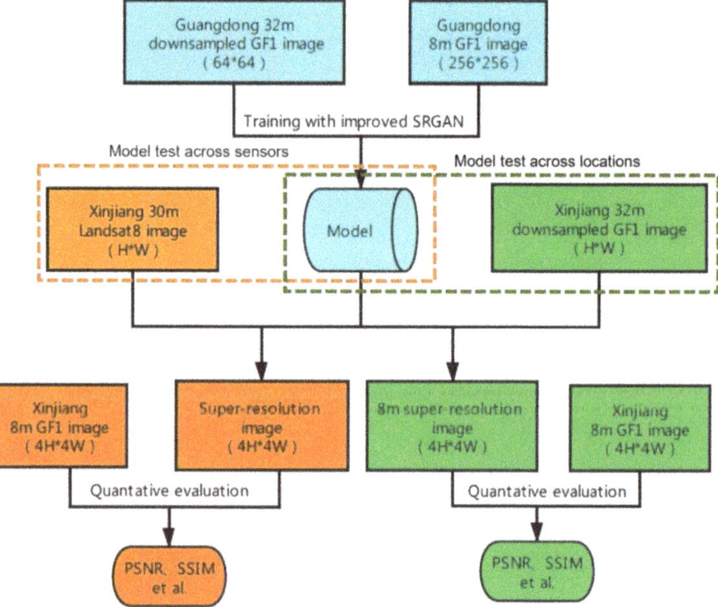

Figure 1. The workflow of the experiment.

First, we used the ISRGAN to train the Guangdong GF 1 data and obtained the super-resolution model. At the same time, we divided the test set into three parts, namely test dataset 1 (GF 1 data in Guangdong province), test dataset 2 (GF 1 data in Xinjiang province) and test dataset 3 (Landsat 8 data in Xinjiang province).

For the test of whether the model had generalization ability across locations in the green part, we tested dataset 1 and dataset 2 and obtained the PSNR and SSIM, respectively, in order to conduct a t-test to determine whether the model had generalization ability across locations. For the test of whether the model had generalization ability across sensors in the orange part, we tested dataset 2 and dataset 3 and obtained the PSNR and SSIM, respectively, in order to conduct a t-test to determine whether the model had generalization ability across sensors.

2.2. SRGAN Review

The GAN is a deep learning model proposed by Goodfellow et al. [30] in 2014. The structure of the GAN is inspired by the two-person zero-sum game in game theory. The framework consists of a generator (G) and a discriminator (D), where the generator (G) learns the distribution of real sample data and generates new sample data, and the discriminator (D) is a binary classifier used to distinguish whether the data is from real samples or generated samples. The structure diagram of the GAN is shown in Figure 2. The optimization process of the GAN is a minimax problem. When the generator and discriminator reach the Nash equilibrium, the optimization process is completed.

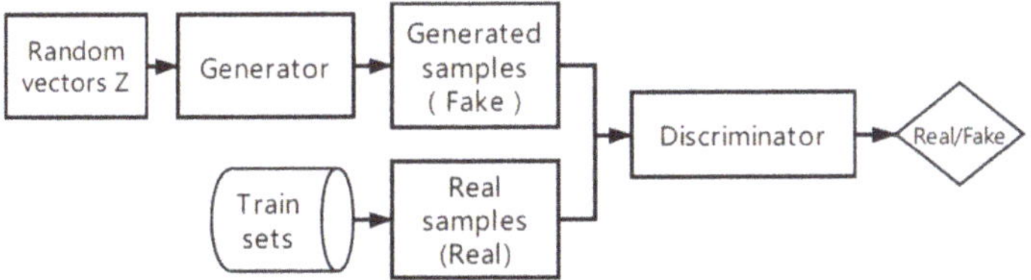

Figure 2. The typical structure of generative adversarial nets (GANs) adapted from [30].

In machine learning, GANs have become a hot research direction. At present, the field of computer vision has become the most widely studied and applied field of GANs, which have broad applications.

The SRGAN is a super-resolution network structure proposed by Christian Ledig [29] in a paper published at the 2017 CVPR conference, which brings the effect of super-resolution to a new height. The SRGAN is trained based on the GAN network, which consists of a generator and a discriminator. The generators use a ResNet structure [31], the former part of the network is connected with several residual blocks, each containing two 3 × 3 convolution layers, which is followed by the batch normalization layer, which is activated with the ReLu function. In the latter part, two subpixel network modules are added to increase the size so that the generator can learn high-resolution image details in the front layer during the training process and improve the image resolution later, so as to achieve the purpose of reducing computing resources. The discriminator adopts the vgg-19 network structure [32], including eight convolution layers, where the LeakyReLu function is used as the activation function for the hidden layer, and finally, the probability of the predicted image coming from the real high-resolution and generated high-resolution image is obtained by using the full connection layer and sigmoid activation function.

The main innovation of the SRGAN is to propose an optimization algorithm of perceived loss based on a neural network, which is to replace the mean square error (MSE) content loss with the loss calculated based on the vgg-19 network feature map. The original pixel-level MSE loss calculation (L_{mse}^{SR}) and the characteristic loss calculation ($L_{VGG/i,j}^{SR}$) based on the vgg-19 network are shown in Equations (1) and (2), respectively

$$L_{mse}^{SR} = \frac{1}{r^2 WH} \sum_{x=1}^{rW} \sum_{y=1}^{rH} (I_{x,y}^{HR} - G_{\theta G}(I^{LR})_{x,y})^2 \tag{1}$$

$$L_{VGG/i,j}^{SR} = \frac{1}{W_{i,j}H_{i,j}} \sum_{x=1}^{W_{i,j}} \sum_{y=1}^{H_{i,j}} (\varnothing_{i,j}(I^{HR})_{x,y} - \varnothing_{i,j}(G_{\theta G}(I^{LR}))_{x,y})^2 \tag{2}$$

where represents the ratio between the spatial resolution of a high- and low-resolution image, W and H respectively represent the pixel numbers of the width and height of the low-resolution image, $W_{i,j}$ and $H_{i,j}$ represent the dimension of the corresponding feature map in the vgg-19 network, $\varnothing_{i,j}$ represents the feature map obtained before the jth convolution before the ith max-pooling layer within the vgg-19 network, $I_{x,y}$ represents the gray value of the layer map at the point (x, y), and $G_{\theta G}$ represents the reconstructed image.

2.3. The Improved SRGAN

2.3.1. Problems in SRGAN

Due to the phenomena of gradient disappearance and mode collapse in the SRGAN training process, the model does not have a good generalization ability for remote sensing image super-resolution across locations and sensors. Next, we analyze the nature of this phenomenon.

The first phenomenon, gradient disappearance, can be explained as follows:

In the training process of the SRGAN, we want the discriminator to be strong enough to distinguish the samples well, and we want to give a result of 1 for the real high-resolution sample image and a result of 0 for the generated high-resolution sample image. Therefore, the discriminator loss function of the original SRGAN is shown in Equation (3):

$$maxV(G,D) = E_{x \sim P_r}[logD(HR)] + E_{x \sim P_g}[log(1 - D(G(LR)))] \tag{3}$$

where $V(G,D)$ represents the difference between the ground true high-resolution (HR) image and the model-generated image, HR and LR represent high-resolution images and low-resolution images, respectively, G and D represent the generator and discriminator, respectively, P_r and P_g represent the distribution of the real HR image and the generated super-resolution (SR) image, and E represents the expectation.

The generator hopes that the image generated by itself can be marked as 1 by the discriminator, so the adversarial loss function of the generator is:

$$minV(G,D) = E_{x \sim P_g}[log(1 - D(G(LR)))] \tag{4}$$

Therefore, the loss function of the SRGAN is shown in Equation (5):

$$\min_G \max_D V(G,D) = E_{x \sim P_r}[logD(HR)] + E_{x \sim P_g}[log(1 - D(G(LR)))] \tag{5}$$

The training process of the SRGAN is divided into two steps. The first step is to fix the generator and train the discriminator. For Equation (3):

$$
\begin{aligned}
V(G,D) \quad &= E_{x \sim P_r}[logD(HR)] + E_{x \sim P_g}[log(1 - D(G(LR)))] \\
&= \int_{HR} P_r log(D(HR))d(HR) + \int_{LR} P_g log(1 - D(G(LR)))d(LR) \\
&= \int_{HR} (P_r log(D(HR)) + P_g log(1 \\
&\quad -D(HR)))d(HR)
\end{aligned}
\tag{6}
$$

In order to optimize the discriminator's ability to distinguish data sources, Equation (6) is maximized as follows

$$D_G^* = \frac{P_r}{P_r + P_g} \tag{7}$$

where D_G^* is the optimal discriminator.

The second step is to fix the discriminator trained in the first step to optimize the generator. For Equation (5), we substitute Equation (7) into it. The new form of $V(G,D)$ as shown in Equation (8):

$$
\begin{aligned}
V(G,D) \quad &= E_{x \sim P_r}\left[logD_G^*(HR)\right] + E_{x \sim P_g}\left[log\left(1 - D_G^*(G(LR))\right)\right] \\
&= E_{x \sim P_r}\left[log\frac{P_r}{P_r + P_g}\right] + E_{x \sim P_g}\left[log\frac{P_g}{P_r + P_g}\right] \\
&= \int_{HR} P_r log\frac{P_r}{P_r + P_g}d(HR) + \int_{LR} P_g log\frac{P_g}{P_r + P_g}d(LR) \\
&= -2log2 + \int_{HR} P_r log\frac{P_r}{\frac{P_r + P_g}{2}}d(HR) + \int_{LR} P_g log\frac{P_g}{\frac{P_r + P_g}{2}}d(LR)
\end{aligned}
\tag{8}
$$

In addition, the Kullback-Leibler (KL) divergence and Jensen-Shannon Divergence (JSD) can be expressed as Equations (9) and (10)

$$KL(P_1\|P_2) = E_{x\sim P_1}log\frac{P_1}{P_2} = \int P_1 log\frac{P_1}{P_2}dx \tag{9}$$

$$JSD(P_1\|P_2) = \frac{1}{2}(KL(P_1\|\frac{P_1+P_2}{2}) + KL(P_1\|\frac{P_1+P_2}{2})) \tag{10}$$

where P_1 and P_2 represent the two probability distributions.

By substituting Equations (9) and (10) into Equation (8), we can get the final form of $V(G,D)$

$$\begin{aligned} V(G,D) &= -2log2 + KL(P_r\|\frac{P_r+P_g}{2}) + KL(P_g\|\frac{P_r+P_g}{2}) \\ &= -2log2 + 2JSD(P_r\|P_g) \end{aligned} \tag{11}$$

where $JSD(P_r\|P_g)$ represents the JSD divergence of the P_r and P_g.

As can be seen from Equation (11), only if $P_r = P_g$, $V(G, D)$ reaches the minimum value is the generation effect the best. In practice, the generated distribution can only be infinitely close to the real distribution, but the two can never overlap completely. According to Equation (11), only when the real distribution and the generated distribution overlap completely is the $V(G,D)$ equal to $-2log2$, otherwise, it always equals 0. Therefore, when using the gradient descent method, the generator cannot get any gradient information, which means it faces the problem of gradient disappearance.

The second phenomenon, mode collapse, can be explained as follows:

The generator loss function can also be written as:

$$E_{x\sim P_g}[-\log(D(G(LR)))] \tag{12}$$

Due to the KL divergence, P_r and P_g can be transformed into the form containing D^*:

$$\begin{aligned} KL(P_g\|P_r) &= E_{x\sim P_g}\left[log\frac{P_g}{P_r}\right] = E_{x\sim P_g}\left[log\frac{\frac{P_g}{P_g+P_r}}{\frac{P_r}{P_g+P_r}}\right] = E_{x\sim P_g}\left[log\frac{1-D_G^*}{D_G^*}\right] \\ &= E_{x\sim P_g}log\left[1-D_G^*\right] - E_{x\sim P_g}logD_G^* \end{aligned} \tag{13}$$

From Equations (12) and (13), it can be concluded that the loss function is equivalent to:

$$\begin{aligned} E_{x\sim P_g}\left[-\log\left(D_G^*(G(LR))\right)\right] &= KL(P_g\|P_r) - E_{x\sim P_g}log\left[1-D_G^*(G(LR))\right] \\ &= KL(P_g\|P_r) - 2JS(P_r\|P_r) + 2log2 + E_{x\sim P_r}logD_G^*(HR) \end{aligned} \tag{14}$$

Since only the first two terms depend on the generator (G), the final loss function of the generator is equivalent to minimizing the following function:

$$F = KL(P_r\|P_g) - 2JS(P_r\|P_g) \tag{15}$$

According to Equation (15), in the process of training the discriminator, the KL divergence should be reduced while the JS divergence should be increased, resulting in unstable training. At the same time, due to the asymmetry of the KL divergence, the following phenomenon occurs: the generator generates an unreal image, and the punishment is relatively high, so the generator generates an image similar to the real image, with a lower penalty. The result of this phenomenon is that the generator tends to produce similar images, which is called the mode collapse problem.

The two problems stated above led to the unstable performance of the SRGAN in the training process, which led to a poor generalization ability for remote sensing image super-resolution across locations and sensors.

2.3.2. Improved SRGAN

Inspired by the idea of WGANs [33], we modified the loss function and the partial structure of the network in view of the weak generalization ability caused by the instability of the SRGAN in the training process and proposed the ISRGAN. In this paper, the Wasserstein distance is used to replace the KL divergence and JS divergence. The calculation of the Wasserstein distance is shown in Equation (16)

$$V(G,D) = W\big(P_r, P_g\big) = \inf_{\gamma \sim \prod(P_r, P_g)} E_{(x,y)\sim\gamma}\big[\|x - y\|\big] \tag{16}$$

where $W\big(P_r, P_g\big)$ represents the Wasserstein distance of the P_r and P_g, γ represents the joint distribution of the real HR image and the generated image, $\prod(P_r, P_g)$ represents the set of all possible joint probability distributions of the P_r and P_g, $\|x - y\|$ represents the distance between a pair of samples x and y, $(x, y) \sim \gamma$ represents how much "mass" must be transported from x to y in order to transform the distributions of the P_r into the distribution of the P_g, and the marginal distributions of x and y are the P_r and P_g, respectively, and inf represents the lower bound that we can take on the expectation of all possible joint distributions.

However, when learning the image distribution, the random variable has thousands of dimensions, and it is difficult to solve directly by solving the linear programming problem. Therefore, we convert it into the dual form

$$V(G,D) = W\big(P_r, P_g\big) = \frac{1}{K} \sup_{\|f\|_L \leq K} E_{x\sim P_r}[f(x)] - E_{x\sim P_g}[f(x)] \tag{17}$$

where $\|f\|_L \leq K$ means that f is a K-Lipschitz function, the K-Lipschitz function is defined as, for $K > 0$, $\|f(x_1) - f(x_2)\| \leq K\|x_1 - x_2\|$ and K is the Lipschitz constant of f, and sup represents the upper bound of the expectation. Therefore, based on the SRGAN structure and the ideal of the WGAN, this paper modifies the network structure, and the loss function during training can be summarized as follows:

(1) Sigmoid was removed from the last layer of the discriminator to transform the classification problem into a regression problem;
(2) The loss functions of the generator and discriminator were not logarithmic;
(3) During the training process, after updating the discriminator parameters, its absolute value was truncated to no more than a fixed constant;
(4) The convolution kernel of the last layer of the generator was changed from 1×1 to 9×9.

3. Experiments

3.1. Study Area

The study area selected included some areas of the Guangdong–Hong Kong–Macao greater bay area and Yuli county, Xinjiang. The Guangdong–Hong Kong—Macao greater bay area is located in the pearl river delta of Guangdong, which is the fourth largest bay area in the world after the New York bay area, the San Francisco bay area, and the Tokyo bay area. The economic foundation of this region is solid, the development potential is huge, and it has a pivotal status; Yuli county, located in the middle of Xinjiang, is an important transportation hub in Xinjiang. The area is particularly rich in mineral resources and tourist resources and is known as the "back garden" of Korla. The two research areas are representative of the study of the super-resolution of remote sensing images and subsequent ground-object extraction. Figure 3 shows the image coverage of GF 1 in the study area.

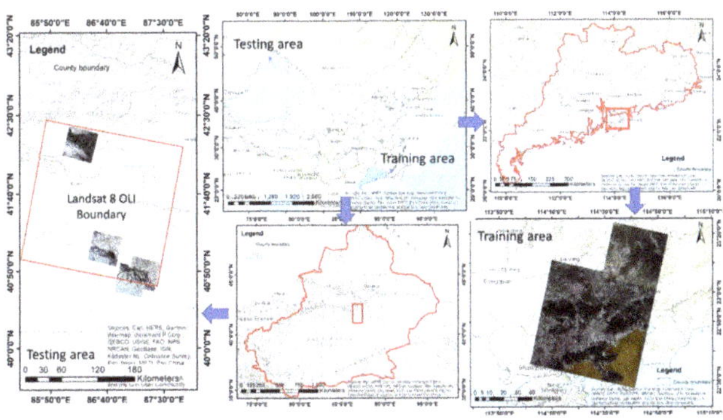

Figure 3. Study area and the image coverage of GF 1.

3.2. Data and Datasets

The experimental data included the GF 1 satellite data (Land observation satellite data service platform: http://218.247.138.119:7777/DSSPlatform/index.html) and the Landsat 8 satellite data (the USGS Earth Explorer: https://earthexplorer.usgs.gov/). The GF 1 satellite was successfully launched from the Jiuquan Satellite Launch Center on April 26, 2013, and is the first satellite from a major Chinese project using a high-resolution earth observation system. The GF 1 satellite is equipped with two 2-m panchromatic resolution cameras, two 8-m resolution multispectral cameras, and four 16-m resolution multispectral wide-width cameras, among which PMS sensor multispectral cameras contain four bands: blue, green, red, and near-infrared. Landsat 8 was launched by NASA on February 11, 2013, and is equipped with two sensors: the Operational Land Imager (OLI) and the Thermal Infrared Sensor (TIRS) thermal infrared sensor, among which the OLI sensor multispectral cameras contain nine bands: coastal, blue (B), green (G), red (R), near-infrared (NIR), short wave infrared 1 (SWIR1), and short wave infrared 2 (SWIR2). The details are shown in Tables 2 and 3.

Table 2. The parameters of GF 1.

Parameters	2m-PAN Sensor/8m-MS Sensor		16m-MS Sensor
	Panchromatic	0.45-0.90um	
		0.45-0.52um	0.45-0.52um
Spectral range	Multispectral	0.52-0.59um	0.52-0.59um
		0.63-0.69um	0.63-0.69um
		0.77-0.89um	0.77-0.89um
Spatial resolution	Panchromatic	2m	16m
	Multispectral	8m	
Scale range	60km (2 cameras)		800km (4 cameras)
Revisit cycle (side swing)	4 days		
Revisit cycle (non-side swing)	41 days		4 days

Table 3. The parameters of Landsat 8.

Sensor	Band	Spectral Range/um	Spatial Resolution/m
OLI	1-COASTAL	0.43-0.45	30
	2-Blue	0.45-0.51	30
	3-Green	0.53-0.59	30
	4-Red	0.64-0.67	30
	5-NIR	0.85-0.88	30
	6-SWIR1	1.57-1.65	30
	7-SWIR2	2.11-2.29	30
	8-PAN	0.50-0.68	15
	9-Cirrus	1.36-1.38	30
TIRS	10-TIR	10.60-11.19	100
	11-TIR	11.50-12.51	100

Table 4 shows the data details used in this paper. Considering the difference in the spectral range between the two kinds of satellite data in each channel, the experimental data selected in this paper only includes three RGB bands.

Table 4. The information of used data.

Location	Sensor	Latitude and Longitude	Time	Filename
Guangdong	GF1-PMS1	E114.2, N22.7	20171211	GF1_PMS1_E114.2_N22.7_20171211_L1A0002840075
Guangdong	GF1-PMS1	E114.3, N23.0	20171211	GF1_PMS1_E114.3_N23.0_20171211_L1A0002840076
Guangdong	GF1-PMS2	E114.6, N22.7	20171211	GF1_PMS2_E114.6_N22.7_20171211_L1A0002840310
Guangdong	GF1-PMS2	E114.6, N22.9	20171211	GF1_PMS2_E114.6_N22.9_20171211_L1A0002840309
Guangdong	GF1-PMS2	E114.7, N23.2	20171211	GF1_PMS2_E114.7_N23.2_20171211_L1A0002840307
Xinjiang	GF1-PMS1	E86.3, N42.2	20180901	GF1_PMS1_E86.3_N42.2_20180901_L1A0003427463
Xinjiang	GF1-PMS1	E86.6, N41.1	20171028	GF1_PMS1_E86.6_N41.1_20171028_L1A0002715402
Xinjiang	GF1-PMS1	E87.2, N40.8	20171130	GF1_PMS1_E87.2_N40.8_20171130_L1A0002807153
Xinjiang	GF1-PMS2	E87.0, N40.7	20170917	GF1_PMS2_E87.0_N40.7_20170917_L1A0002605364
Xinjiang	Landsat-8 OLI	E86.2, N41.7	20171229	LC08_L1TP_143031_20171229_20180103_01_T1

When doing the experiment, we need to make sure that the same ratio of the high-resolution image and low-resolution image pixel must be strictly 1:4, while the GF 1 PMS sensor data and Landsat satellite's OLI sensor data is not a strict 1:4 relation. In order to guarantee the feasibility of the experiment, we used the GF1 data and cut it into images of 256×256 size, while obtaining images of 64×64 size through cubic subsampling. Finally, we selected 3200 images of 256×256 size and their subsamples for training as high-resolution and low-resolution images, respectively.

3.3. Network Parameter Setting and Training

The proportion factor of the high-resolution image (HR) and low-resolution image (LR) used in the experiment was $\times 4$, among which the LR images were obtained by sampling the HR image four times, using the nearest-neighbor method in Python. During training, the batch size was set to 16, and the training process was divided into two steps. In the first step, the ResNet [31] was trained to obtain the mean square error (MSE) between the generated high-resolution image and the real high-resolution image, namely the traditional pixel-based loss, and the learning rate was initialized to 10^{-4}, training 100 epochs in total. In the second step, we used the model trained in the first step as the initialization of the generator. Using pretreatment based on pixel losses can make the method based on the GAN achieve a better effect. The reason for this can be summarized as follows: 1) The high-resolution image generated by preprocessing is a relatively good image for the discriminator, so it pays more attention to texture details in the following training process. 2) It is better to avoid the generator reaching the local optimization. The initialization learning rate of the generator training was 10^{-4}, and it was reduced to 1/2 in each 250 iterations, training 500 epochs in total.

We used the RMSProp optimization algorithm to update the generator and discriminator alternately until the model converged. The model was implemented using the Tensorflow framework (Google Inc. https://www.tensorflow.org) and was trained on four NVIDIA GeForce GTX TITAN X GPUs. The code is modified based on the SRGAN code, which can be freely downloaded from the GitHub website (https://github.com/tensorlayer/srgan).

3.4. Assessment

In this paper, we adopted the peak signal-to-noise ratio (PSNR) [34] and structural similarity index measurement (SSIM) [35] as the evaluation indexes for the experimental results. Usually, after an image is compressed, the image spectrum will change, so the output image will be somewhat different from the original image. In order to measure the quality of the processed image, the PSNR is usually used to measure whether a processor meets the expected requirements and is calculated as follows

$$PSNR = 10 \times log_{10}\left(\frac{MAX^2}{MSE}\right) \tag{18}$$

where the MSE is calculated as shown in Equation (19)

$$MSE = \frac{\sum_{n=1}^{N}(I^n - P^n)}{N} \tag{19}$$

where I^n refers to the gray value of the nth pixel of the original image, and P^n refers to the gray value of the nth pixel after processing. The unit of the peak signal-to-noise ratio (PSNR) is dB, and the higher the value, the better the image quality.

SSIM is a method used to measure the subjective experience quality of television, film, or other digital images and video. This method was first proposed by the image and video engineering laboratory of the University of Texas at Austin and then developed in cooperation with New York University. The SSIM algorithm is used to test the similarity of two images, and its measurement or prediction of image quality is based on uncompressed or undistorted images as a reference. The model measures image similarity in brightness, contrast, and structure. Its calculation is shown in Equation (20)

$$SSIM(X, Y) = \frac{(2u_X u_Y + C_1)(2\sigma_{XY} + C_2)}{\left(u_X^2 + u_Y^2 + C_1\right)\left(\delta_X^2 + \delta_Y^2 + C_2\right)} \tag{20}$$

where u_X and u_Y represent the means of the gray values of image X and image Y, respectively, and δ_X^2, δ_Y^2 and σ_{XY} represent the variances of the gray values of image X and image Y, respectively.

Generally, $C_1 = (K_1 * L)^2, C_2 = (K_2 * L)^2, K_1 = 0.01, K_2 = 0.03$ and L is the maximum image value. The range of SSIM is (0,1); the larger its value, the less image distortion there is.

4. Results

4.1. High Spatial Resolution Image of ISRGAN across Locations and Sensors

Based on the ISRGAN super-resolution training model, we tested the ISRGAN on a GF 1 dataset from Guangdong, a GF 1 dataset in Xinjiang, and a Landsat 8 dataset in Xinjiang. Some test results are shown in Figures 4–6.

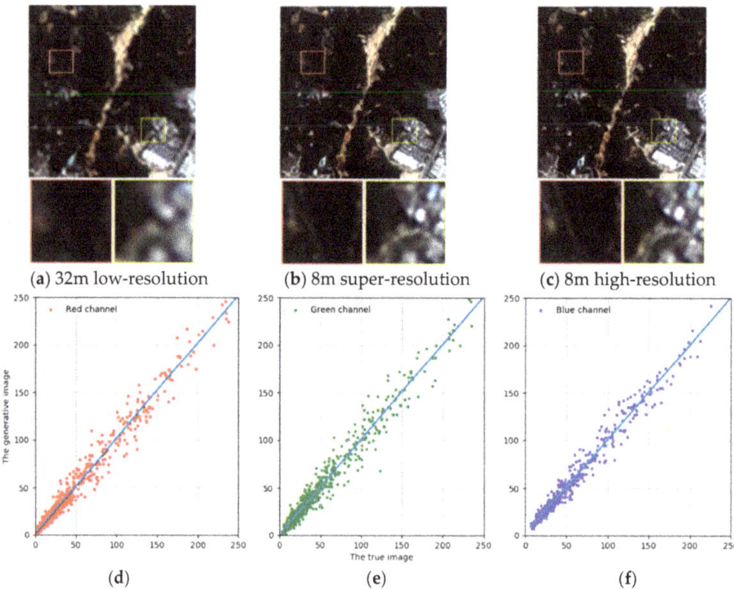

(a) 32m low-resolution   **(b)** 8m super-resolution   **(c)** 8m high-resolution

(d)   **(e)**   **(f)**

Figure 4. The predicted super-resolution image for our ISRGAN model (**b**) on a GF 1 dataset in Guangdong, compared to the input image (**a**) and the ground truth (**c**). Figures (**d**), (**e**), and (**f**) are 1:1 plots of the Digital Number (DN) value in the red, green and blue bands compared to the ground truth (**c**), with slopes of 1.0063, 1.0032, and 0.9955, respectively.

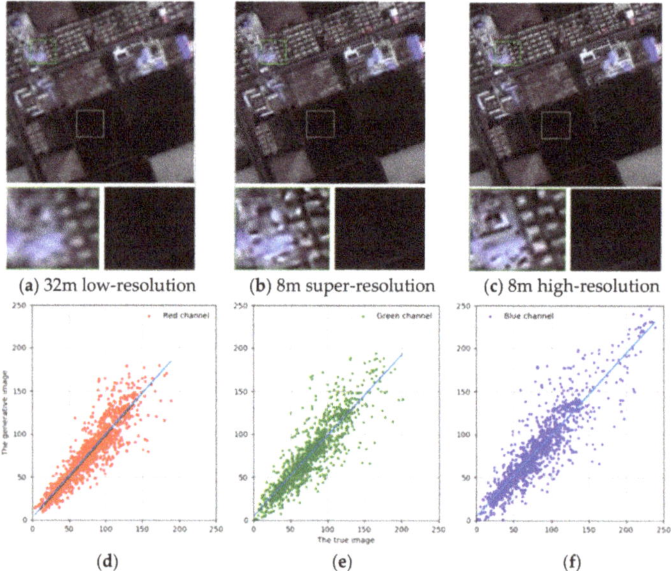

(a) 32m low-resolution   **(b)** 8m super-resolution   **(c)** 8m high-resolution

(d)   **(e)**   **(f)**

Figure 5. Cross-location comparison: the predicted super-resolution image for our ISRGAN model (**b**) on a GF 1 dataset in Xinjiang, compared to the input image (**a**) and the ground truth (**c**). Figures (**d**), (**e**), and (**f**) are 1:1 plots of the DN value in the red, green and blue bands compared to the ground truth (**c**), with slopes of 0.9658, 0.9378 and 0.9485, respectively.

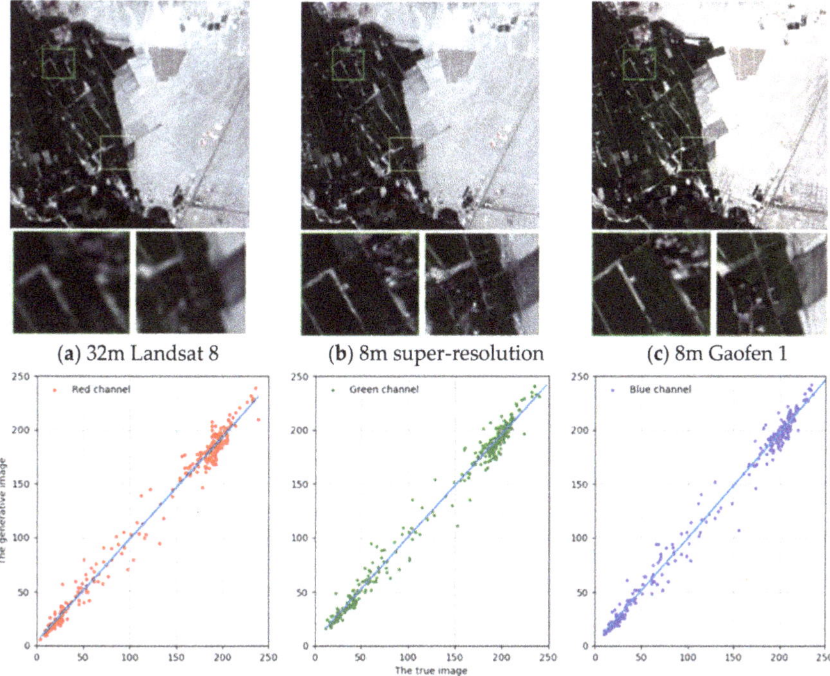

(**a**) 32m Landsat 8   (**b**) 8m super-resolution   (**c**) 8m Gaofen 1

Figure 6. Cross-sensor and -location comparison: the predicted super-resolution image for our ISRGAN model (**b**) on a Landsat 8 dataset in Xinjiang, compared to the input image (**a**) and the ground truth (**c**). Figures (**d**), (**e**), and (**f**) are 1:1 plots of the DN value in the red, green and blue bands compared to the ground truth (**c**), with slopes of 0.9527, 0.9564 and 0.9760, respectively.

As can be seen from the examples, the slope of the statistical pictures of the image after super-resolution and the real image in the gray value of the three bands is close to 1, so this model maintains the spectral information of the original image.

In addition, in order to verify the generalization ability of the super-resolution model across locations and sensors, we defined the following: if the evaluation indexes on two data sets follow a normal distribution and the mean value is not significantly different, then the model is approximately considered to have the same property on two datasets. Therefore, when verifying the generalization ability of the model across locations, we conducted t-tests on the evaluation indexes of the Guangdong (GF 1) dataset and the Xinjiang (GF 1) dataset. When verifying the generalization ability of the model across sensors, we also conducted a t-test on the evaluation indexes of the Xinjiang (GF 1) dataset and the Xinjiang (Landsat) dataset by using the R software with the "car" package (https://www.rdocumentation.org/packages/car/versions/3.0-3), for which the confidence was 95%. The test results are shown in Tables 5 and 6, respectively.

Table 5. T-Test of the model across locations.

		Levene Test of Variance Sig	T-Test of the Mean Sig(2-tailed)
SSIM	Variances are equal	0.510	1.000
	Variances are not equal		1.000
PSNR	Variances are equal	0.033	0.621
	Variances are not equal		0.625

Table 6. T-Test of the model across sensors.

		Levene Test of Variance Sig	T-Test of the Mean Sig(2-tailed)
SSIM	Variances are equal	0.011	0.824
	Variances are not equal		0.826
PSNR	Variances are equal	0.197	0.000
	Variances are not equal		0.000

In Tables 5 and 6, first, we judge whether there is a significant difference between the two groups of data, which is the Levene test of the variance equation. If the Sig parameter values are greater than 0.05, there is no significant difference in variance. After judging the variance, a t-test was performed on the mean value. Similarly, if the Sig (2-tailed) index value is greater than 0.05, there is no significant difference in the mean value of the two groups of data, which means that there is no significant difference between the two groups of data. As can be seen from Table 5, there is no significant difference in the mean values of the two groups of data in the PSNR and SSIM. Therefore, the model has generalization ability across locations. As can be seen from Table 6, there is no significant difference between the two groups of data in the SSIM, while there is a significant difference between the two groups of data in the PSNR. Since the PSNR only considers the gray value of pixels between the two groups of images, while the SSIM comprehensively considers the brightness, contrast, structure, and other information between the two groups of images, the SSIM has generalization ability across sensors.

4.2. Compare ISRGAN with NE, SCSR, and SRGAN

In this paper, the super-resolution methods we compared include the NE and SCSR methods, and the SRGAN. According to the super-resolution results, we calculated the evaluation indexes between them and the original images. In this paper, it was difficult to obtain the Landsat 8 satellite data and GF 1 satellite data at the same time and in the same scene, and the corresponding number of pixels was not consistent, considering the subsequent consistency of the computational criteria in validating whether the model has generalization ability across locations and sensors. Therefore, all the reference images of the quantitative calculation in this paper were the original images before super-resolution. The calculation method was to reduce the sampling of the image after super-resolution and then calculate the quantification index between the image after super-resolution and the original image. Figure 7 shows the partial comparison results of this paper based on the NE and SCSR methods, the SRGAN, and the ISRGAN in the three test sets.

We counted the evaluation indexes of the test data on the three test sets with three super-resolution algorithms and calculated their means. The results are shown in Table 7:

As can be seen from Table 7, in the horizontal comparison of the three super-resolution algorithms, the ISRGAN algorithm in this paper is significantly superior to the other three methods in the PSNR and SSIM, so it has certain advantages.

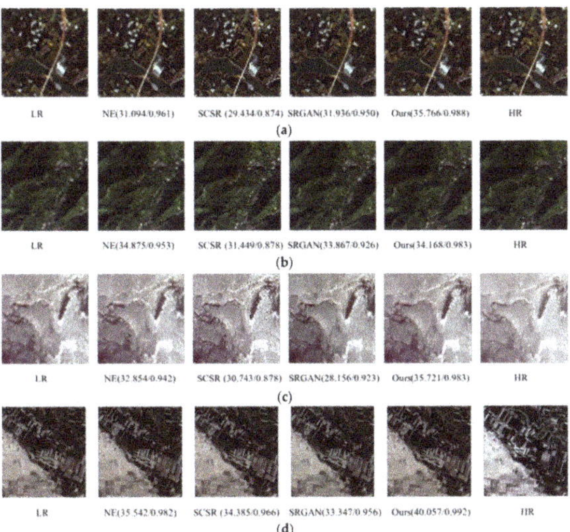

Figure 7. Comparison of the output of our ISRGAN model (the fifth column) to other models (NE, SCSR, SRGAN). Figures (**a**) and (**b**) are the results of the GF 1 dataset in Guangdong, (**c**) is the result of the cross-location testing in the GF 1 dataset in Xinjiang, and (**d**) is the result of Landsat 8 in Xinjiang. The ground truth is shown as HR. The low-resolution inputs are marked as LR.

Table 7. Image quality comparison of different super-resolution algorithms.

Data Sets	Index	NE	Super-Resolution Algorithms		
			SCSR	SRGAN	Ours
Guangdong	PSNR	33.6349	31.1207	32.1658	36.3744
(GF 1)	SSIM	0.9600	0.8854	0.9375	0.9884
Xinjiang	PSNR	32.0343	30.1631	31.3782	35.8160
(GF 1)	SSIM	0.9608	0.9001	0.9524	0.9885
Xinjiang	PSNR	35.0010	33.639	32.8203	38.0922
(Landsat 8)	SSIM	0.9818	0.9654	0.9488	0.9879

4.3. The Improvement of Land Cover Classification after ISRGAN Super-Resolution

The purpose of image super-resolution is to make better use of the advantages of high spatial resolution images and improve the accuracy of target recognition [36], classification [37], and change detection [38,39]. Taking the land cover classification and ground feature extraction as examples, the changes in the land use classification and ground feature extraction in the Landsat images before and after super-resolution were compared and analyzed.

The classification area selected was in the area at the border of Guangzhou and Shenzhen, where the feature categories are rich and densely distributed. It is beneficial to fully take advantage of high spatial resolution images in the land use classification and feature extraction. As shown in Figure 8, the original Landsat 8 image of the super-resolution application sample area was obtained on February 7, 2016.

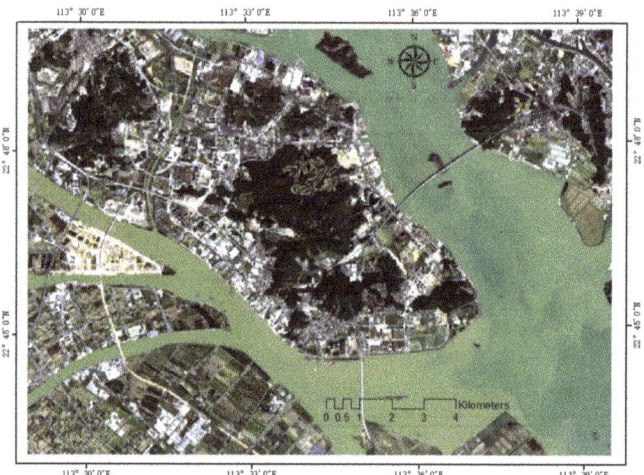

Figure 8. Landsat 8 image of the demonstration area.

For the classification of features, in order to avoid the interference of human factors on the classification results, we used the K-means algorithm to classify the before and after images of super-resolution, and the classification results are shown in Figure 9.

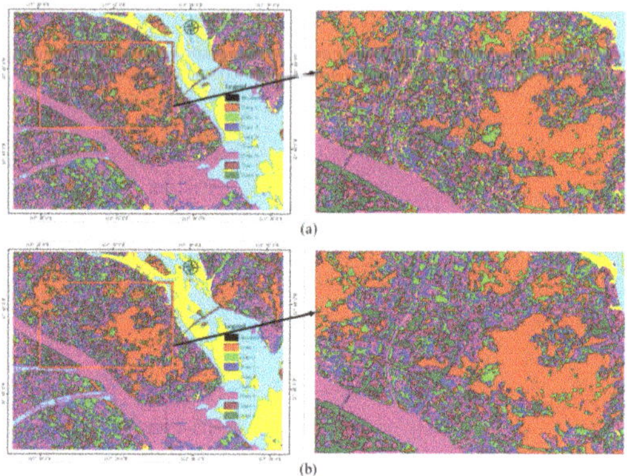

Figure 9. Comparison of the classification results between the images before and after super-resolution. Figure (**a**) is the classification result on the image before super-resolution and Figure (**b**) is the classification result on the image after super-resolution.

By comparing the classification results before and after the super-resolution images, we can see that the extraction effect of the image after super-resolution is better than that before super-resolution in the areas where the features are densely distributed and texture details are not obvious, such as roads and impervious surfaces. From the overall extraction effect, the result of the image after the super-resolution classification is better than that of the former in boundary and separability. Therefore, the image after super-resolution is broadly applicable in the classification of remote sensing images.

In addition, in order to better reflect the superiority of the image after super-resolution on the extraction of impervious surfaces, we used the Support Vector Machine (SVM) algorithm to extract impervious surfaces based on images before and after super-resolution. Similarly, in order to minimize the influence of human factors on the extraction results, we randomly selected the same set of training samples and test samples on Google Earth, converted them into the area of interest at the corresponding resolution, and then classified and extracted them. The extraction results of the impervious surfaces before and after super-resolution are shown in Figure 10.

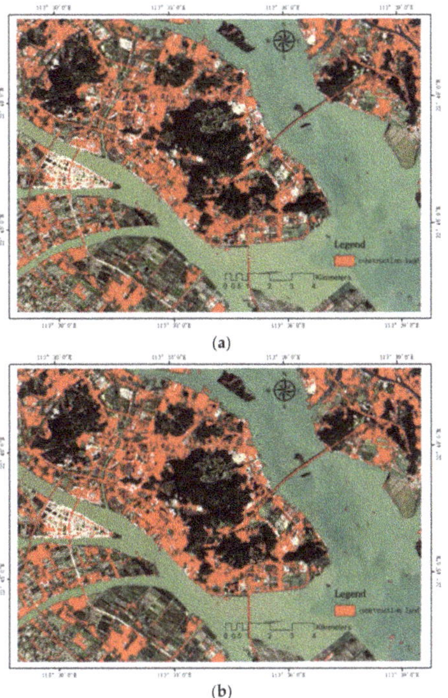

(a)

(b)

Figure 10. Comparison of the extraction results of impervious surfaces between the images before and after super-resolution (Figure (**a**) is the extraction result on the image before super-resolution and Figure (**b**) is the extraction result on the image after super-resolution).

In order to quantitatively verify the improvement of the extraction accuracy of impervious surfaces based on the image after super-resolution, we tested the same set of test samples selected above, which included 122 impervious surface sample points and 71 non-impervious surface sample points. The confusion matrix of the impervious surface extraction before and after super-resolution is shown in Tables 8 and 9. The overall accuracy of the impervious surface extraction before super-resolution was 70.1%, and the Kappa coefficient was 0.419. After super-resolution, the overall extraction accuracy of impervious surfaces was 86.1%, and the Kappa coefficient was 0.720. The quantitative results show that the extraction accuracy after super-resolution was nearly 15% higher than that before super-resolution.

Table 8. Confusion matrix of construction on the image before super-resolution.

Class	Impervious Surface	Others
Impervious Surface	61.48%	15.28%
Others	38.52%	84.71%

Table 9. Confusion matrix of construction on the image after super-resolution.

Class	Impervious Surface	Others
Impervious Surface	79.51%	2.78%
Others	20.49%	97.22%

5. Discussion

In comparison with the NE method, SCSR method, and the SRGAN, the ISRGAN shows better performance of generalization in the cross-location and -sensor tests. However, like many other super-resolution algorithms, the pseudo-textures can still be seen in the output after super-resolution, and the bell effect near the edges still needs to be improved. The edge enhancement algorithms can be further applied to recover the edge details, especially in the high spatial variation area.

Meanwhile, the scale ratio can lead to a dramatic change in the visual satisfaction of the model output. For most super-resolution algorithms, the perfect scale ratio is about 2:4. This means you can get a satisfying prediction when you down-scale an image with 30-m resolution down to 8-m rather than to 1-m. By increasing this ratio, the output of the image texture could show more random pseudo-textures and lead to a more serious bell effect. The fundamental reason for this is that the learning-based super-resolution algorithms try to recover the nonlinear point spread function (PSF) from a large number of available samples. However, when the ratio increases, the image details lost when crossing the different scales could be more and more complex and harder to capture by one universal PSF based on limited samples. The process of recovering the image details by the super-resolution algorithms is ill-posed, since the number of pixels needed to be predicted always needs to be larger than the number of known low spatial resolution pixels. Other than the large-scale ratio, the other possible way to recover the image details is by using image fusion technology, such as the spatial and temporal adaptive reflectance fusion model (STARFM) [40], enhanced STARFM (ESTARFM) [41] and the U-STFM model [42], which is basically "borrowing" the detailed image texture from the high spatial resolution reference image rather than predicting it. However, the consequences are when the discrepancy between the reference image and the input image goes large or the land cover changes rapidly, and thus the fusion-based method can fail to predict these changes.

In addition, due to the fact that GANs have two networks, which are named the generator and discriminator networks, more parameters than a convolutional neural network need to be optimized during training, so a long training time is needed. Currently, Muyang Li [43] has proposed the GAN compression method, which greatly shortens the training time. Through a large number of experiments, the computations in pix2pix [44], CycleGAN [45], and GauGAN [46] with this method were reduced from 1/9 to 1/21 without losing the fidelity of the generated image in the meantime. Therefore, the GAN compression method can be applied to the super-resolution network to effectively reduce the training time of the model.

6. Conclusions

Based on the super-resolution algorithm of the generated adversarial network in the computer vision field, this paper aimed to solve the problems of gradient disappearance and mode collapse that exist in the training of the generated adversarial network itself. Combined with the method of minimizing the Wasserstein distance proposed in the WGAN, we modified the original super-resolution network (SRGAN) and proposed the ISRGAN. Then we applied it to the super-resolution of remote sensing influence and drew the following conclusions:

(1) The ISRGAN super-resolution network proposed in this paper is applied to the super-resolution of remote sensing images, and the results obtained are better than other super-resolution methods, such as the neighbor-embedding method, the sparse representation method, and the SRGAN;

(2) In order to realize the one-time training and repeated use of the super-resolution model, we directly applied the model trained with the Guangdong GF 1 image data to the Xinjiang GF 1 image data and conducted a t-test on the evaluation indexes of the super-resolution results of the two datasets, which verified the generalization ability of the super-resolution model across locations;

(3) In order to combine the advantages of the domestic GF 1 image data and the Landsat 8 data, the super-resolution model trained on the GF 1 data was directly applied to the Landsat 8 data, and a t-test was performed on the evaluation indexes of the super-resolution results, verifying the generalization ability of the super-resolution model across sensors;

(4) Taking the land use classification and ground-object extraction as examples, we compared the classification and extraction accuracy of the images before and after super-resolution. The K-means clustering algorithm was adopted for the land use classification, and the SVM algorithm was used for the ground-object extraction. The results show that the visual effect and accuracy of the images after super-resolution are improved in the classification and ground-object extraction, indicating that the super-resolution of remote sensing images is of great application value in resource development, environmental monitoring, disaster research, and global change analysis.

Author Contributions: Conceptualization, Y.X., S.G., and J.C.; Formal analysis, Y.X.; Methodology, Y.X. and X.D.; Project administration, J.C.; Software, X.Z.; Supervision, S.G., X.D., and L.S.; Validation, W.X.; Visualization, X.Z.; Writing – original draft, Y.X.; Writing – review and editing, Y.X., S.G., J.C., X.D., and L.S. Authors have read and agreed to the published version of the manuscript.

Funding: This research was supported by the National Key Research and Development Program of China (Project No. 2017YFB0504203). Natural science foundation of China project (41801358, 41801360, 41771403, 41601212). Fundamental Research Foundation of Shenzhen Technology and Innovation Council (JCYJ20170818155853672),

Acknowledgments: The authors thank Y. Shen and C. Ling from SIAT and reviewers for suggestions.

References

1. Lim, H.S.; MatJafri, M.Z.; Abdullah, K. High spatial resolution land cover mapping using remotely sensed image. *Modern Appl. Sci.* **2009**, *3*, 82–91. [CrossRef]

2. Mou, L.; Zhu, X.; Vakalopoulou, M.; Karantzalos, K.; Paragios, N.; Le Saux, B.; Moser, G.; Tuia, D. Multitemporal Very High Resolution From Space: Outcome of the 2016 IEEE GRSS Data Fusion Contest. *IEEE J. Sel. Top. Appl. Earth Obs. Remote Sens.* **2017**, *10*, 3435–3447. [CrossRef]

3. Schmitt, M.; Zhu, X.X. Data Fusion and Remote Sensing: An ever-growing relationship. *IEEE Geosci. Remote Sens. Mag.* **2016**, *4*, 6–23. [CrossRef]

4. Yang, D.Q.; Li, Z.M.; Xia, Y.T.; Chen, Z.Z. Remote Sensing Image Super-resolution: Challenges and Approaches. In Proceedings of the 2015 IEEE International Conference on Digital Signal Processing (DSP), Singapore, 21–24 July 2015; pp. 196–200.

5. Luo, Q.; Shao, X.; Peng, L.; Wang, Y.; Wang, L. Super-resolution imaging in remote sensing. *Satell. Data Compress. Commun. Process. XI* **2015**, *9501*, 950108.

6. Zhang, X.G. A new kind of super-resolution reconstruction algorithm based on the ICM and the nearest neighbor interpolation. *Adv. Sci. Through Comput.* **2008**, 344–346.

7. Zhang, X.G. A New Kind of Super-Resolution Reconstruction Algorithm Based on the ICM and the Bilinear Interpolation. In Proceedings of the 2008 International Seminar on Future BioMedical Information Engineering, Wuhan, China, 18 December 2008; pp. 183–186.

8. Zhang, X.G. A New Kind of Super-Resolution Reconstruction Algorithm Based on the ICM and the Bicubic Interpolation. In Proceedings of the 2008 International Symposium on Intelligent Information Technology Application Workshops, Shanghai, China, 21–22 December 2008; pp. 817–820.

9. Hardie, R.C.; Barnard, K.J.; Armstrong, E.E. Joint MAP registration and high-resolution image estimation using a sequence of undersampled images. *IEEE Trans. Image Process.* **1997**, *6*, 1621–1633. [CrossRef]

10. Tian, J.; Ma, K.K. Stochastic super-resolution image reconstruction. *J. Vis. Commun. Image Represent.* **2010**, *21*, 232–244. [CrossRef]

11. Kim, K.I.; Kwon, Y. Single-image super-resolution using sparse regression and natural image prior. *IEEE Trans. Pattern Anal. Mach. Intell.* **2010**, *32*, 1127–1133.

12. Kursun, O.; Favorov, O. Super-resolution by unsupervised learning of high level features in natural images. In Proceedings of the 6th World Multi-Conference on Systemics, Cybernetics and Informatics (SCI 2002)/8th International Conference on Information Systems Analysis and Synthesis (ISAS 2002), Orlando, FL, USA, 14–18 July 2002; pp. 178–183.

13. Begin, I.; Ferrie, F.P. Blind super-resolution using a learning-based approach. In Proceedings of the 17th International Conference on Pattern Recognition, Cambridge, UK, 23–26 August 2004; Volume 2, pp. 85–89.

14. Joshi, M.V.; Chaudhuri, S.; Panuganti, R. A learning-based method for image super-resolution from zoomed observations. *IEEE Trans. Syst. Man Cybern. Part B Cybern.* **2005**, *35*, 527–537. [CrossRef]

15. Chan, T.M.; Zhang, J.P. An improved super-resolution with manifold learning and histogram matching. In *International Conference on Biometrics*; Springer: Berlin/Heidelberg, Germany, 2006; pp. 756–762.

16. Rajaram, S.; Das Gupta, M.; Petrovic, N.; Huang, T.S. Learning-based nonparametric image super-resolution. *EURASIP J. Appl. Signal Process.* **2006**, 51306. [CrossRef]

17. Kim, C.; Choi, K.; Ra, J.B. Improvement on Learning-Based Super-Resolution by Adopting Residual Information and Patch Reliability. In Proceedings of the 2009 16th IEEE International Conference on Image Processing, Cairo, Egypt, 7–10 November 2009; pp. 1197–1200.

18. Yang, M.C.; Chu, C.T.; Wang, Y.C.F. Learning Sparse Image Representation with Support Vector Regression for Single-Image Super-Resolution. In Proceedings of the 2010 IEEE International Conference on Image Processing, Hong Kong, China, 26–29 September 2010; pp. 1973–1976.

19. Zhang, J.; Zhao, C.; Xiong, R.Q.; Ma, S.W.; Zhao, D.B. Image Super-Resolution Via Dual-Dictionary Learning and Sparse Representation. In Proceedings of the 2012 IEEE International Symposium on Circuits and Systems (ISCAS), Seoul, Korea, 20–23 May 2012; pp. 1688–1691.

20. Dong, C.; Loy CC, G.; He, K.M.; Tang, X.O. Learning a Deep Convolutional Network for Image Super-Resolution. In *European Conference on Computer Vision*; Springer: Cham, Switzerland, 2014; pp. 184–199.

21. Chang, H.; Yeung, D.Y.; Xiong, Y. Super-resolution through neighbor embedding. In Proceedings of the 2004 IEEE Computer Society Conference on Computer Vision and Pattern Recognition, Washington, DC, USA, 27 June–2 July 2004; Volume 1, pp. 275–282.

22. Yang, J.; Wright, J.; Huang, T.S.; Ma, Y. Image super-resolution via sparse representation. *IEEE Trans. Image Process.* **2010**, *19*, 2861–2873. [CrossRef] [PubMed]

23. Miskin, J.; MacKay, D.C. Ensemble Learning for Blind Image Separation and Deconvolution. In *Advances in Independent Component Analysis*; Springer: London, UK, 2000; pp. 123–141.

24. Guo, S.; Sun, B.; Zhang, H.K.; Liu, J.; Chen, J.; Wang, J.; Jiang, X.; Yang, Y. MODIS ocean color product downscaling via spatio-temporal fusion and regression: The case of chlorophyll-a in coastal waters. *Int. J. Appl. Earth Obs. Geoinf.* **2018**, *73*, 340–361. [CrossRef]

25. Dong, C.; Loy, C.C.; He, K. Image Super-Resolution Using Deep Convolutional Networks. *IEEE Trans. Pattern Anal. Mach. Intell.* **2016**, *38*, 295–307. [CrossRef] [PubMed]

26. Lu, Y.; Qin, X.S. A coupled K-nearest neighbour and Bayesian neural network model for daily rainfall downscaling. *Int. J. Climatol.* **2014**, *34*, 3221–3236. [CrossRef]

27. Takeda, H.; Farsiu, S.; Milanfar, P. Kernel regression for image processing and reconstruction. *IEEE Trans. Image Process.* **2007**, *16*, 349–366. [CrossRef] [PubMed]

28. Zhang, H.; Huang, B. Support vector regression-based downscaling for intercalibration of multiresolution satellite images. *IEEE Trans. Geosci. Remote Sens.* **2013**, *51*, 1114–1123. [CrossRef]

29. Ledig, C.; Theis, L.; Huszar, F.; Caballero, J.; Cunningham, A.; Acosta, A.; Aitken, A.; Tejani, A.; Totz, J.; Wang, Z.H.; et al. Photo-Realistic Single Image Super-Resolution Using a Generative Adversarial Network. In Proceedings of the IEEE Conference on Computer Vision and Pattern Recognition, Honolulu, HI, USA, 21–26 July 2017; pp. 105–114.

30. Goodfellow, I.J.; Pouget-Abadie, J.; Mirza, M.; Xu, B.; Warde-Farley, D.; Ozair, S.; Courville, A.; Bengio, Y. Generative Adversarial Nets. In *Proceedings of the NIPS'14: 27th International Conference on Neural Information Processing Systems*; MIT Press: Cambridge, MA, USA, 2014; pp. 2672–2680.

31. He, K.; Zhang, X.; Ren, S.; Sun, J. Deep residual learning for image recognition. In Proceedings of the 2016 IEEE Conference on Computer Vision and Pattern Recognition (CVPR), Las Vegas, NV, USA, 26 June–1 July 2016; pp. 770–778.

32. Simonyan, K.; Zisserman, A. Very Deep Convolutional Networks for Large-Scale Image Recognition. *arXiv* **2014**, arXiv:1409.1556.

33. Arjovsky, M.; Chintala, S.; Bottou, L. Wasserstein GAN. *arXiv* **2017**, arXiv:1701.07875.

34. Steele, R. Peak Signal-to-Noise Ratio Formulas for Multistage Delta Modulation with Rc-Shaped Gaussian Input Signals. *Bell Syst. Tech. J.* **1982**, *61*, 347–362. [CrossRef]

35. Wang, Z.; Bovik, A.C.; Sheikh, H.R.; Simoncelli, E.P. Image quality assessment: From error visibility to structural similarity. *IEEE Trans. Image Process.* **2004**, *13*, 600–612. [CrossRef]

36. Coulter, L.L.; Stow, D.A. Monitoring habitat preserves in southern California using high spatial resolution multispectral imagery. *Environ. Monit. Assess.* **2009**, *152*, 343–356. [CrossRef] [PubMed]

37. Chang, C.W.; Shi, C.H.; Liew, S.C.; Kwoh, L.K. Land Cover Classification of Very High Spatial Resolution Satelite Imagery. In Proceedings of the 2013 IEEE International Geoscience and Remote Sensing Symposium—IGARSS, Melbourne, Australia, 21–26 July 2013; pp. 2685–2687.

38. Lu, D.; Hetrick, S.; Moran, E.; Li, G. Detection of urban expansion in an urban-rural landscape with multitemporal QuickBird images. *J. Appl. Remote Sens.* **2010**, *4*, 041880. [CrossRef] [PubMed]

39. Coppin, P.; Jonckheere, I.; Nackaerts, K.; Muys, B.; Lambin, E. Digital change detection methods in ecosystem monitoring: A review. *Int. J. Remote Sens.* **2004**, *25*, 1565–1596. [CrossRef]

40. Gao, F.; Masek, J.; Schwaller, M.; Hall, F. On the Blending of the Landsat and MODIS Surface Reflectance: Predicting Daily Landsat Surface Reflectance. *IEEE Trans. Geosci. Remote Sens.* **2006**, *44*, 2207–2218.

41. Zhu, X.; Chen, J.; Gao, F.; Chen, X.; Masek, J.G. An enhanced spatial and temporal adaptive reflectance fusion model for complex heterogeneous regions. *Remote Sens. Environ.* **2010**, *114*, 2610–2623. [CrossRef]

42. Huang, B.; Zhang, H. Spatio-temporal reflectance fusion via unmixing: Accounting for both phenological and land-cover changes. *Int. J. Remote Sens.* **2014**, *35*, 6213–6233. [CrossRef]

43. Li, M.; Lin, J.; Ding, Y.; Liu, Z.; Zhu, J.Y.; Han, S. GAN Compression: Efficient Architectures for Interactive Conditional GANs. *arXiv* **2020**, arXiv:2003.08936.

44. Isola, P.; Zhu, J.Y.; Zhou, T.; Efros, A.A. Image-to-Image Translation with Conditional Adversarial Networks. In Proceedings of the 2017 IEEE Conference on Computer Vision and Pattern Recognition (CVPR), Honolulu, HI, USA, 21–26 July 2017; pp. 5967–5976.

45. Zhu, J.Y.; Park, T.; Isola, P.; Efros, A.A. Unpaired Image-to-Image Translation using Cycle-Consistent Adversarial Networks Jun-Yan. In Proceedings of the IEEE International Conference on Computer Vision, Venice, Italy, 22–29 October 2017; pp. 2223–2232.

46. Park, T.; Liu, M.Y.; Wang, T.C.; Zhu, J.Y. Semantic image synthesis with spatially-adaptive normalization. In Proceedings of the IEEE Conference on Computer Vision and Pattern Recognition, Long Beach, CA, USA, 16–20 June 2019; pp. 2332–2341.

Article

Design of Feedforward Neural Networks in the Classification of Hyperspectral Imagery Using Superstructural Optimization

Hasan Sildir [1], Erdal Aydin [2] and Taskin Kavzoglu [3,*

[1] Chemical Engineering, Gebze Technical University, 41400 Gebze, Kocaeli, Turkey; hasansildir@gtu.edu.tr
[2] Chemical Engineering, Bogazici University, 34342 Bebek, Istanbul, Turkey; erdal.aydin@boun.edu.tr
[3] Geomatics Engineering, Gebze Technical University, 41400 Gebze, Kocaeli, Turkey
* Correspondence: kavzoglu@gtu.edu.tr

Received: 5 February 2020; Accepted: 13 March 2020; Published: 16 March 2020

Abstract: Artificial Neural Networks (ANNs) have been used in a wide range of applications for complex datasets with their flexible mathematical architecture. The flexibility is favored by the introduction of a higher number of connections and variables, in general. However, over-parameterization of the ANN equations and the existence of redundant input variables usually result in poor test performance. This paper proposes a superstructure-based mixed-integer nonlinear programming method for optimal structural design including neuron number selection, pruning, and input selection for multilayer perceptron (MLP) ANNs. In addition, this method uses statistical measures such as the parameter covariance matrix in order to increase the test performance while permitting reduced training performance. The suggested approach was implemented on two public hyperspectral datasets (with 10% and 50% sampling ratios), namely Indian Pines and Pavia University, for the classification problem. The test results revealed promising performances compared to the standard fully connected neural networks in terms of the estimated overall and individual class accuracies. With the application of the proposed superstructural optimization, fully connected networks were pruned by over 60% in terms of the total number of connections, resulting in an increase of 4% for the 10% sampling ratio and a 1% decrease for the 50% sampling ratio. Moreover, over 20% of the spectral bands in the Indian Pines data and 30% in the Pavia University data were found statistically insignificant, and they were thus removed from the MLP networks. As a result, the proposed method was found effective in optimizing the architectural design with high generalization capabilities, particularly for fewer numbers of samples. The analysis of the eliminated spectral bands revealed that the proposed algorithm mostly removed the bands adjacent to the pre-eliminated noisy bands and highly correlated bands carrying similar information.

Keywords: artificial neural networks; classification; superstructure optimization; mixed-inter nonlinear programming; hyperspectral images

---

1. Introduction

Since the introduction of perceptron by Rosenblatt in 1958 [1], numerous studies in almost all scientific fields have been conducted to apply neural network models and test their performances. Starting with the first pioneering study of Benediktsson et al. [2], artificial neural networks (ANNs) have been extensively used in remote sensing fields, frequently for the supervised classification of remotely sensed images in the production of thematic maps [3–6]. Historical development reveals that ANNs were initially applied for comparative studies with conventional classifiers (e.g., maximum likelihood classifier), and later with other machine learning algorithms (e.g., support vector machines, random forest) for a wide range of problems [7–11]. In the last decade, new and advanced satellite sensors were launched, producing a vast amount of data repeatedly, at a higher number of bands. Both spatial and spectral resolutions of the sensors have increased; thus, the selection of the most appropriate data as inputs, known as feature selection, has become a more critical issue, particularly for neural networks. For this purpose, the pruning of neural networks has been suggested as an alternative to existing statistical methods [12–15].

The topology of Multi-Layer Perceptron (MLP) networks includes three types of layers called input, hidden, and output layers, each consisting of fully interconnected processing nodes, except that there are no interconnections between the nodes within the same layer. These networks typically have one input layer, one or more hidden layers, and one output layer. The input layer nodes correspond to individual data sources, which can be either spectral bands or other sources of data. The output nodes correspond to the desired classes of information, such as land use/land cover (LULC) classes in classification. Hidden layers are required for computational purposes. The values at each node are estimated through the summation of the multiplications between previous node values and weights of the links connected to that node. Since the nodes on input and output layers are usually pre-defined, except for the feature selection case where some irrelevant or highly correlated inputs are eliminated, the number of hidden layers and their nodes are the unknown hyper-parameters in the network, the choice of which directly affects the performance and generalization capabilities of the network. Several heuristics and formulations have been suggested in literature to estimate the optimal size for the hidden layer(s), but there is no universally accepted method that exists for estimating the optimal number of hidden layer nodes for a particular problem [16–19]. The use of ANNs in remote sensing has been reviewed by several studies, including [17,19–21]. Furthermore, the limitations and crucial issues in the application of neural networks have been discussed in [17,19,21,22].

Several approaches or methods exist in literature for the construction of optimal network architecture in addition to the heuristics mentioned above. These methods can be categorized as exhaustive search algorithms, also known as brute-force, constructive, pruning, and a combination of these methods. In the brute-force approach, after many small network architectures are formed and trained, the best smallest architecture producing the lowest error level or the highest accuracy for the dataset is selected. This approach is computationally expensive since many networks must be trained to obtain a solution [22,23]. Constructive methods start with a small network and add new hidden nodes to the network after each epoch if the training error or the proposed accuracy is not at the acceptable level. On the other hand, pruning methods work opposite to the constructive methods, in that a large network is selected and unimportant or ineffective links and/or hidden layer nodes are removed. Thus, overfitting to the training data can be avoided. These methods have the advantages of both small and large networks. For a start, the user has to determine the initial large network structure for the problem and the stopping criterion to end the training process. It was reported that training a large network and then pruning it is advantageous and favorable compared to that of training a small network [17,24]. There are also hybrid methods, also known as growing and pruning methods, that can both add and remove hidden layer units [25–28]. These methods are less popular due to the training of small networks suffering from the noisy fitness evaluation problem, and they are likely to be stuck into a local minimum together with a longer training time requirement.

The design of a neural network is not a simple task. The number of nodes in the hidden layer(s) should be large enough for the correct representation of the problem, but at the same time low enough to have adequate generalization capabilities [29]. The optimum number of hidden layer nodes depends on various factors including the numbers of input and output units, the number of training cases, the complexity of the classification to be learned, the level of noise in the data, the network architecture, the nature of the hidden unit activation function, the training algorithm, and regularization [30]. It is impractical to state that neural network topology is minimal and optimal since the optimality criteria actually varies for each problem under consideration [31]. If the network is too small, it cannot learn from the data, resulting in a high training error, which is a characteristic of underfitting. Small networks can have better generalization capabilities, but there is a risk of not learning the problem under consideration due to the insufficient number of processing elements [23,32]. On the other hand, if the network is too large, a well-known overfitting problem occurs. In other words, it becomes over-specific to the training data and likely to fail with the test data, producing lower classification accuracies. However, large networks have better fault tolerance [33]. Ideally, a close correspondence between training and testing errors is desired [34]. From the above argument, it can be concluded that a large network should be preferred to a small one since underfitting is a more serious issue than overfitting as it can be avoided using training strategies and pruning techniques by downsizing the network wisely. The optimum structure for a neural network should be large enough to learn the underlying characteristics of the problem and small enough to generalize for other datasets [17,32]. The motivation in this study is to not only remove some interconnections or eliminate some hidden layer neurons to improve generalization capabilities, but also to reduce the dimension of the input layer by eliminating the least effective and correlated spectral bands, and thus achieve improved performance. This is particularly important for the processing of hyperspectral images that comprise many correlated and sometimes irrelevant spectral bands for the problem under consideration.

Sildir and Aydin [35] suggested using a mixed-integer programming method in order to optimally and simultaneously design and train ANNs via superstructure optimization and parameter identifiability. In this study, a similar superstructure-based optimization technique is proposed for the classification of two benchmark hyperspectral images. The first essential part of the suggested method is to set up the superstructure formulation where inputs, number of neurons, and connections between inputs, hidden neurons, and outputs are all binary decision variables. At the same time, standard ANN parameters, e.g., connection weights, can take continuous values. This strong formulation brings about a mixed-integer program, usually, a nonlinear one (MINLP), which has to be solved with respect to a certain design metric. As a result, 'redundant' input variables, neurons, and connections for larger datasets are eliminated automatically.

In addition to the superstructure formulation, we also suggest integrating the use of statistical measures, namely parameter uncertainty for the purpose of enhancing the prediction performance of ANNs. This statistical approach takes the covariance of ANN parameters into account and integrates the measure with the objective function of the training algorithm. To the best of authors' knowledge, this paper is the first application of such an optimal and robust ANN algorithm addressing the classification of remotely sensed imagery. In addition to this novel application concept, extra linking constraints are added to this newer formulation that forces the optimization algorithm not to iterate for the continuous variables when certain binary variables are equal to zero, which in turn decreases the computational load of the resulting mixed-integer type ANN related problems significantly.

2. Test Sites and Datasets

For the experiments, aimed to show the effectiveness of the proposed optimization algorithm, two well-known hyperspectral datasets that are widely used in literature to test new algorithms and approaches were employed in this study. The effectiveness of the proposed method was investigated with different sampling ratios using 10% and 50% of the ground reference data.

2.1. The Indian Pines Dataset

The Indian Pines scene recorded by the Airborne Visible/Infrared Imaging Spectrometer (AVIRIS) sensor on June 12, 1992, was used in this study. The image and its ground reference data are made available by Purdue University (https://purr.purdue.edu/publications/1947/1). The dataset, covering a 2.9 by 2.9 km (145 by 145 pixels) agriculture dominated land in Tippecanoe County of Indiana, USA, has 220 spectral bands at 20 m spatial resolution (Figure 1). Twenty spectral bands (104–108, 150–163, 220) comprising the region of water absorption were removed from the dataset. The ground reference dataset including 16 LULC classes was collected through a field study in June 1992 [36]. The Indian Pines dataset has been employed in many publications to test and compare the performances of various algorithms [37,38]. The dataset is regarded as a challenging one for classification problems because of three major reasons. Firstly, the crops in the study site (mainly corn and soybeans) were very early in their growth cycle (about 5% canopy cover). Secondly, the imagery has a moderate spatial resolution of 20 m, resulting in a high number of mixed pixels. Lastly, the number of reference samples for the 16 LULC classes varies greatly among the classes, ranging from 20 samples to 2,455, which is regarded as an imbalanced dataset. Because of the availability of the limited number of samples for each LULC class, many researchers either combined the particular class types into a single one or avoid using some of the classes (e.g., oats, alfalfa, stone-steel towers). Considering that 20 pixels of the oats class must be divided into training and testing in the application that makes the learning process theoretically challenging, this class is left out in further processes. Thus, the Indian Pines dataset with 15 LULC classes, which are shown in Table 1, was considered in this study.

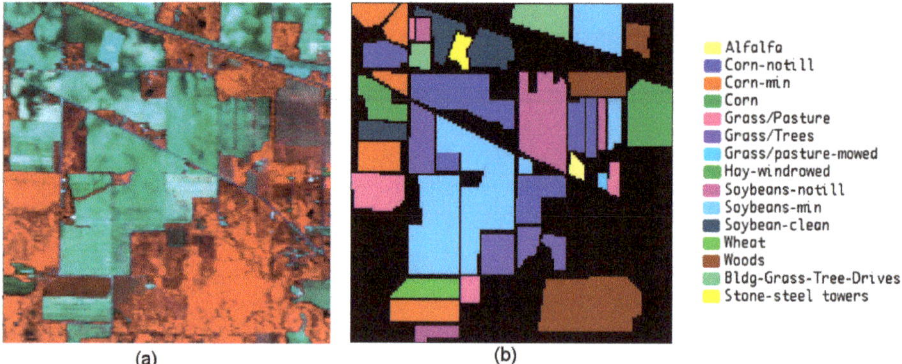

Figure 1. (**a**) Three-band color composite of Indian Pines Airborne Visible/Infrared Imaging Spectrometer (AVIRIS) hyperspectral image, and (**b**) ground reference data.

Table 1. Descriptions of the classes for Indian Pines data.

Index	Description	Number of Samples
O_1	Alfalfa	54
O_2	Corn-notill	1434
O_3	Corn-min	834
O_4	Corn	234
O_5	Grass-pasture	497
O_6	Grass-trees	747
O_7	Grass-pasture-mowed	26
O_8	Hay-windrowed	489
O_9	Soybean-notill	968
O_{10}	Soybean-min	2468
O_{11}	Soybean-clean	614
O_{12}	Wheat	212
O_{13}	Woods	1294
O_{14}	Buildings-Grass-Trees-Drives	380
O_{15}	Stone-Steel towers	95

2.2. The Pavia University Dataset

The Pavia University hyperspectral image was acquired with a Reflective Optics Spectrographic Image System (ROSIS) sensor during a flight campaign over Pavia, northern Italy. The ROSIS optical sensor provides images at a spectral range from 0.43 to 0.86µm with 115 bands. Twelve bands that were noisy or impacted by water absorption were removed from the dataset and the remaining 103 bands were employed in this study. The dataset captured over the Engineering School of the Pavia

University has 610 × 340 pixels with a spatial resolution of 1.3 m. The Pavia University dataset has ground truth maps of 9 classes and 42,776 labeled samples. The image and the ground reference data are shown in Figure 2 and details about the samples of all classes are given in Table 2.

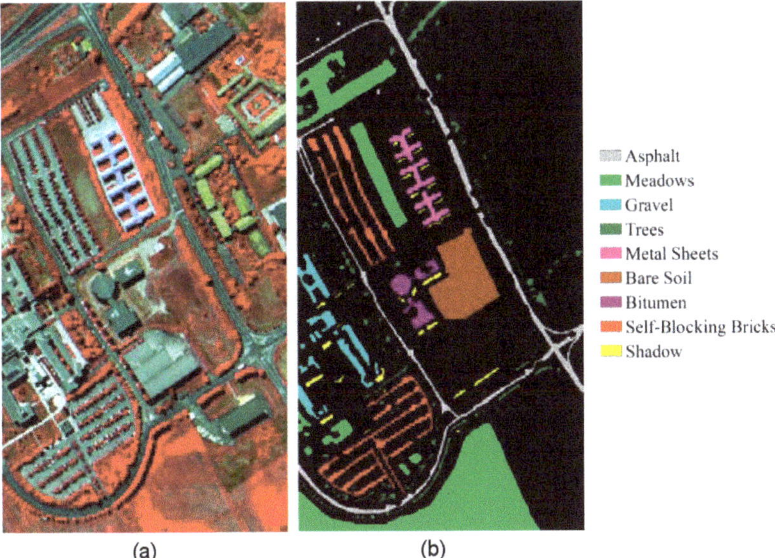

Figure 2. (**a**) Three-band color composite of University of Pavia Reflective Optics Spectrographic Image System (ROSIS) hyperspectral image, and (**b**) ground reference data.

Table 2. Descriptions of the classes for Pavia University data.

Index	Description	Number of Samples
O_1	Asphalt	6631
O_2	Meadows	18,649
O_3	Gravel	2099
O_4	Trees	3064
O_5	Painted metal sheets	1345
O_6	Bare soil	5029
O_7	Bitumen	1330
O_8	Self-blocking bricks	3682
O_9	Shadows	947

3. Optimal ANN Structure Detection and Training Methodology

Typical ANN structures usually contain a single hidden layer in addition to input and output layers containing identity activation functions. All those layers are fully connected in a traditional sense. The expression for a typical fully connected ANN (FC-ANN) is given by:

$$y = f_1(A \cdot f_2(B \cdot u + C) + D) \tag{1}$$

where A, B, C, and D are continuous weights with proper dimensions; f_1 and f_2 are output and hidden layer activation functions, respectively. For classification problems, the selected output activation function usually used for normalization. The softmax function is a typical example among other alternatives [39]. Note that the output activation function also calculates the individual probabilities for the classification problems, whereas the hidden layer activation function is not necessarily limited to normalization.

For an FC-ANN, the weights are traditionally assigned as non-zero in order to represent the connections among the neural network variables and layers. Those weights are estimated in the training by nonlinear optimization through the solution of:

$$Min_{A,B,C,D} \sum_{i=1}^{N} \|f_1(A \cdot f_2(B \cdot u_i + C) + D) - y_i\| \tag{2}$$

where N is the number of training samples; y_i is the ith sample vector; u_i is the ith input vector. Note that Equation (2) might also include additional box constraints to either reduce the search space for the training of ANNs or for specifically tailoring the training formulation.

As mentioned above, the solution of Equation (2) is usually obtained via programming a non-linear optimization problem (NLP). This solution delivers the FC-ANN weights (continuous variables), which minimize the training error without considering parameter identifiability issues, architecture orientation, and overfitting. In theory, as the number of decision variables and connections increases, the ANN training formulation should generate more flexibility, which in turn enhances the representative nature of ANNs on more complex datasets. The numbers of outputs, inputs, and hidden layer neurons together represent the number of decision variables. Traditionally, the structural hyper-parameters including the number of neurons, contained layers with the neurons, and the activation functions are manually tuned after trial and error. In addition to the structural parameters, the selection of proper input variables is another vital decision that is not included in (2) explicitly. However, it should be noted that complex and large datasets contain a significant amount of correlation and redundancy, especially in the big data era. On the other hand, it should be mentioned that deep neural networks including dropout layers can easily deal with the overfitting issues in a sequential manner. Yet, using deep neural nets is not in the scope of this paper. The integration of the proposed novel structure detection and training algorithm with the deep neural networks, which can be carried out without a loss of generality, is left for a future study.

Once the number of neurons lifts up, the dimension of continuous variables increases proportionally, and more connections are introduced in FC-ANNs. As a result, FC-ANN architecture becomes more challenging to train. Moreover, the optimal estimation of those parameters suffers from identifiability issues when the ANN architecture is poorly designed, or the training data do not contain statistically significant information [40–43].

The covariance matrix of the continuous ANN parameters have been adopted as a measure of identifiability in previous studies ([44]) and is used as a statistical metric in this study, for the elimination of the ANN variables including the number of neurons, connections, and input variables. Once the sum of the elements of the covariance matrix has a higher numerical value, the accompanying uncertainty in that estimated parameter leads to much larger prediction bounds due to the prorogation of uncertainty ([45]). In addition, a significant amount of computational power might be required for the training of ANNs, since there are many combinations of parameter values resulting in similar training performances.

Sildir and Aydin ([35]) proposed an MINLP formulation that realizes the optimal training of ANNs via superstructure modifications and parameter identifiability. They showed the contribution of the proposed formulation on regression problems. Results showed that the suggested method increases the predictive capabilities of ANNs with a significant reduction in the ANN superstructure compared to FC-ANNs. The MINLP formulation introduces additional binary variables to the traditional ANN equations in order to detect the optimal ANN architecture and to favor the optimal determination of input variables, hidden neurons, and connections for larger datasets among a maximum ANN structure. The modified one hidden layer ANN output equation is given as follows:

$$y = f_1\big(\big(A \circ A_{binary}\big) \cdot diag\big(N_{binary}\big) \cdot f_2\big(\big(B \circ B_{binary}\big) \cdot diag\big(U_{binary}\big) \cdot u + C\big) + D\big) \tag{3}$$

where $\circ$ is the Hadamard product operator; A_{binary} and B_{binary} are matrices with binary values representing the existence of connections. The existence of a particular connection is defined by the binary variable $A_{binary,ij}$. A_{ij} is the continuous weight parameter of the connection between the jth neuron and the ith output and can be non-zero only if the connection is decided to exist after solving the training optimization problem. Similarly, B_{ij} represents the connection between the input and corresponding neurons. In practice, once a particular column of B_{ij} is zero, then the jth input does not deliver information to the hidden layer and thus to the outputs as a result of feed-forward design. N_{binary} and U_{binary} are the binary vectors defining the existence of the neuron and input, respectively. For instance, if a particular element of U_{binary} is zero, it makes the corresponding column of B_{binary} zero, eliminating all the connections from the particular input; thus, the corresponding input is eliminated. These rules are realized via the introduction of extra linking constraints to the formulation, and the resulting problem exhibits a strong mixed-integer program formulation. The training optimization problem is given by:

$$Min_{A,A_{binary},B,B_{binary},C,D,N_{binary},U_{binary}} \gamma \sum diag\big(cov_p\big) + F$$
$$s.t.$$
$$F = \sum_{i=1}^{N} \|f_1\big(\big(A \circ A_{binary}\big) \cdot diag\big(N_{binary}\big) \cdot f_2\big(\big(B \circ B_{binary}\big) \cdot diag\big(U_{binary}\big) \cdot u_i + C\big) + D\big) - y_i\|$$
$$A_{binary,ij} \leq N_{binary,j} \tag{4}$$
$$B_{binary,ij} \leq U_{binary,j}$$
$$-A_{LB} \times A_{binary,j} \leq A_{i,j} \leq A_{UB} \times A_{binary,j}$$
$$-B_{LB} \times B_{binary,j} \leq B_{i,j} \leq B_{UB} \times B_{binary,j}$$
$$A_{binary}, B_{binary}, N_{binary}, U_{binary} \in \{0, 1\}$$

where γ is the tuning parameter for the multi-objective optimization; A_{LB} and A_{UB} are lower and upper bounds on A respectively; B_{LB} and B_{UB} are lower and upper bounds on B respectively ([35]). cov_p, which is a measure of parameter identifiability in this formulation, is the covariance matrix of the estimated ANN weights. Intuitively, diagonal elements of this refer to the variances of the corresponding weight. In theory, those values would increase significantly when overfitting occurs.

The problem given in Equation (4) is a relatively large scale and non-convex MINLP, which is quite challenging to solve to the global optimum. There are various efficient commercial solvers utilizing branch and bound ([46]), generalized benders decomposition ([47]), and outer approximation methods ([48]) for solving convex MINLPs. Nevertheless, solving non-convex MINLPs to global optimality is still an open research area and is not in the scope of this study. We should also mention that both the training and testing performances of the ANNs can be increased dramatically when a global solution algorithm is implemented to solve the problem given in Equation (4).

In this work, the adaptive, hybrid evolutionary algorithm suggested in [35] is used to solve the non-convex MINLP program given by (4). This method decomposes the original MINLP into integer programming (IP) and nonlinear programming (NLP) problems ([49–51]). IPs only include integer (or binary) decision variables that can be adjusted during optimization, whereas NLPs only involve continuous decision variables. For detailed information about the aforementioned optimization problems and their solution methods, we refer the reader to [52]. The IP stands on the outer loop and is solved via the genetic algorithm-based IP solver of Matlab while the inner loop NLP is solved by an interior point-based open source nonlinear programming solver IPOPT ([53]). Two problems are solved sequentially until the tolerance value of the original problem objective value or the maximum wall clock time is reached. This quasi decomposition feature is usually beneficial for solving large-scale problems.

It should be noted that all the experiments in this study were carried out using our in-house programs in Matlab software (v.2019b). The fully connected network (FC) was trained using the Matlab Neural Net Toolbox, implementing a standard back-propagation algorithm for training. A pseudo-algorithm for the mentioned optimal ANN structure detection and training approach is shown in Table 3. Also, a simplified diagram of the problem solution is shown in Figure 3.

Table 3. Pseudo algorithm adopted in this study for superstructure optimization.

Begin
Start with an initial guess and calculate the objective function
While (t<Maximum Wall Cock Time) or (Stopping Criterion)
Assign binary decision variables
Update linking constraints
While (Iteration number < Criterion) or (Stopping Criterion)
Update continuous decision variables
End While
Calculate the covariance matrix of parameters
Calculate the objective function
If (Objective function is improved)
Update binary decision variables (e.g., the ANN structure)
End If
End While

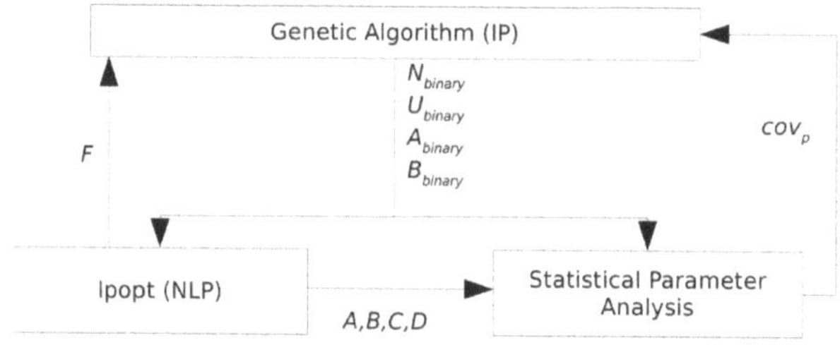

Figure 3. The proposed solution algorithm Problem (4).

There are efficient duality-based decomposition algorithms, which have proven to be very powerful for solving non-convex MINLP problems to global optimality. Nevertheless, these methods often require the NLP to be solved to global optimality, which is a challenging task for highly nonlinear relations (e.g., the tanh function of ANNs), and they demand high computational power. Unless the NLP converges to global optima, the decomposition algorithm may converge to an infeasible point or even diverge. On the other hand, these requirements do not usually apply to adaptive black-box optimization methods, with the possible drawback of converging to local optima. As mentioned above, the solution of the suggested ANN training problem to global optimality is not in the scope of this work and is left to a future study.

4. Results

The optimization problem given in Equation (4) was solved for the two public datasets considered in this study. The ANN architecture obtained from Equation (4) is called the optimal superstructure ANN (designated as OS hereafter) whose performance is compared to the fully connected ANN (designated as FC hereafter) to show the contribution of the current approach. Unlike FC, OS contains a significantly smaller number of neurons and connections, produced by eliminating the least effective or redundant hidden neurons, interconnections, and input variables. In order to test the effect of sample sizes used in the training process, 10% and 50% samples of the whole dataset were employed in the processing of FC and OS neural networks. For the Indian Pines dataset, 1082 training samples for the 10% sampling ratio and 5173 training samples for the 50% sampling ratio using 200 spectral bands as inputs were considered for the prediction of 15 LULC classes.

Figure 4 represents the remaining connections within the network with the white color representing a non-zero value, and thus existing connections, and the black color showing the removed connections for the network trained with approximately 10% sampling ratio. Whilst Figure 4a shows the connections between input and hidden layers, Figure 4b shows the connections between hidden and output layers. It can be noticed easily that no hidden layer node was removed from the network; thus, only the connections were removed by the proposed method. The final structure of the network was estimated as 158-10-15, indicating that 42 inputs (i.e., spectral bands) that have no connection to any hidden neuron were eliminated, represented by a black column in Figure 4a. On the other hand, 1271 of 2,150 connections, representing 61% of the total connections, were also removed from the network to simplify the network and improve its generalization capabilities.

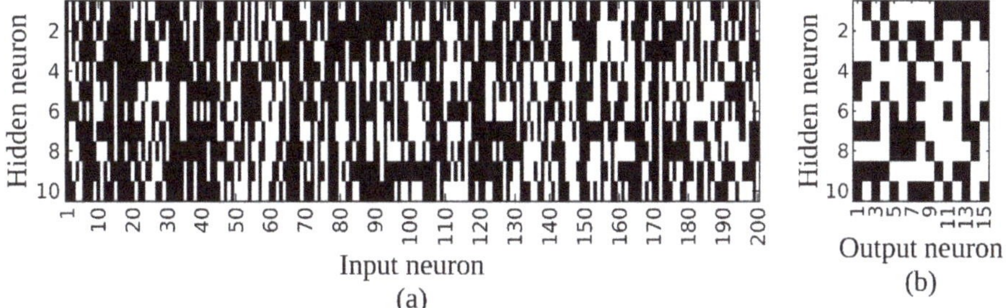

Figure 4. Eliminated (black) and remaining connections (white) for Indian Pines (**a**) between input and hidden neurons, and (**b**) between hidden and output neurons for 10% sampling ratio.

For the 50% sampling ratio, an optimal network superstructure with dimensions of 147-9-15 was calculated through the proposed approach, resulting in a significant reduction compared to the fully connected network of 200-10-15. The result of the process is given in Figure 5, showing the ultimate connections in the network between input and hidden layers, and hidden and output layers, respectively.

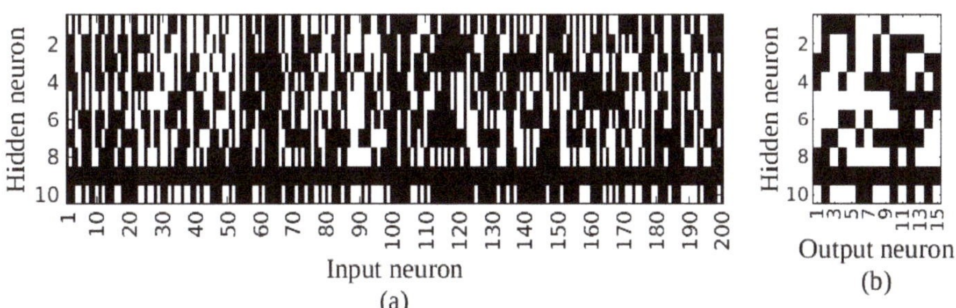

Figure 5. Eliminated (black) and remaining connections (white) for Indian Pines (**a**) between input and hidden neurons, and (**b**) between hidden and output neurons for 50% sampling ratio.

Note that, due to the linking constraints in Equation (4), the connections to and from a neuron are eliminated once a particular neuron is eliminated. In that case, the connection to and from the hidden neuron nine was removed, shown as a black row in Figure 5a,b. Therefore, it can be said that there is no information flow through the corresponding neuron. Similarly, 53 inputs that have no connection to any hidden neuron were eliminated, represented by a black column in Figure 5a. As a result, a considerable number of connections were removed from the network. To be more specific, 1413 of 2150 connections (i.e., almost 66% of the total connections) were removed from the network.

In order to show the position of the eliminated inputs (i.e., spectral bands), mean spectral signatures of 15 LULC classes in the Indian Pines dataset were extracted from the ground reference and the eliminated 53 bands for the 50% sampling ratio were depicted on the figure with vertical lines for further analysis (Figure 6). Perhaps the most striking result is that the proposed method removed the spectral bands adjacent to the previously eliminated noisy bands from the original datasets. It was also noticed that the algorithm detected some spectral ranges (e.g., 764–898 nm, 1004–1071 nm, 1205–1322 nm, 1591–1660 nm) as more beneficial compared to the others for discriminating the LULC classes. However, the spectral bands at the ranges of 918–1004 nm and 1501–1591 nm that indicate similar reflectance measures with the remaining ones were eliminated. Therefore, it can be concluded that the proposed algorithm removed the bands carrying similar information by considering the change or trend in the spectral curves. It is clear from the figure that most of the vegetation types have similar spectral signatures, but they have a varying range of reflectances at blue, green, near-infrared, and shortwave infrared (~1500–1700 nm) regions. The distinct spectral signature of stone-steel towers class can be also noticed.

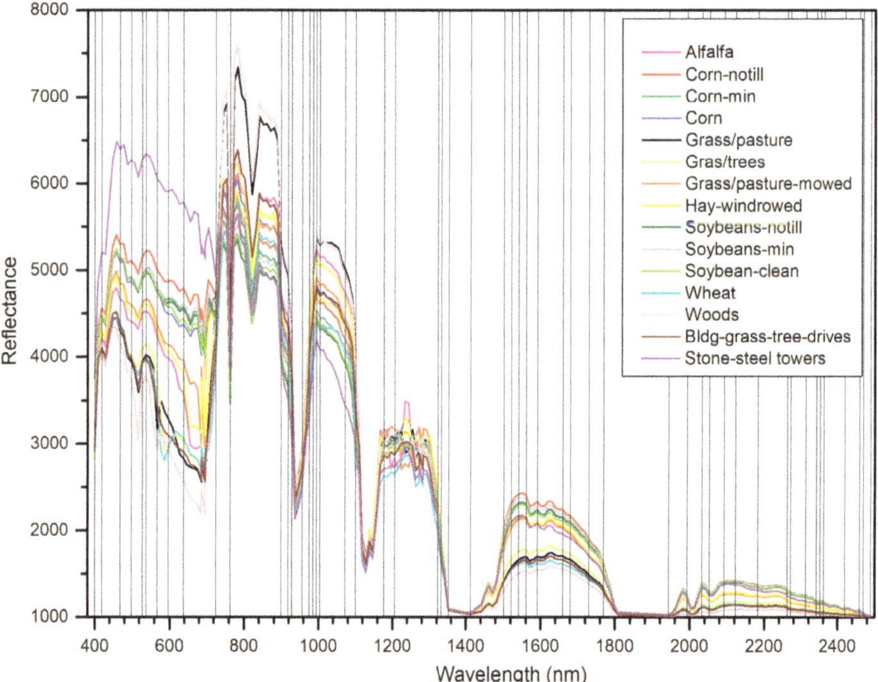

Figure 6. Spectral signatures of land use/land cover (LULC) classes in Indian Pines ground reference data and eliminated spectral bands (vertical lines) by the proposed algorithm for 50% sampling case.

For the analyses of the OS and FC networks using the test datasets, individual and overall accuracy measures were calculated (Table 4). While the F-score measure indicating the harmonic mean of user's and producer's accuracies was estimated for individual class accuracy assessment, overall accuracy (OA), Kappa, and weighted Kappa coefficients were used to evaluate the accuracy of the thematic maps. When the 10% sampling strategy was employed, the total number of connections in the network decreased from 2,150 to 879, indicating a 61% shrinkage. Although the network was highly compressed, the overall accuracy increased by about 4%, Kappa and weighted Kappa coefficients increased by about 5%. The performance of the FC network dropped, which is obviously a result of the occurrence of overfitting (over 99% overall accuracy on the training data). In the case of OS, the network was prevented from overfitting to training data. On the other hand, for the 50% sampling case, the overall accuracy decreased from 83.80% to 82.72%, indicating only a 1% decrease in the classification performance by decreasing the size of the network by about 66%. Similar results were calculated for Kappa and weighted Kappa coefficients.

Table 4. Classification accuracies obtained by different training sample sizes for the Indian Pines hyperspectral dataset using fully connected (FC) and optimal superstructure (OS) networks.

Class	10% of Samples			50% of Samples		
	Train Pixels	F-Score (FC)	F-Score (OS)	Train Pixels	F-Score (FC)	F-Score (OS)
Alfalfa	22	48.48	64.10	27	77.55	87.27
Corn-notill	144	68.81	74.94	717	78.84	79.77
Corn-min	84	54.30	57.47	417	74.64	70.79
Corn	24	47.72	53.14	117	59.23	64.80
Grass-pasture	50	74.40	76.48	248	88.29	83.86
Grass–trees	75	88.08	87.03	373	92.95	91.72
Grass-pasture-mowed	19	9.92	40.00	13	81.48	92.31
Hay-windrowed	49	90.02	95.34	245	95.35	97.75
Soybean-notill	97	67.00	74.31	484	79.84	79.79
Soybean-min	247	74.84	76.08	1234	84.34	80.97
Soybean-clean	62	54.70	65.03	307	82.93	78.42
Wheat	22	82.90	85.86	106	93.90	92.73
Woods	130	88.54	89.72	647	92.03	92.73
Bldg–Grass–Trees–Drives	38	45.41	62.15	190	67.91	69.54
Stone-Steel towers	19	74.59	81.44	48	95.92	91.11
Overall Acc. (%)		71.98	76.37		83.80	82.72
Kappa		0.680	0.730		0.815	0.802
Weighted kappa		0.715	0.753		0.848	0.850

When individual class accuracies estimated for each class were analyzed, some important results were obtained. Firstly, both FC and OS networks produced highly accurate results for some classes, namely grass–trees, hay-windrowed, wheat, woods, and stone-steel towers. However, networks performed poorly for two particular classes, namely corn, and building–grass–trees–drives. The corn class was mostly confused with other corn related classes (i.e., corn-notill and corn-min). The confusion was severe for the fully connected network, producing a 9.92% F-score value for grass-pasture-mowed class, which clearly shows failure in the delineation of this particular cover type. The negative effects of limited and imbalanced data can be easily seen from this class since individual class accuracy varies by about 30% for the 10% sampling case and 11% for the 50% sampling case. The building–grass–trees–drives class covering buildings and their surrounding pervious and impervious features were mainly mixed with the woods class that resulted in a decrease in classification accuracy. Thematic maps produced for the whole dataset using FC and OS networks trained with 50% of whole samples are shown in Figure 7. Misclassified pixels, particularly for the corn related ones, can be easily observed from the comparison of the thematic maps.

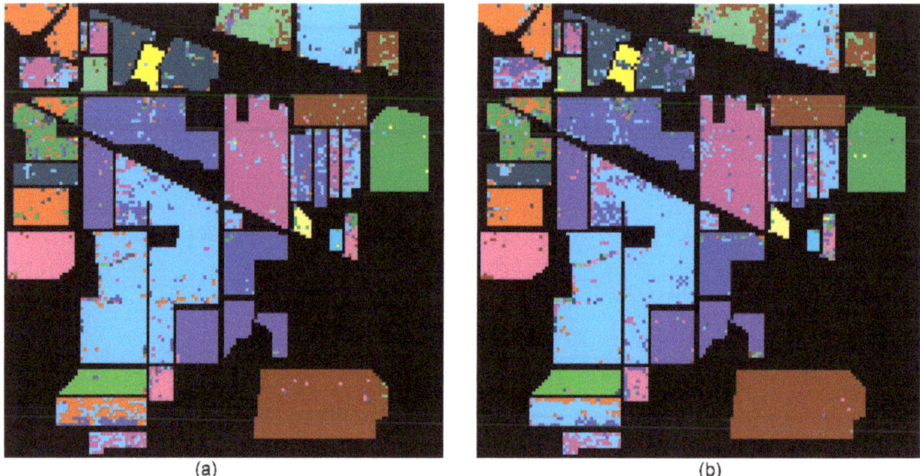

(a)   (b)

Figure 7. Classification results using (**a**) FC and (**b**) OS networks with 50% of the samples for the Indian Pines dataset.

For the Pavia University dataset, a fully connected network of 103-10-9 was optimized throughout the training process to learn the characteristics of the nine LULC classes from 103 spectral bands, and networks of 69-10-9 and 69-7-9 were found optimal in terms of its size and performance for the 10% and 50% sampling ratios, respectively. For the 10% sampling ratio, 697 of 1120 connections were removed from the network, showing a 58% shrinkage. For the 50% sampling ratio, 718 of 1120 links were removed from the network, indicating a 64% shrinkage in the network. For both sampling cases, 34 inputs were removed, indicating a feature selection rate of 33%. In other words, the fully connected network was trimmed by an average of 61%, and 33% of the spectral bands were disregarded as a result of the input selection process. The eliminated and remaining network connections for the 10% sampling ratio were shown in Figure 8. It can be noticed that a comparably smaller number of connections were removed between hidden and output layers, and none of the hidden layer nodes were removed by the proposed algorithm.

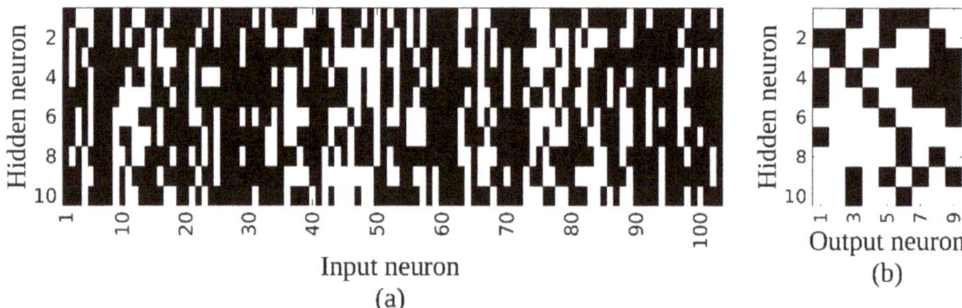

Input neuron   Output neuron
(a)   (b)

Figure 8. Eliminated (black) and remaining connections (white) for Pavia University (**a**) between input and hidden neurons, and (**b**) between hidden and output neurons for 10% sampling ratio.

Figure 9 shows the final network connections between the layers for the case of the 50% sampling ratio. The removal of three hidden neurons, namely seven, eight, and nine can be easily noticed from the figure (black horizontal lines). Similar to the results produced for the 10% sampling ratio, a smaller number of connections were removed between hidden and output layers.

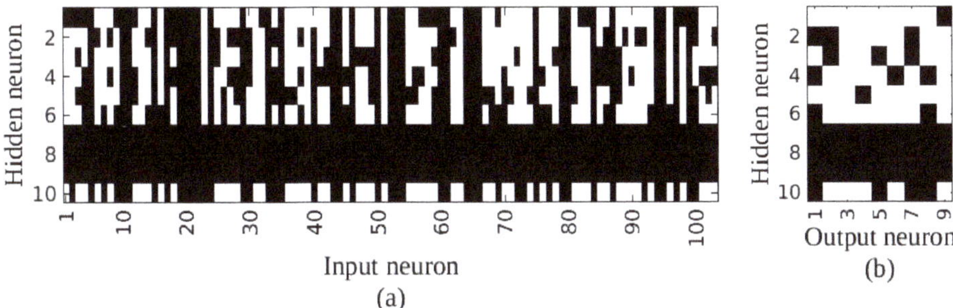

Figure 9. Eliminated (black) and remaining connections (white) for Pavia University (**a**) between input and hidden neurons, and (**b**) between hidden and output neurons for 50% sampling ratio.

For a clear explanation of the eliminated spectral bands, mean spectral signatures of the classes in Pavia University data were obtained from the ground reference and the location of the eliminated 34 spectral bands for the 50% sampling ratio were shown on the same figure with vertical lines (Figure 10). Similar results with the Indian Pines dataset were observed for the elimination of spectral bands as the highly correlated neighboring bands introducing similar reflectance values were mostly removed from the dataset. The method determines the spectral regions of 573–607, 704–742, and 793–822 nm as discriminating ones for the delineation of the characteristics of the LULC classes. In addition, it removed the spectral bands in the ranges of 468–527 and 607–653 nm. It can be observed that green and red-edge bands were mainly selected for the modeling of the problem. Spectral signature curves also revealed that there were high resemblances between bitumen and asphalt classes, also between gravel and self-blocking bricks classes. Metal sheets and shadow classes had distinct spectral reflectances compared to the other classes. On the other hand, a typical vegetation curve was observed for trees and meadows.

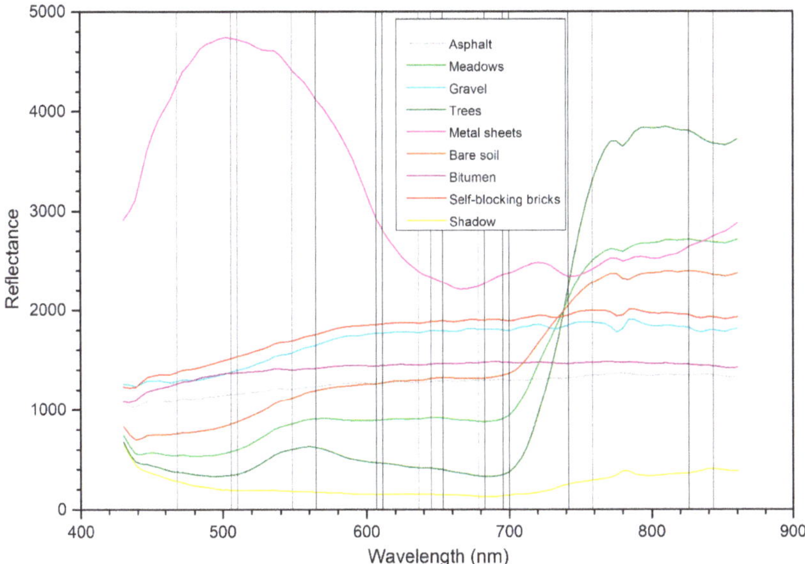

Figure 10. Spectral signatures of LULC classes in Pavia University ground reference data and eliminated spectral bands (vertical lines) by the proposed algorithm for 50% sampling case.

After the training stage for the FC and OS networks, the test data including the rest of the ground reference data for Pavia University were introduced to those networks, and corresponding network performances were presented in Table 5. With the 10% sampling ratio, the overall accuracy of 87.26%, and a Kappa coefficient of 0.830 were obtained with the optimal superstructure network (OS) while overall accuracy of 84.63% and a Kappa coefficient of 0.796 was achieved by the fully connected network (FC). This clearly shows the robustness of the proposed method, producing about 4% improvement in classification accuracy. With the 50% sampling ratio, the overall accuracy of 89.21% and the Kappa coefficient of 0.856 was obtained with the optimal superstructure network (OS) while the fully connected network (FC) achieved an overall accuracy of 90.76% and Kappa coefficient of 0.877. The accuracy decrease was about 1% for overall accuracy. From these results, it can be stated that the proposed method performs well for a fewer number of samples.

Table 5. Classification accuracies obtained by different training sample sizes for the University of Pavia hyperspectral dataset using FC and OS networks.

Class	10% of Samples			50% of Samples		
	Train Pixels	F-Score (FC)	F-Score (OS)	Train Pixels	F-Score (FC)	F-Score (OS)
Asphalt	668	80.24	86.84	3276	90.47	88.56
Meadows	1872	93.18	93.68	9366	95.38	95.07
Gravel	188	65.31	70.87	1032	78.57	72.31
Trees	306	90.28	91.00	1520	93.13	93.48
Painted metal sheets	135	99.34	98.84	670	99.12	99.34
Bare soil	513	75.28	76.47	2529	83.27	80.99
Bitumen	132	62.38	72.23	660	84.66	76.77
Self-blocking bricks	373	70.43	75.03	1860	80.94	77.32
Shadows	91	84.69	91.23	475	94.04	90.85
Overall Acc. (%)		84.63	87.26		90.76	89.21
Kappa		0.796	0.830		0.877	0.856
Weighted kappa		0.813	0.864		0.867	0.871

When the individual class accuracies measured by the F-score accuracy measure were analyzed, it was noticed that the lowest accuracies were estimated for the gravel class that was mainly confused with the self-blocking bricks for both FC and OS networks. Similarly, bitumen pixels were confused with asphalt pixels. This is certainly related to the spectral similarity of the corresponding classes that can be easily observed from the mean spectral reflectance curves (spectral signatures) given in Figure 10. The highest individual class accuracy was achieved for the metal sheets class (over 99%), which has a distinct spectral signature compared to the other classes. The trained networks using the 50% sampling ratio were applied to the whole image to produce the thematic maps of the study, which is presented in Figure 11. Confusion in the class definition for the above-mentioned classes can be observed clearly from the figure. The mixture of gravel and self-blocking bricks pixels is quite obvious in the thematic map produced with the OS network (Figure 11b). Moreover, misclassified pixels within the meadows and asphalt fields are in the form of "salt-and-pepper" noise.

Performances of the FC and OS networks for both datasets were summarized in Table 6. For the considered datasets, no hidden neuron was removed from the networks when limited training data (only 10% of the whole datasets) were considered. When the 50% sampling ratio was employed in the training phase, one hidden neuron was eliminated from the FC network for the Indian Pines data and three hidden neurons were removed for the Pavia University data. Smaller networks were found sufficient to learn the underlying characteristics of the LULC classes, and for both cases, the initial networks were trimmed by about 60% in terms of the total number of links, which can be regarded as a success of the proposed algorithm. In addition, a considerable number of inputs (i.e., spectral bands) were removed from the datasets, achieving even better classification performances (about 4% overall accuracy difference). With the implementation of the proposed superstructure optimization, the networks avoided overfitting, thus producing higher classification accuracies for the limited training data (i.e., 10% sampling ratio). It should be mentioned that the FC networks had low generalization capabilities, producing very high accuracy for the training data but comparatively lower accuracies for the test data. The obtained results are promising for the proposed algorithm, being a good alternative to feature selection methods, especially the statistical ones. $\gamma \sum diag(cov_p)$ values in the table indicate the level of overfitting that occurred in the training process. The computed values were much higher for the fully connected networks, particularly the one calculated for the Indian Pines data. Finally, it should be also mentioned that the multiply accumulates (MACS) are directly proportional to the number of connections; therefore, they can be estimated from Table 6.

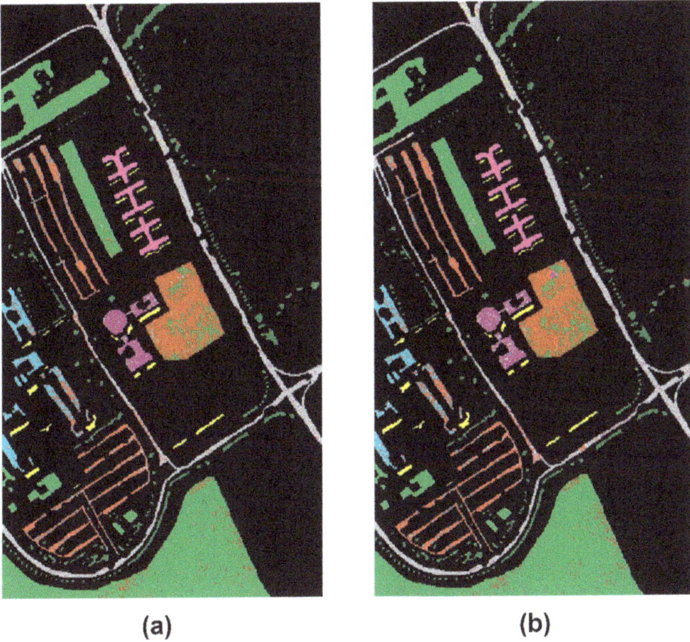

Figure 11. Classification results using (**a**) FC and (**b**) OS networks with 50% of the samples for the Pavia University dataset.

The comparison of the CPU times of different sampling ratios for the two datasets is given in Table 7. All the results were obtained using an Intel Core i5-6400 CPU 2.7 GHz 4 core 16 Gb RAM machine using a Linux operating system. It was observed that the CPU times of the proposed training method were larger than the FCs because of the MINLP programs. MINLPs are known to be NP-hard and cannot be solved in polynomial time, whereas standard training algorithms (NLPs) are P-only types. Therefore, the required computational time can be much higher for the proposed method. On the other hand, reduced and optimal ANN structures should result in faster CPU times since the number of required multiplication operations is much lower, which might also be a beneficial feature for testing larger ANNs, e.g., deep neural networks.

Table 6. Performance comparison and network architecture for FC and OS for 10% and 50% samples. Note that OA indicates overall accuracy.

	Indian Pines				Pavia University			
	10% Sample		50% Sample		10% Sample		50% Sample	
	FC	OS	FC	OS	FC	OS	FC	OS
OA (training)	99.91	97.04	90.37	86.45	99.53	92.10	92.06	90.23
OA (test)	71.98	76.37	83.80	82.72	84.63	87.26	90.76	89.21
$\gamma \sum diag(cov_p)$	6.57	0.08	9.80	0.11	2.34	0.04	1.01	0.03
Number of hidden neurons	10	10	10	9	10	10	10	7
Number of inputs	200	158	200	147	103	69	103	69
Number of connections	2150	879	2150	737	1120	423	1120	402

Table 7. CPU time comparison for FC and OS for 10% and 50% sampling ratios.

	Indian Pines				Pavia University			
	10% Sampling		50% Sampling		10% Sampling		50% Sampling	
	FC	OS	FC	OS	FC	OS	FC	OS
Training (s)	41.5	7213	65.3	10823	40.2	5418	63.7	9036
Test (s)	0.015	0.010	0.007	0.005	0.05	0.03	0.02	0.01

5. Conclusions

This study investigates the optimal training of multi-layer perceptrons through formulating and solving a mixed-integer non-linear optimization problem, delivering a significant reduction in the number of network connections, neurons, and input variables. It differs from the other methods proposed in literature as it introduces both a strong and general mixed-integer programming method for optimal structural design and allows automatic and simultaneous design and training. This feature is particularly advantageous since the presence of redundant inputs and connections decreases the prediction performance of the ANNs and increases the computational load for training. Furthermore, classical input selection (i.e., feature selection) and pruning methods, including dropout layers into deep neural networks, usually require many sequential iterations between design and training instead of automatic and simultaneous design and training. Two classification case studies with two sampling ratios (10% and 50% sampling ratios), namely Indian Pines and Pavia University datasets, were considered as benchmark test sites, and the results showed that optimal ANN structures contained a significantly lower number of inputs, connections, and neuron numbers. To be more specific, although about 60% of the network connections and 25% of the inputs (i.e., spectral bands) were removed by the proposed algorithm, superior classification performances (~4% in terms of overall accuracy) were achieved with the estimated optimal superstructure for the case of limited training samples (10% of the whole samples). It was observed that the method eliminated the least effective and correlated spectral bands that have an insignificant or trivial contribution to the delineation of the LULC characteristics. To the best of authors' knowledge, this paper is the first application of such an automatic and optimal design and training method for MLP type neural networks for classification problems. Finally, the method presented in this work can be applied using global optimization algorithms for further enhancement in terms of the prediction performance of ANNs. Moreover, reduced and optimal ANN structures should result in faster CPU times, since the number of required multiplication operations is much smaller, which could be a beneficial feature for testing larger ANNs, e.g., deep neural networks.

Author Contributions: Conceptualization, T.K., H.S. and E.A.; methodology, T.K., H.S. and E.A.; formal analysis, T.K., H.S. and E.A.; investigation, T.K. and H.S.; data curation, H.S. and E.A.; writing—original draft preparation,

T.K., H.S. and E.A.; writing—review and editing, T.K., H.S. and E.A.; project administration, T.K.; All authors have read and agreed to the published version of the manuscript.

Funding: This research received no external funding.

Acknowledgments: The authors would like to thank P. Gamba from the University of Pavia, Italy, and D. Landgrebe from Purdue University for providing the Pavia University and Indian Pines datasets, respectively.

Conflicts of Interest: The authors declare no conflict of interest.

References

1. Rosenblatt, F. The perceptron: A probabilistic model for information storage and organization in the brain. *Psychol. Rev.* **1958**, *65*, 386–408. [CrossRef] [PubMed]
2. Benediktsson, J.A.; Swain, P.H.; Ersoy, O.K. Neural network approaches versus statistical methods in classification of multisource remote sensing data. *IEEE Trans. Geosci. Remote Sens.* **1990**, *28*, 540–552. [CrossRef]
3. Serpico, S.B.; Bruzzone, L.; Roli, F. An experimental comparison of neural and statistical non-parametric algorithms for supervised classification of remote-sensing images. *Pattern Recognit. Lett.* **1996**, *17*, 1331–1341. [CrossRef]
4. Lu, D.; Weng, Q. A survey of image classification methods and techniques for improving classification performance. *Int. J. Remote Sens.* **2007**, *28*, 823–870. [CrossRef]
5. Yuan, H.; Van Der Wiele, C.F.; Khorram, S. An automated artificial neural network system for land use/land cover classification from Landsat TM imagery. *Remote Sens.* **2009**, *1*, 243–265. [CrossRef]
6. Taravat, A.; Proud, S.; Peronaci, S.; Del Frate, F.; Oppelt, N. Multilayer perceptron neural networks model for Meteosat second generation SEVIRI daytime cloud masking. *Remote Sens.* **2015**, *7*, 1529–1539. [CrossRef]
7. Paola, J.D.; Schowengerdt, R.A. A review and analysis of backpropagation neural networks for classification of remotely-sensed multi-spectral imagery. *Int. J. Remote Sens.* **1995**, *16*, 3033–3058. [CrossRef]
8. Bruzzone, L.; Conese, C.; Maselli, F.; Roli, F. Multisource classification of complex rural areas by statistical and neural-network approaches. *Photogramm. Eng. Remote Sens.* **1997**, *63*, 523–533.
9. Sunar Erbek, F.; Özkan, C.; Taberner, M. Comparison of maximum likelihood classification method with supervised artificial neural network algorithms for land use activities. *Int. J. Remote Sens.* **2004**, *25*, 1733–1748. [CrossRef]
10. Kavzoglu, T.; Reis, S. Performance analysis of maximum likelihood and artificial neural network classifiers for training sets with mixed pixels. *GISci. Remote Sens.* **2008**, *45*, 330–342. [CrossRef]
11. Mahdianpari, M.; Salehi, B.; Rezaee, M.; Mohammadimanesh, F.; Zhang, Y. Very deep convolutional neural networks for complex land cover mapping using multispectral remote sensing imagery. *Remote Sens.* **2018**, *10*, 1119. [CrossRef]
12. Kavzoglu, T.; Mather, P.M. The role of feature selection in artificial neural network applications. *Int. J. Remote Sens.* **2002**, *23*, 2919–2937. [CrossRef]
13. Ledesma, S.; Cerda, G.; Aviña, G.; Hernández, D.; Torres, M. Feature selection using artificial neural networks. In *Lecture Notes in Computer Science (Including Subseries Lecture Notes in Artificial Intelligence and Lecture Notes in Bioinformatics)*; Springer: Berlin/Heidelberg, Germany, 2008; Volume 5317, pp. 351–359.
14. Roy, D.; Murty, K.S.R.; Mohan, C.K. Feature selection using Deep Neural Networks. In Proceedings of the International Joint Conference on Neural Networks, Killarney, Ireland, 12–17 July 2015.
15. Deraeve, J.; Alexander, W.H. Fast, accurate, and stable feature selection using neural networks. *Neuroinformatics* **2018**, *16*, 253–268. [CrossRef] [PubMed]
16. Kavzoglu, T. An Investigation of the Design and Use of Feed-Forward Artificial Neural Networks. Ph.D. Thesis, The University of Nottingham, Nottingham, UK, 2001.
17. Kavzoglu, T.; Mather, P.M. The use of backpropagating artificial neural networks in land cover classification. *Int. J. Remote Sens.* **2003**, *24*, 4907–4938. [CrossRef]
18. Stathakis, D.; Kanellopoulos, I. Global optimization versus deterministic pruning for the classification of remotely sensed imagery. *Photogramm. Eng. Remote Sens.* **2008**, *74*, 1259–1265. [CrossRef]
19. Mas, J.F.; Flores, J.J. The application of artificial neural networks to the analysis of remotely sensed data. *Int. J. Remote Sens.* **2008**, *29*, 617–663. [CrossRef]

20. Atkinson, P.M.; Tatnall, A.R.L. Introduction neural networks in remote sensing. *Int. J. Remote Sens.* **1997**, *18*, 699–709. [CrossRef]

21. Suliman, A.; Zhang, Y. A Review on back-propagation neural networks in the application of remote sensing image classification. *J. Earth Sci. Eng.* **2015**, *5*, 52–65.

22. Stathakis, D. How many hidden layers and nodes? *Int. J. Remote Sens.* **2009**, *30*, 2133–2147. [CrossRef]

23. Augasta, M.; Kathirvalavakumar, T. Pruning algorithms of neural networks—A comparative study. *Open Comput. Sci.* **2013**, *3*, 105–115. [CrossRef]

24. Castellano, G.; Fanelli, A.M.; Pelillo, M. An iterative pruning algorithm for feedforward neural networks. *IEEE Trans. Neural Netw.* **1997**, *8*, 519–531. [CrossRef] [PubMed]

25. Paetz, J. Reducing the number of neurons in radial basis function networks with dynamic decay adjustment. *Neurocomputing* **2004**, *62*, 79–91. [CrossRef]

26. Narasimha, P.L.; Delashmit, W.H.; Manry, M.T.; Li, J.; Maldonado, F. An integrated growing-pruning method for feedforward network training. *Neurocomputing* **2008**, *71*, 2831–2847. [CrossRef]

27. Zanchettin, C.; Ludermir, T.B. Hybrid optimization technique for artificial neural networks design. In Proceedings of the ICEIS 2009—11th International Conference on Enterprise Information Systems, Milan, Italy, 6–10 May 2009; pp. 242–247.

28. Gan, M.; Peng, H.; Dong, X.P. A hybrid algorithm to optimize RBF network architecture and parameters for nonlinear time series prediction. *Appl. Math. Model.* **2012**, *36*, 2911–2919. [CrossRef]

29. Kavzoglu, T.; Mather, P.M. Pruning artificial neural networks: An example using land cover classification of multi-sensor images. *Int. J. Remote Sens.* **1999**, *20*, 2761–2785. [CrossRef]

30. Sarle, W. Neural Network FAQ. Available online: Ftp://ftp.sas.com/pub/neural/FAQ.html (accessed on 8 January 2020).

31. Thimm, G.; Fiesler, E. *Pruning of Neural Networks*. IDIAP Research Report: IDIAP-RR 97-03. 1997, pp. 1–17. Available online: https://publications.idiap.ch/downloads/reports/1997/rr97-03.pdf (accessed on 8 January 2020).

32. Reed, R. Pruning algorithms—A survey. *IEEE Trans. Neural Netw.* **1993**, *4*, 740–747. [CrossRef]

33. Emmerson, M.D.; Damper, R.I. Determining and improving the fault tolerance of multilayer perceptrons in a pattern-recognition application. *IEEE Trans. Neural Netw.* **1993**, *4*, 788–793. [CrossRef]

34. Kimes, D.S.; Nelson, R.F.; Manry, M.T.; Fung, A.K. Review article: Attributes of neural networks for extracting continuous vegetation variables from optical and radar measurements. *Int. J. Remote Sens.* **1998**, *19*, 2639–2663. [CrossRef]

35. Sildir, H.; Aydin, E. Optimal Artificial Neural Network Design and Training: Input Selection and Architecture. Submitted. **2019**, 1–9.

36. Jackson, Q.Z.; Landgrebe, D. Design of an Adaptive Classification Procedure for the Analysis of High-Dimensional Data with Limited Training Samples. Ph.D. Thesis, School of Electrical & Computer Engineering, Purdue University, West Lafayette, IN, USA, 2001; 137p.

37. Kavzoglu, T.; Tonbul, H.; Yildiz Erdemir, M.; Colkesen, I. Dimensionality reduction and classification of hyperspectral images using object-based image analysis. *J. Indian Soc. Remote Sens.* **2018**, *46*, 1297–1306. [CrossRef]

38. Maxwell, A.E.; Warner, T.A.; Fang, F. Implementation of machine-learning classification in remote sensing: An applied review. *Int. J. Remote Sens.* **2018**, *39*, 2784–2817. [CrossRef]

39. De Brébisson, A.; Vincent, P. An exploration of softmax alternatives belonging to the spherical loss family. In Proceedings of the 4th International Conference on Learning Representations, ICLR 2016—Conference Track Proceedings, San Juan, PR, USA, 2–4 May 2016.

40. McLean, K.A.P.; McAuley, K.B. Mathematical modelling of chemical processes - obtaining the best model predictions and parameter estimates using identifiability and estimability procedures. *Can. J. Chem. Eng.* **2012**, *90*, 351–366. [CrossRef]

41. Dua, V. A mixed-integer programming approach for optimal configuration of artificial neural networks. *Chem. Eng. Res. Des.* **2010**, *88*, 55–60. [CrossRef]

42. Dua, V. Optimal configuration of artificial neural networks. *Comput. Aided Chem. Eng.* **2006**, *21*, 1599–1604.

43. Kavzoglu, T. Determining optimum structure for artificial neural networks. In Proceedings of the 25th Annual Technical Conference and Exhibition of the Remote Sensing Society (Earth Observation: From Data to Information), Cardiff, UK, 8–10 September 1999; pp. 675–682.

44. Lin, Z.; Zou, Q.; Ward, E.S.; Ober, R.J. Cramer-Rao lower bound for parameter estimation in nonlinear systems. *IEEE Signal Process. Lett.* **2005**, *12*, 855–858.
45. Tellinghuisen, J. Statistical error propagation. *J. Phys. Chem. A* **2001**, *105*, 3917–3921. [CrossRef]
46. Lawler, E.L.; Wood, D.E. Branch-and-bound methods: A survey. *Oper. Res.* **1966**, *14*, 699–719. [CrossRef]
47. Geoffrion, A.M. Generalized Benders decomposition. *J. Optim. Theory Appl.* **1972**, *10*, 237–260. [CrossRef]
48. Duran, M.A.; Grossmann, I.E. An outer-approximation algorithm for a class of mixed-integer nonlinear programs. *Math. Program.* **1986**, *36*, 307–339. [CrossRef]
49. Chen, X.; Li, Z.; Yang, J.; Shao, Z.; Zhu, L. Nested tabu search (TS) and sequential quadratic programming (SQP) method, combined with adaptive model reformulation for heat exchanger network synthesis (HENS). *Ind. Eng. Chem. Res.* **2008**, *47*, 2320–2330. [CrossRef]
50. Pintarič, Z.N.; Kravanja, Z. The two-level strategy for MINLP synthesis of process flowsheets under uncertainty. *Comput. Chem. Eng.* **2000**, *24*, 195–201. [CrossRef]
51. Chen, X.; Li, Z.; Wan, W.; Zhu, L.; Shao, Z. A master-slave solving method with adaptive model reformulation technique for water network synthesis using MINLP. *Sep. Purif. Technol.* **2012**, *98*, 516–530. [CrossRef]
52. Biegler, L.T.; Grossmann, I.E. Retrospective on optimization. *Comput. Chem. Eng.* **2004**, *28*, 1169–1192. [CrossRef]
53. Biegler, L.T. Large-scale nonlinear programming: An integrating framework for enterprise-wide dynamic optimization. *Comput. Aided Chem. Eng.* **2007**, *24*, 575–582.

Article

Deep Quadruplet Network for Hyperspectral Image Classification with a Small Number of Samples

Chengye Zhang [1,2,3], Jun Yue [2,*] and Qiming Qin [2]

[1] State Key Laboratory of Coal Resources and Safe Mining, China University of Mining & Technology (Beijing), Beijing 100083, China; czhang@cumtb.edu.cn
[2] Institute of Remote Sensing and Geographic Information System, School of Earth and Space Sciences, Peking University, Beijing 100871, China; qmqin@pku.edu.cn
[3] College of Geoscience and Surveying Engineering, China University of Mining & Technology (Beijing), Beijing 100083, China
[*] Correspondence: jyue@pku.edu.cn

Received: 17 January 2020; Accepted: 13 February 2020; Published: 15 February 2020

Abstract: This study proposes a deep quadruplet network (DQN) for hyperspectral image classification given the limitation of having a small number of samples. A quadruplet network is designed, which makes use of a new quadruplet loss function in order to learn a feature space where the distances between samples from the same class are shortened, while those from a different class are enlarged. A deep 3-D convolutional neural network (CNN) with characteristics of both dense convolution and dilated convolution is then employed and embedded in the quadruplet network to extract spatial-spectral features. Finally, the nearest neighbor (NN) classifier is used to accomplish the classification in the learned feature space. The results show that the proposed network can learn a feature space and is able to undertake hyperspectral image classification using only a limited number of samples. The main highlights of the study include: (1) The proposed approach was found to have high overall accuracy and can be classified as state-of-the-art; (2) Results of the ablation study suggest that all the modules of the proposed approach are effective in improving accuracy and that the proposed quadruplet loss contributes the most; (3) Time-analysis shows the proposed methodology has a similar level of time consumption as compared with existing methods.

Keywords: deep learning; hyperspectral image classification; few-shot learning; quadruplet loss; dense network; dilated convolutional network

---

1. Introduction

A hyperspectral image covers hundreds of bands with high spectral resolution and provides a detailed spectral curve for each pixel [1,2]. Both the spatial and the spectral information are gathered in a hyperspectral image. Hyperspectral image classification is aimed at identifying the specific class (i.e., label) for each pixel (for example, cropland, lake, river, grassland, forest, mineral rocks, building, and roads). As the first step in many hyperspectral remote sensing applications, image classification is vital in the fields of agricultural statistics, disaster reduction, mineral exploration, and environmental monitoring.

In recent decades, many methods have been proposed for the hyperspectral image classification, such as spectral angle mapper (SAM), mixture tuned matched filtering (MTMF), spectral feature fitting (SFF) [3,4], neural network (NN) [5], support vector machine (SVM) [6,7], and random forest (RF) [8,9]. SAM, MTMF, and SFF are heavily influenced by anthropogenic decision-making, while NN, SVM, and RF are gradually becoming more dependent on new machine learning methods. Since the concept of deep learning was introduced into hyperspectral image classification for the first time [10], deep neural network has been gaining popularity and has triggered global research interest in establishing deep learning models for hyperspectral image classification [11–14]. In particular, some deep-learning methods have been proposed by combining spectral and spatial features to improve classification accuracy [15–21].

However, the aforementioned methods still require substantial improvements in hyperspectral image classification, especially under the condition of small-samples. For supervised classification of remotely sensed images, the training samples are usually acquired by two methods: (1) from field surveys and (2) directly from images with higher resolution. In particular, higher classification accuracy is usually acquired from training samples collected by field surveys. However, compared with laboratory work, field survey is costly, complicated, and time-consuming, which can significantly restrict the number of training samples. A small dataset of training samples can substantially diminish accuracy in hyperspectral image classification. Moreover, hyperspectral images suffer more from data redundancy in the spectral dimension compared with multi-spectral images, which creates additional difficulties for classification.

Few-shot learning involves solving the problem using a limited number of samples and has been used for various applications such as image segmentation, image caption, object recognition, and face identification. [22–25]. Given the limited accuracy due to having only a few labeled samples per class, few-shot learning usually trains the model based on a well-labeled dataset, and the model is then generalized into new classes [26]. A metric learning strategy is usually adopted to learn the features of the object and distinguish based on the absolute distance between samples [27]. In recent years, several few-shot learning methods have been proposed for hyperspectral image classification, e.g., DFSL (deep few-shot learning) [28,29]. However, absolute distance ignores the relationship between inter-class and intra-class and limits classification accuracy. The use of relative distance, based on widening inter-class distance and shortening the intra-class distance, has been proposed in lieu of absolute distance [30]. Proposing new methods that account for the relative relationship between inter-class and intra-class is therefore crucial in improving the accuracy of hyperspectral image classification with a limited number of samples.

This study proposes a deep quadruplet network (DQN) for hyperspectral image classification with a small number of samples. To improve the accuracy, we designed a quadruplet network, in particular, a new quadruplet loss function, and a deep 3-D CNN with double branches consisting of dense convolution and dilated convolution.

2. Materials and Methods

2.1. Data

2.1.1. Training Data

The training data used in this study are four well-known public hyperspectral datasets: "Houston", "Chikusei", "KSC", and "Botswana" [28]. The details of the four hyperspectral datasets used in training are presented in Table 1.

Table 1. The details of the four hyperspectral datasets for training networks [28].

Dataset Name	Houston	Chikusei	KSC	Botswana
Location	Houston	Chikusei	Florida	Botswana
Height	349	2517	512	1476
Width	1905	2335	614	256
Bands	144	128	176	145
Spectral Range (nm)	380–1050	363–1018	400–2500	400–2500
Spatial Resolution (m)	2.5	2.5	18	30
Num. of Classes	31	19	13	14

2.1.2. Testing Data

The testing data used in this study are three widely-known public hyperspectral datasets: "Salinas", "Indian Pines" (IP), and "University of Pavia" (UP). The details of the three hyperspectral datasets used in testing are summarized in Table 2. The ground-truth maps of the three hyperspectral datasets are shown in Figures 1–3.

Table 2. The details of the three hyperspectral datasets for testing networks [28].

Dataset Name	Salinas	IP	UP
Location	California	Indiana	Pavia
Height	512	145	610
Width	217	145	340
Bands	204	200	103
Spectral Range (nm)	400–2500	400–2500	430–860
Spatial Resolution (m)	3.7	20	1.3
Num. of Classes	16	16	9

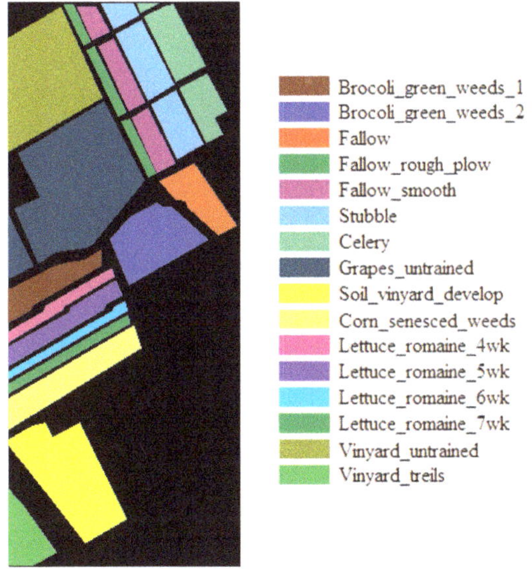

Figure 1. Ground-truth map of the Salinas dataset.

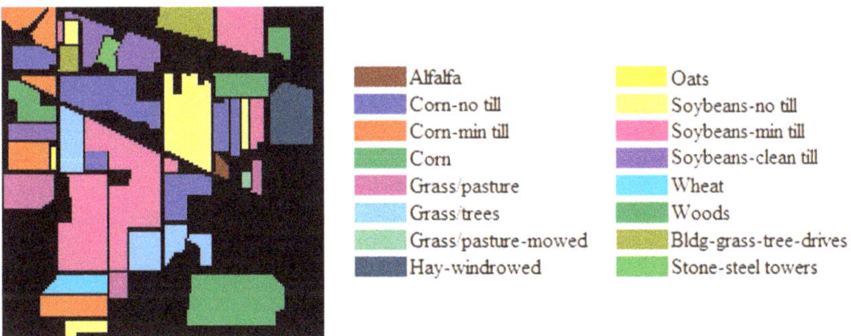

Figure 2. Ground-truth map of the Indian Pines (IP) dataset.

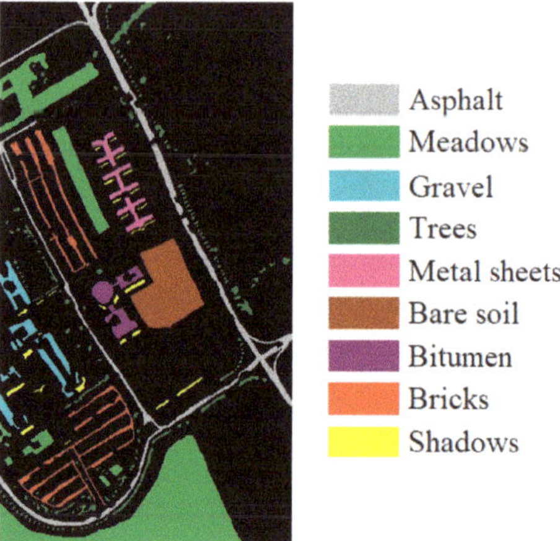

Figure 3. Ground-truth map of the UP dataset.

The training datasets "Houston" and "Chikusei" can be acquired from the following websites, respectively: "http://hyperspectral.ee.uh.edu/2egf4tg8hial13gt/2013_DFTC.zip" and "http://park.itc. u-tokyo.ac.jp/sal/hyperdata/Hyperspec_Chikusei_MATLAB.zip". The training datasets "KSC" and "Botswana" and all the testing datasets can be acquired from the website "http://www.ehu.eus/ccwintco/ index.php?title=Hyperspectral_Remote_Sensing_Scenes".

2.2. Structure of the Proposed Method

The structure of the proposed methodology in this study is shown in Figure 4. A deep quadruplet network is trained to learn a feature space. The testing data is transferred to the learned feature space to extract features. The classification is accomplished using the Euclidean distance and the nearest neighbor (NN) classifier.

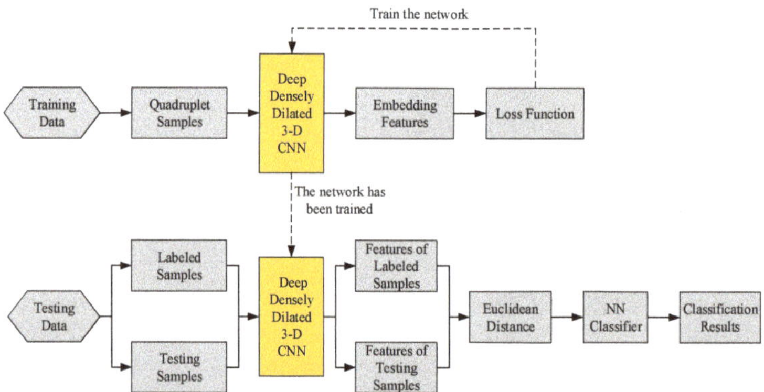

Figure 4. The structure of the proposed methodology.

2.3. Quadruplet Learning

Metric learning refers to the transfer of input data from the original space R^F into a new feature space R^D (i.e., $f_\theta: R^F \rightarrow R^D$). F and D refer to the dimension of the original space and the new space, respectively, and θ is the learnable parameter. In the new feature space R^D, samples from the same class are expected to be closer than those from different classes so that the classification can be finished in R^D using nearest neighbor classifier. Several networks have been developed to accomplish this task, including the siamese network, triplet network [31], and quadruplet network [32].

In a siamese network, a contrastive loss function is designed to train the network to distinguish between pairs of samples from the same class and those from different classes. The designed loss function limits the samples within the same class and enlarges the samples from the different classes. However, for classification purposes, the feature space learned by the siamese network is inferior to that of the triplet network. In addition, siamese networks are sensitive to calibration in order to contextualize similarity vs. dissimilarity [31]. The loss function for a siamese network is:

$$L_s = \frac{1}{N_s} \sum_{i=1}^{N_s} d(x_a{}^{(i)}, x_p{}^{(i)}) \tag{1}$$

where $x_a{}^{(i)}$ and $x_p{}^{(i)}$ are two samples from the same class, which has been transferred by f_θ: $\mathrm{R}^F \to \mathrm{R}^D$; N_s is the number of siamese pairs; and $d\,(\cdot)$ is the Euclidean distance of two elements.

Triplet network refers to training based on the use of many triplets. A triplet contains three different samples $(x_a{}^{(i)}, x_p{}^{(i)}, x_n{}^{(i)})$, where $x_a{}^{(i)}$ and $x_p{}^{(i)}$ are two samples from the same class (i.e., positive pairs), while $x_a{}^{(i)}$ and $x_n{}^{(i)}$ are samples from different classes (i.e., negative pairs). Each sample in a triplet has been transferred by f_θ: $\mathrm{R}^F \to \mathrm{R}^D$. The loss function for the triplet network is given by [32]:

$$L_t = \frac{1}{N_t} \sum_{i=1}^{N_t} \left(d(x_a{}^{(i)}, x_p{}^{(i)}) - d(x_a{}^{(i)}, x_n{}^{(i)}) + \gamma \right)_+ \tag{2}$$

where γ is the value of the margin set that segregates the positive pairs with the negative pairs; N_t is the number of triplets; and $(z)_+ = \max(0, z)$. The first term is intended to shorten the distance between two samples from the same class, while the second term is designed to enlarge the distance between two samples from different classes. For the loss function in triplet networks, each positive pair and negative pair share a given sample (i.e., $x_a{}^{(i)}$), which compels triplet networks to focus more on obtaining the correct ranks for the pair distances. In other words, the triplet loss only considers the relative distances of the positive and negative pairs, which results in poor generalization for the triplet network and difficulty applying in tracking tasks [32].

The quadruplet loss (QL) [30] introduces a different negative pair into the triplet loss. The quadruplet loss function contains four different samples: $(x_a{}^{(i)}, x_p{}^{(i)}, x_{n1}{}^{(i)}, x_{n2}{}^{(i)})$, where $x_a{}^{(i)}$ and $x_p{}^{(i)}$ are samples from a same class while $x_{n1}{}^{(i)}$ and $x_{n2}{}^{(i)}$ are samples from another two classes. All the samples have been transferred to the featured space by f_θ: $\mathrm{R}^F \to \mathrm{R}^D$. The quadruplet loss is given by the equation:

$$L_q = \frac{1}{N_q} \sum_{i=1}^{N_q} \left((d(x_a{}^{(i)}, x_p{}^{(i)}) - d(x_a{}^{(i)}, x_{n1}{}^{(i)}) + \gamma)_+ + (d(x_a{}^{(i)}, x_p{}^{(i)}) - d(x_{n1}{}^{(i)}, x_{n2}{}^{(i)}) + \beta)_+ \right) \tag{3}$$

where γ and β are the margins for the two terms; and N_q is the number of quadruplets. The first term in quadruplet loss is the same as that in the triplet loss (Equations (2) and (3)). The second term constrains the intra-class distances to be smaller than the inter-class distances [30]. However, the loss function in Equation (3) usually performs poorly because the number of quadruplets and quadruplet pairs would grow rapidly when the dataset gets more extensive. Moreover, most samples are not so useful towards adequately training the network and can overwhelm the relevant hard-learning samples, leading to the poor performance of the network [32].

Hence, this study designed a new quadruplet loss function, as shown in Equation (4):

$$L_{nq} = \frac{1}{N_{nq}} \sum_{i=1}^{N_{nq}} \left(d(x_a{}^{(i)}, x_p{}^{(i)}) - d(x_m{}^{(i)}, x_n{}^{(i)}) + \gamma \right)_+ \tag{4}$$

where $x_p^{(i)}$ is the farthest sample to the reference $x_a^{(i)}$ in the same class; $x_m^{(i)}$ and $x_n^{(i)}$ are the closest negative pairs in the whole batch; N_{nq} is the number of quadruplets in the new loss function; and γ is the value of the margin. Each sample in Equation (4) has been transferred by f_θ: $R^F \rightarrow R^D$. The conceptual diagram of the quadruplet network, as proposed in this study, is presented in Figure 5. The proposed loss function compensates for the shortcomings of Equations (2)–(4). The procedure for batch training using the proposed loss function is shown in Table 3, where $T = \{t_1, t_2, t_3, \dots, t_s\}$ is the training dataset for this batch, and s is the number of labeled samples. In a batch (shown in Table 3), $N_{nq} = s$, and t_j or t_k represents a sample in the dataset T. (t_j, t_k) represents a pair of samples, C (t_j) is the class label of the sample t_j, and α is the learning rate. The variables t_a, t_p, t_m, and t_n are the quadruplets before the deep network, while x_a, x_p, x_m, and x_n are the corresponding quadruplets after the deep network.

Table 3. The procedure for training a batch.

Input: The training dataset for this batch $T=\{t_1, t_2, t_3, \dots, t_s\}$
The initialized or updated learnable parameter θ

For all pairs of samples (t_j, t_k) in T **Do**
   Calculate the Euclidean distance $d\ (f_\theta\ (t_j), f_\theta\ (t_k))$
End For
Set the loss $L_{nq} = 0$
Set $(m, n) = \underset{(j,k)}{\operatorname{argmin}} d(f_\theta(t_j), f_\theta(t_k))$, under the condition C$(t_j) \neq$ C(t_k)
$x_m = f_\theta\ (t_m)$, $x_n = f_\theta\ (t_n)$. (Transfer t_m, t_n to x_m, x_n by the network f_θ: $R^F \rightarrow R^D$)
For t_a in the dataset T **DO**
   $x_a = f_\theta\ (t_a)$. (Transfer t_a to x_a by the network f_θ: $R^F \rightarrow R^D$)
   Set $p = \underset{(j)}{\operatorname{argmax}} d(f_\theta(t_j), x_a)$, under the condition C$(t_j) =$ C(t_a)
   $x_p = f_\theta\ (t_p)$. (Transfer t_p to x_p by the network f_θ: $R^F \rightarrow R^D$)
   Update: $L_{nq} = L_{nq} + \frac{1}{N_{nq}}(d(x_a, x_p) - d(x_m, x_n) + \gamma)_+$
End For
Update: $\theta = \theta - \alpha \nabla_\theta L_{nq}$
Output: θ, L_{nq}

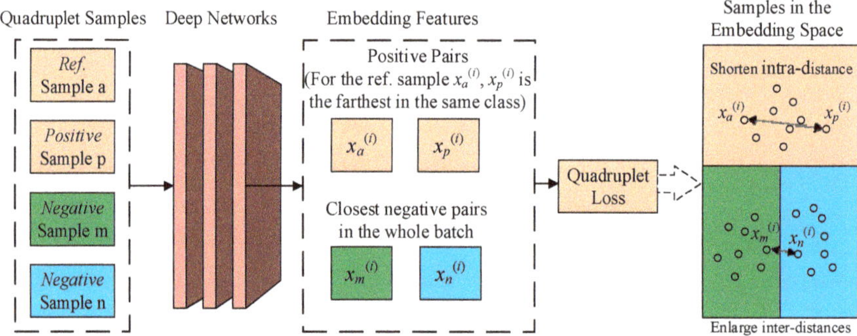

Figure 5. The concept of the quadruplet network proposed in this study.

2.4. Deep Dense Dilated 3-D CNN

2.4.1. Deep Network Framework

As shown in Figure 6, the overall framework of the proposed deep dense dilated 3-D CNN contains two branches: a dense CNN and a dilated CNN.

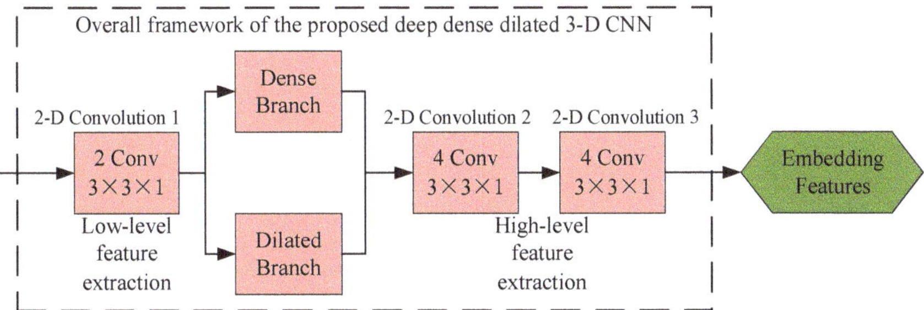

Figure 6. The overall framework of the proposed deep dense dilated 3-D convolutional neural network (CNN).

2.4.2. Dense CNN

The dense CNN block consists of five convolutional layers (see Figure 7a). "Conv" is a convolutional operation with a $3 \times 3 \times 3$ kernel, while "2 Conv" represents a convolutional layer with two kernels (i.e., two convolutional operations). For a normal CNN block with five layers, there are five connections (a connection is between a layer and its subsequent layer) [33] (see Figure 7b). However, as the network becomes more and more deep, the problem with a normal CNN is that the features contained in the input can vanish after it passes through many layers until it reaches the end [33]. So instead of using the normal CNN, a dense CNN was used in this study. Aside from the preserving the five connections from the normal CNN block, the dense CNN provides six other connections: three connections are between the 1st layer and the 3rd, 4th, and 5th layers; two connections are between the 2nd Conv and the 4th and 5th Conv; and one connection is between the 3rd Conv and the 5th Conv. Figure 7a shows the operation at a connection point in the dense CNN, and "$\oplus$" represents the sum of all the imported connected lines.

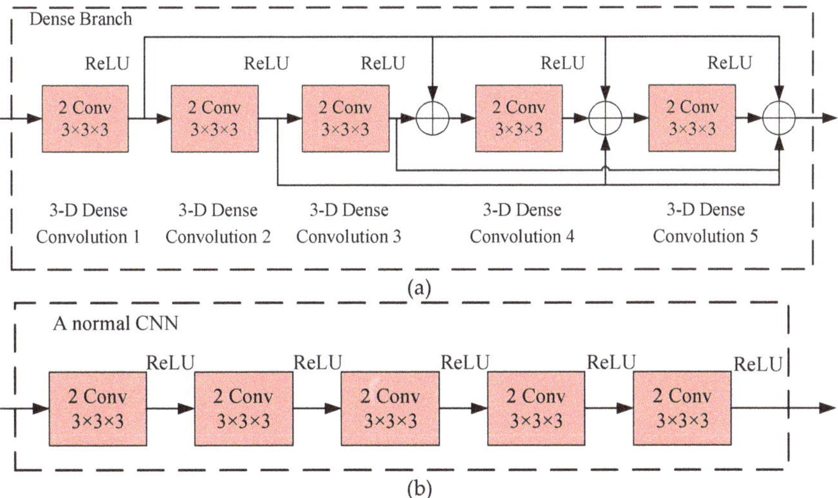

Figure 7. The dense CNN block in this study and a normal CNN block. (**a**) The structure of the dense CNN block with 5 layers in this study; (**b**) a normal CNN block with 5 layers.

The use of the dense CNN in this study alleviates the problem regarding information vanishing when passing through numerous layers and make full use of the features extracted by all the layers.

2.4.3. Dilated CNN

For normal convolutional operation, the convolutional kernel covers an image area using the same size (Figure 8a). A normal CNN employed in image classification represents the image using many tiny feature scenes, resulting in obscure spatial structures [34]. Moreover, the spatial acuity and details that are lost are almost impossible to restore through upsampling and training. Hence, the image classification accuracy is limited using the normal CNN, especially for images requiring detailed scene understanding [34]. Dilated convolution is an operation in which the convolutional kernel covers an image area with a bigger size (Figure 8b). For example, a 3 × 3 convolutional kernel only covers a 3 × 3 image area in normal convolutional operation (Figure 8a), but a 3 × 3 convolutional kernel can enlarge the receptive field to 5 × 5 (Figure 8b), or even bigger field. The dilated CNN represents the image features on a bigger scale and alleviates the disadvantage of data redundancy in a hyperspectral dataset without increasing the network's depth or complexity [34]. The structure of the dilated branch in this study is shown in Figure 9.

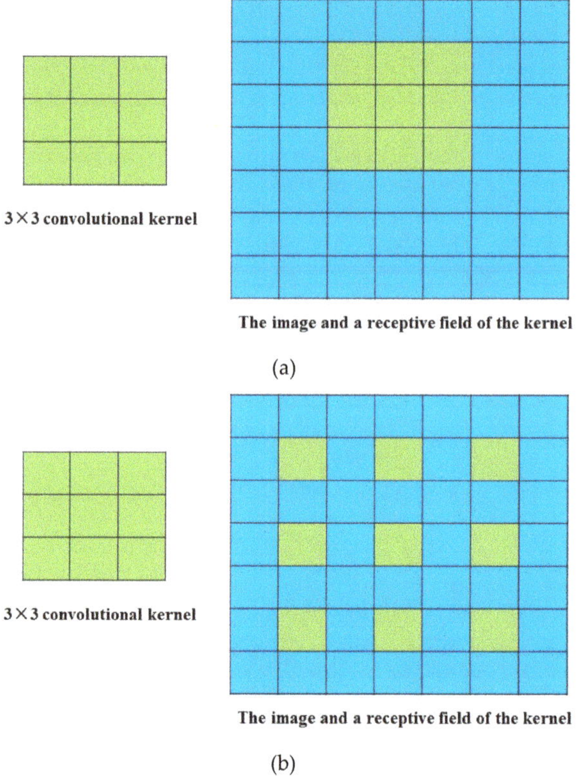

Figure 8. The convolutional operation in a normal and dilated CNN. (**a**) The convolutional operation in a normal CNN using a 3 × 3 convolutional kernel; (**b**) the convolutional operation in a dilated CNN using a 3 × 3 convolutional kernel.

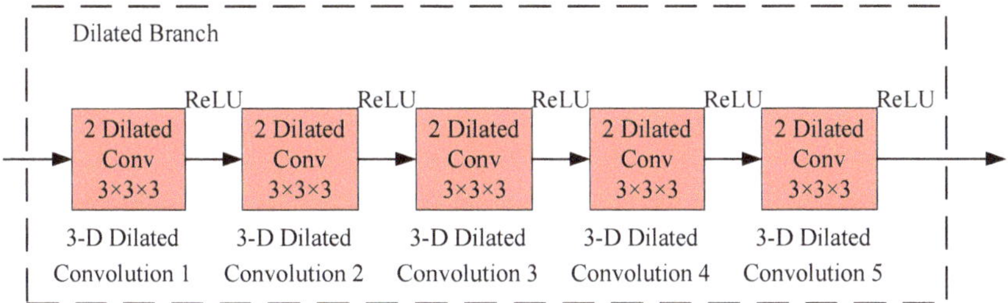

Figure 9. The structure of the dilated branch in this study.

In both the dense CNN block and the dilated CNN block, there is an operation called ReLU. ReLU is an activation function defined as:

$$f(x)=\max(0, x) \tag{5}$$

2.5. Nearest Neighbor (NN) for Classification

In this study, an embedded feature space is learned after training the proposed deep quadruplet CNN using the training data. For the testing data, the supervised samples and the samples to be classified are transferred to the embedded feature space by the trained deep quadruplet CNN. The classification is completed using the average Euclidean distance between the supervised samples and the samples to be classified using the nearest neighbor classifier.

2.6. Parameters Setting

Details of the network architecture for the proposed deep quadruplet network are summarized in Table 4. The layer names in Table 4 correspond with the blocks in Figure 6, Figure 7, and Figure 9. N is the band number of the hyperspectral dataset. The N bands are selected by graph representation based band selection (GRBS) [35]. "ceil ($N/2$)" is the ceiling function, which is equal to the rounded-up integer of $N/2$. The learning rate α for optimizing DQN is set to be 10^{-3} with a weight decay of 10^{-4} and a momentum of 0.9. The sensitivity of the margin γ to the classification accuracy was tested with 15 supervised samples per class. The overall accuracy (OA) for all three testing datasets is shown in Figure 10. The value of γ was set to be 0.4, which refers to the best accuracy obtained for all the three testing datasets, as presented in Figure 10.

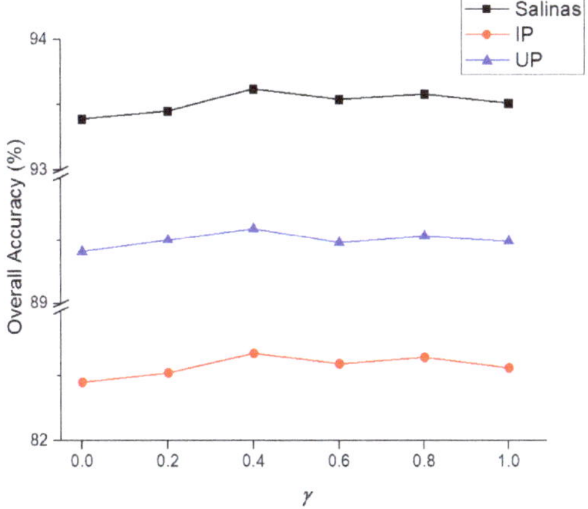

Figure 10. The overall accuracy with different γ for all the three testing datasets.

Table 4. The network architecture details of the proposed deep quadruplet network.

Layer Name	Input Layer	Filter Size	Padding	Output Shape
Input	/	/	/	$9 \times 9 \times N$
2-D Convolution 1	Input	$3 \times 3 \times 1 \times 2$	Yes	$9 \times 9 \times N \times 2$
3-D Dense Convolution 1	2-D Convolution 1	$3 \times 3 \times 3 \times 2$	Yes	$9 \times 9 \times N \times 2$
3-D Dense Convolution 2	3-D Dense Convolution 1	$3 \times 3 \times 3 \times 2$	Yes	$9 \times 9 \times N \times 2$
3-D Dense Convolution 3	3-D Dense Convolution 2	$3 \times 3 \times 3 \times 2$	Yes	$9 \times 9 \times N \times 2$
3-D Dense Shortcut 1	3-D Dense Convolution 1&3	3-D Dense Convolution 1 + 3-D Dense Convolution 3		$9 \times 9 \times N \times 2$
3-D Dense Convolution 4	3-D Dense Shortcut 1	$3 \times 3 \times 3 \times 2$	Yes	$9 \times 9 \times N \times 2$
3-D Dense Shortcut 2	3-D Dense Convolution 1&2&4	3-D Dense Convolution 1 + 3-D Dense Convolution 2 + 3-D Dense Convolution 4		$9 \times 9 \times N \times 2$
3-D Dense Convolution 5	3-D Dense Shortcut 2	$3 \times 3 \times 3 \times 2$	Yes	$9 \times 9 \times N \times 2$
3-D Dense Shortcut 3	3-D Dense Convolution 1&2&3&5	3-D Dense Convolution 1 + 3-D Dense Convolution 2 + 3-D Dense Convolution 3 + 3-D Dense Convolution 5		$9 \times 9 \times N \times 2$
Max Pooling 1	3-D Dense Shortcut 3	$2 \times 2 \times 2$	No	$5 \times 5 \times \text{ceil}\,(N/2) \times 2$
3-D Dilated Convolution 1	2-D Convolution 1	3-D Dilated $3 \times 3 \times 3 \times 2$	Yes	$9 \times 9 \times N \times 2$
3-D Dilated Convolution 2	3-D Dilated Convolution 1	3-D Dilated $3 \times 3 \times 3 \times 2$	Yes	$9 \times 9 \times N \times 2$
3-D Dilated Convolution 3	3-D Dilated Convolution 2	3-D Dilated $3 \times 3 \times 3 \times 2$	Yes	$9 \times 9 \times N \times 2$
3-D Dilated Convolution 4	3-D Dilated Convolution 3	3-D Dilated $3 \times 3 \times 3 \times 2$	Yes	$9 \times 9 \times N \times 2$
3-D Dilated Convolution 5	3-D Dilated Convolution 4	3-D Dilated $3 \times 3 \times 3 \times 2$	Yes	$9 \times 9 \times N \times 2$
Max Pooling 2	3-D Dilated Convolution 5	$2 \times 2 \times 2$	No	$5 \times 5 \times \text{ceil}\,(N/2) \times 2$
Concatenation	Max Pooling 1&2	/	/	$5 \times 5 \times \text{ceil}\,(N/2) \times 4$
2-D Convolution 2	Concatenation	$3 \times 3 \times 1 \times 4$	No	$3 \times 3 \times \text{ceil}\,(N/2) \times 4$
2-D Convolution 3	2-D Convolution 2	$3 \times 3 \times 1 \times 4$	No	$1 \times 1 \times \text{ceil}\,(N/2) \times 4$
Full Connected	2-D Convolution 3	/	/	150

3. Results and Discussion

3.1. Accuracy

For the Salinas and the UP datasets, the number of supervised samples (L) used in the testing experiments was set to 5, 10, 15, 20, and 25. Given the limited total number of labeled samples in the IP dataset, the number of supervised samples (L) used in the testing experiments was set to 5, 10, and 15. For each case of L, the supervised samples are selected randomly and ten runs were performed. The overall accuracy of each run was recorded. The results of the classification using the proposed method (DQN+NN) are presented in Figures 11–13. The overall accuracy of the proposed approach was then compared with other methods, including: SVM [6], LapSVM [36], TSVM [37], SCS^{3}VM [38], SS-LPSVM [39], KNN+SNI [40], MLR+RS [41], SVM+S-CNN, 3D-CNN [19], DFSL+NN, and DFSL+SVM [28]. The average value and the standard deviation (STD) of the overall accuracy in the ten runs for the three testing datasets using the different methods are shown in Tables 5–7. The overall accuracy of 3D-CNN was examined based on the method described by Hamida et al. [19]. The accuracy of the SVM, LapSVM, TSVM, SCS^{3}VM, SS-LPSVM, KNN+SNI, MLR+RS, SVM+S-CNN, DFSL+NN, and DFSL+SVM are derived from the study of Liu et al. [28]. The training datasets and testing datasets in our paper are exactly same with that in Reference [28], which are public and have been widely used for comparing different methods for hyperspectral image classification [28,38–41]. Hence, the comparison of different methods is appropriate.

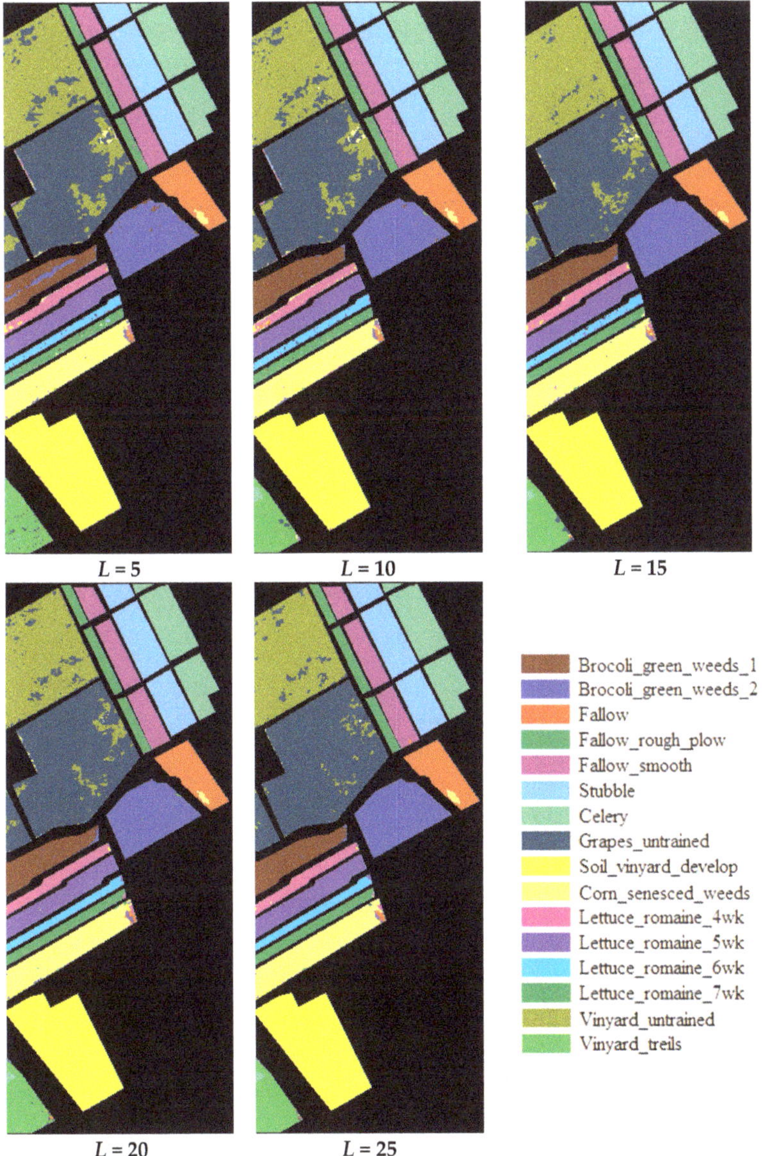

Figure 11. The classification results of the Salinas dataset.

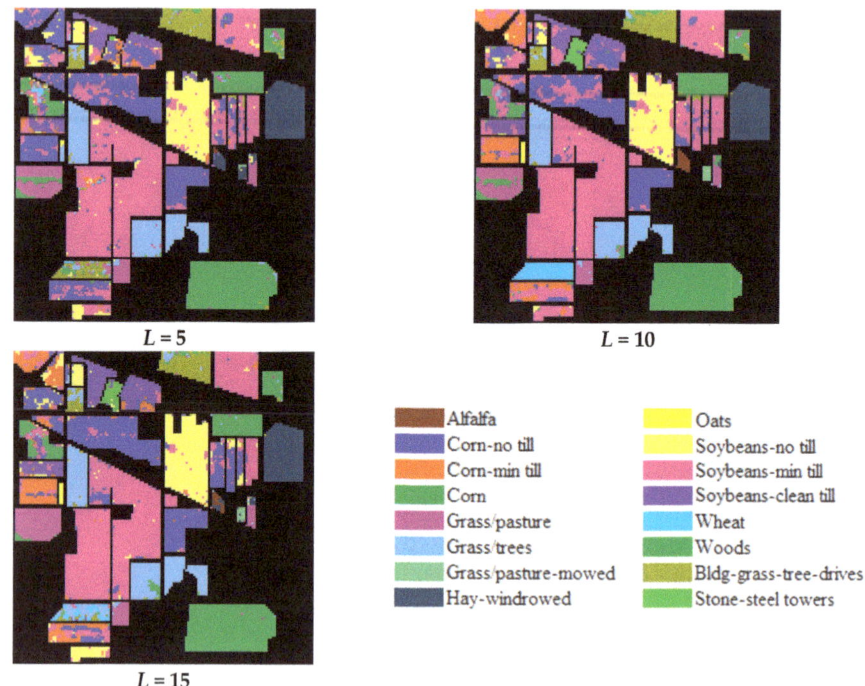

Figure 12. The classification results of the IP dataset.

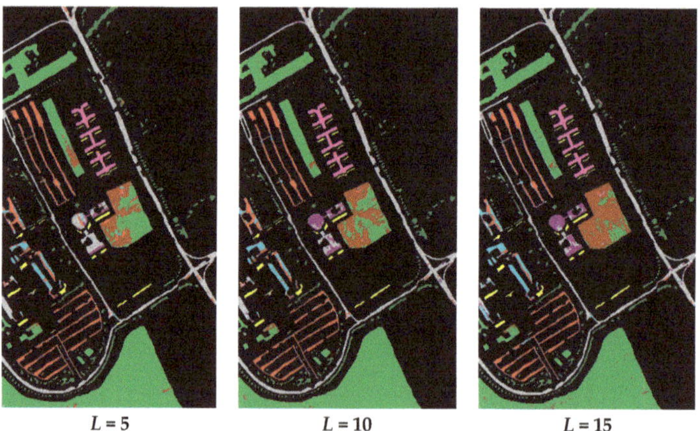

Figure 13. *Cont.*

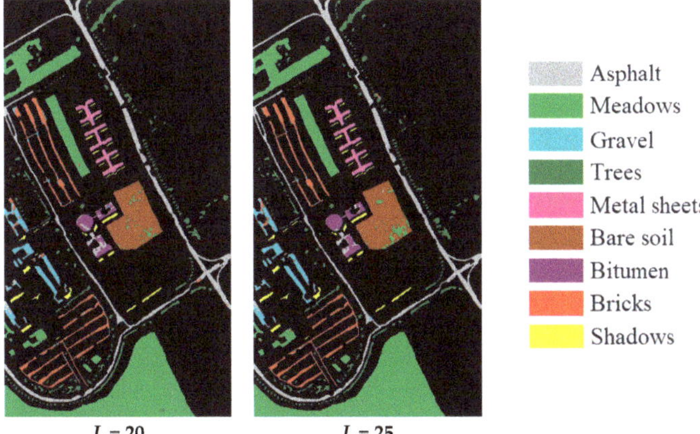

Figure 13. The classification results of the University of Pavia (UP) dataset.

Table 5. The average ± STD overall accuracy (OA, %) with Salinas dataset using different methods (The bold value is the best accuracy in each case).

Method [1]	$L = 5$	$L = 10$	$L = 15$	$L = 20$	$L = 25$
SVM	73.90 ± 1.91	75.62 ± 1.73	79.08 ± 1.45	77.89 ± 1.20	78.05 ± 1.49
LapSVM	75.31 ± 2.31	76.34 ± 1.77	77.93 ± 2.42	79.40 ± 0.73	80.56 ± 1.33
TSVM	60.43 ± 1.40	67.47 ± 1.05	69.12 ± 1.32	71.03 ± 1.78	71.83 ± 1.16
SCS³VM	74.12 ± 2.44	78.49 ± 2.02	81.83 ± 0.93	81.22 ± 1.27	77.08 ± 0.80
SS-LPSVM	86.79 ± 1.75	90.36 ± 1.35	90.86 ± 1.36	91.77 ± 0.96	92.11 ± 1.07
KNN + SNI	80.39 ± 1.58	84.64 ± 1.54	86.94 ± 1.52	88.28 ± 1.49	87.64 ± 1.72
MLR + RS	78.29 ± 2.16	85.03 ± 1.43	87.20 ± 1.74	88.76 ± 1.87	89.42 ± 0.85
SVM + S-CNN	12.66 ± 2.75	50.04 ± 14.34	60.72 ± 4.85	70.30 ± 2.61	71.62 ± 12.05
3D-CNN	85.58 ± 2.18	86.26 ± 1.84	88.01 ± 1.47	88.94 ± 1.38	91.21 ± 1.23
DFSL + NN	88.40 ± 1.54	89.86 ± 1.69	92.15 ± 1.24	92.69 ± 0.98	93.61 ± 0.83
DFSL + SVM	85.58 ± 1.87	89.73 ± 1.24	91.21 ± 1.64	93.42 ± 1.25	94.28 ± 0.80
New Approach	**89.92 ± 1.87**	**91.11 ± 1.50**	**93.66 ± 1.68**	**94.46 ± 1.17**	**95.85 ± 1.14**

[1] The accuracy of SVM, LapSVM, TSVM, SCS³VM, SS-LPSVM, KNN+SNI, MLR+RS, SVM+S-CNN, DFSL+NN, and DFSL+SVM are from Liu et al. [28]. Abbreviations: SVM, support vector machine; LapSVM, laplacian support vector machines; TSVM: transductive support vector machine; SCS³VM: spatial-contextual semi-supervised support vector machine; SS-LPSVM: spatial-spectral label propagation based on the SVM; KNN+SNI: k-nearest neighbor + spatial neighborhood information; MLR+RS: multinomial logistic regression + random selection; CNN: convolutional neural network; S-CNN: Siamese CNN; 3D: 3-dimensions; DFSL: deep few-shot learning; NN: nearest neighbor.

Table 6. The average ± STD overall accuracy (OA, %) with IP dataset for different methods (The bold value is the best accuracy in each case).

Method [1]	$L = 5$	$L = 10$	$L = 15$
SVM	50.23 ± 1.74	55.56 ± 2.04	58.58 ± 0.80
LapSVM	52.31 ± 0.67	56.36 ± 0.71	59.99 ± 0.65
TSVM	62.57 ± 0.23	63.45 ± 0.17	65.42 ± 0.02
SCS³VM	55.42 ± 0.35	60.86 ± 5.08	67.24 ± 0.47
SS-LPSVM	56.95 ± 0.95	64.74 ± 0.39	78.76 ± 0.04
KNN + SNI	56.39 ± 1.03	74.88 ± 0.54	78.92 ± 0.61
MLR + RS	55.38 ± 3.98	69.28 ± 2.63	75.15 ± 1.43
SVM + S-CNN	10.02 ± 1.48	17.71 ± 4.90	44.00 ± 5.73
3D-CNN	63.54 ± 2.72	71.25 ± 1.64	76.25 ± 2.17
DFSL + NN	67.84 ± 1.29	76.49 ± 1.44	78.62 ± 1.59
DFSL + SVM	64.58±2.78	75.53 ± 1.89	79.98 ± 2.23
New Approach	**70.24 ± 1.26**	**78.20 ± 1.64**	**82.65 ± 1.82**

[1] The accuracy of SVM, LapSVM, TSVM, SCS³VM, SS-LPSVM, KNN+SNI, MLR+RS, SVM+S-CNN, DFSL+NN, and DFSL+SVM are from Liu et al. [28].

Table 7. The average ± STD overall accuracy (OA, %) with UP dataset for different methods (The bold value is the best accuracy in each case).

Method [1]	$L = 5$	$L = 10$	$L = 15$	$L = 20$	$L = 25$
SVM	53.73 ± 1.30	61.53 ± 1.14	60.43 ± 0.94	64.89 ± 1.14	68.01 ± 2.62
LapSVM	65.72 ± 0.34	68.26 ± 2.20	68.34 ± 0.29	65.91 ± 0.45	68.88 ± 1.34
TSVM	63.43 ± 1.22	63.73 ± 0.45	68.45 ± 1.07	73.72 ± 0.27	69.96 ± 1.39
SCS^{3}VM	56.76 ± 2.28	64.25 ± 0.40	66.87 ± 0.37	68.24 ± 1.18	69.45 ± 2.19
SS-LPSVM	69.60 ± 2.30	75.88 ± 0.22	80.67 ± 1.21	78.41 ± 0.26	85.56 ± 0.09
KNN + SNI	70.21 ± 1.29	78.97 ± 2.33	82.56 ± 0.51	85.18 ± 0.65	86.26 ± 0.37
MLR + RS	69.73 ± 3.15	80.30 ± 2.54	84.10 ± 1.94	83.52 ± 2.13	87.97 ± 1.69
SVM + S-CNN	23.68 ± 6.34	66.64 ± 2.37	68.35 ± 4.70	78.43 ± 1.93	72.87 ± 7.36
3D-CNN	71.58 ± 3.58	79.63 ± 1.75	83.89 ± 2.93	85.98 ± 1.76	89.56 ± 1.20
DFSL + NN	80.81 ± 3.12	84.79 ± 2.27	86.68 ± 2.61	89.59 ± 1.05	91.11 ± 0.83
DFSL + SVM	72.57 ± 3.93	84.56 ± 1.83	87.23 ± 1.38	90.69 ± 1.29	93.08 ± 0.92
New Approach	**81.03 ± 1.52**	**86.10 ± 0.97**	**89.55 ± 1.28**	**93.11 ± 1.23**	**94.71 ± 1.11**

[1] The accuracy of SVM, LapSVM, TSVM, SCS^{3}VM, SS-LPSVM, KNN+SNI, MLR+RS, SVM+S-CNN, DFSL+NN, and DFSL+SVM are from Liu et al. [28].

Tables 5–7 show that the accuracy of the proposed method with different numbers of supervised samples is better compared to other methods in all three testing datasets. From Tables 5–7, it can be inferred that the proposed method is state-of-the-art for few-shot hyperspectral image classification in terms of classification accuracy.

There are some other results reported in existing publications [28]. The OA could reach 97.81%, 98.35%, and 98.62% for Salinas, IP, and UP dataset, respectively [28]. However, these results are obtained based on 200 supervised samples per class ($L = 200$). It is obvious that more supervised samples lead to better accuracy. Our paper pays attention to the hyperspectral image classification with a small number of samples, so the situation with $L = 200$ is out of our discussion.

Here is an explanation about the generalization of the network from some classes to new classes. Traditional artificial neural network has been successful in data-intensive applications, but it is hard to learn from a limited number of examples. To solve this problem, few-shot learning (FSL) is proposed [42]. Few-shot learning (FSL) is a type of machine learning problems where there is a little supervised information for the target. A strategy is that the network learns prior knowledge, and it can be generalized to new tasks with limited supervised samples. In fact, it is an important and famous branch of machine learning and has been widely used to solve many problems [28,32,42] (e.g., face recognition). The feature space learned from hyperspectral image has been demonstrated the ability to generalize to the new classes [28]. The network trained in this paper is also essentially a generalized feature extractor. The extracted features of supervised samples and samples to be classified are put into the classifier (nearest neighbor) to finish the classification.

Here we make it clear that the L specific supervised samples are not exactly the same as those in [28], due to the randomness of selecting supervised samples. However, the way of selection of supervised samples and comparative analysis in this paper is the same as that in [28,39]. It is common to randomly select supervised samples and conduct several runs of experiments (e.g., 10 runs in [28,39], and also our paper). The purpose is to avoid the occasionality of accuracy in one run.

3.2. Ablation Study

There are three key modules in the proposed methodology: the quadruplet loss, the dense branch, and the dilated branch. To demonstrate the effectiveness of each module, the classification accuracy was calculated when one of the modules is replaced. Simply put, an ablation study was performed. The proposed quadruplet loss was replaced by the siamese loss, the triplet loss, and the original quadruplet loss. The dense branch or dilated branch was replaced using a normal CNN module. In the proposed methodology, when one module is replaced, the other modules are kept the same. The summary of average values ± STD of OA in the ten runs is shown in Tables 8–10 for all three testing datasets.

Table 8. The OA of the Salinas dataset when the modules are replaced.

	$L = 5$	$L = 10$	$L = 15$	$L = 20$	$L = 25$
Method			New Approach		
OA (%)	89.92 ± 1.87	91.11 ± 1.50	93.66 ± 1.68	94.46 ± 1.17	95.85 ± 1.14
Method		The proposed quadruplet loss was replaced by the siamese loss.			
OA (%)	88.93 ± 2.18	90.31 ± 1.88	92.85 ± 1.94	93.68 ± 1.59	94.79 ± 1.48
Method		The proposed quadruplet loss was replaced by the triplet loss.			
OA (%)	89.21 ± 1.96	90.58 ± 1.82	93.01 ± 1.53	93.87 ± 1.46	94.96 ± 1.34
Method		The proposed quadruplet loss was replaced by the original quadruplet loss.			
OA (%)	89.34 ± 2.04	90.62 ± 1.95	93.30 ± 1.62	94.06 ± 1.57	95.14 ± 1.43
Method		The dense branch was replaced by a normal CNN module.			
OA (%)	89.36 ± 2.15	90.75 ± 1.96	93.34 ± 1.91	94.15 ± 1.69	95.25 ± 1.57
Method		The dilated branch was replaced by a normal CNN module.			
OA (%)	89.50 ± 2.36	90.97 ± 2.04	93.52 ± 1.96	94.21 ± 1.54	95.47 ± 1.39

Table 9. The OA of the IP dataset when the when the modules are replaced.

	$L = 5$	$L = 10$	$L = 15$
Method		New Approach	
OA (%)	70.24 ± 1.26	78.20 ± 1.64	82.65 ± 1.82
Method	The proposed quadruplet loss was replaced by the siamese loss.		
OA (%)	68.19 ± 1.59	77.12 ± 1.71	80.35 ± 1.87
Method	The proposed quadruplet loss was replaced by the triplet loss.		
OA (%)	68.58 ± 1.62	77.37 ± 2.09	81.19 ± 1.58
Method	The proposed quadruplet loss was replaced by the original quadruplet loss.		
OA (%)	68.72 ± 1.75	77.62 ± 1.84	81.38 ± 1.57
Method	The dense branch was replaced by a normal CNN module.		
OA (%)	68.71 ± 1.68	77.58 ± 2.07	81.22 ± 1.69
Method	The dilated branch was replaced by a normal CNN module.		
OA (%)	68.85 ± 1.56	77.64 ± 2.14	81.39 ± 1.62

Table 10. The OA of the UP dataset when the modules are replaced.

	$L = 5$	$L = 10$	$L = 15$	$L = 20$	$L = 25$
Method			New Approach		
OA (%)	81.03 ± 1.52	86.10 ± 0.97	89.55 ± 1.28	93.11 ± 1.23	94.71 ± 1.11
Method		The proposed quadruplet loss was replaced by the siamese loss.			
OA (%)	80.07 ± 2.25	84.13 ± 2.08	88.17 ± 1.67	91.09 ± 1.54	93.02 ± 1.28
Method		The proposed quadruplet loss was replaced by the triplet loss.			
OA (%)	80.31 ± 1.87	84.69 ± 1.54	88.53 ± 1.73	91.78 ± 1.70	93.19 ± 1.58
Method		The proposed quadruplet loss was replaced by the original quadruplet loss.			
OA (%)	80.54 ± 1.95	84.88 ± 1.73	88.71 ± 1.76	91.96 ± 1.69	93.27 ± 1.52
Method		The dense branch was replaced by a normal CNN module.			
OA (%)	80.64 ± 2.28	84.94 ± 2.09	88.78 ± 1.89	92.12 ± 1.75	93.35 ± 1.54
Method		The dilated branch was replaced by a normal CNN module.			
OA (%)	80.90 ± 2.11	85.47 ± 1.87	88.96 ± 1.80	92.34 ± 1.67	93.47 ± 1.64

In every method where a module was replaced, the accuracy was lower compared with the proposed methodology (see Tables 8–10). The inclusion of the quadruplet loss, the dense branch, and the dilated branch contributes to improving the accuracy, which was demonstrated by the ablation study. In particular, the decrease in accuracy was most substantial when the quadruplet loss was replaced, which suggests that the designation of the quadruplet loss contributes the most in improving the accuracy in the proposed methodology.

Tables 11–13 shows the average accuracy (AA) and Kappa coefficients of the proposed method, 3D-CNN, and the five experiments in ablation study. There is no enough information in publications for other existing methods. Tables 11–13 suggest that the proposed method obtains satisfying results in terms of average accuracy and Kappa coefficient.

Table 11. The average accuracy (AA, %) and the Kappa coefficient (%) for Salinas dataset.

	$L = 5$	$L = 10$	$L = 15$	$L = 20$	$L = 25$
Method			New Approach		
AA	90.19 ± 1.69	90.75 ± 1.59	93.54 ± 1.79	94.37 ± 1.26	95.58 ± 1.26
Kappa	89.68 ± 1.47	91.06 ± 1.28	93.26 ± 1.59	94.28 ± 1.32	95.46 ± 1.17
Method		The proposed quadruplet loss was replaced by the siamese loss.			
AA	88.86 ± 1.88	90.22 ± 2.02	92.93 ± 1.86	93.79 ± 1.53	94.86 ± 1.36
Kappa	88.72 ± 2.39	89.94±1.97	92.78 ± 2.14	93.59 ± 1.68	94.68 ± 1.54
Method		The proposed quadruplet loss was replaced by the triplet loss.			
AA	89.14 ± 1.87	90.39 ± 1.98	93.16 ± 1.56	93.89 ± 1.50	95.02 ± 1.42
Kappa	89.06 ± 2.16	90.24 ± 2.03	92.86 ± 1.69	93.72 ± 1.54	94.89 ± 1.57
Method		The proposed quadruplet loss was replaced by the original quadruplet loss.			
AA	89.19 ± 1.96	90.44 ± 1.89	93.21 ± 1.59	93.93 ± 1.52	95.11 ± 1.54
Kappa	89.26 ± 2.17	90.32 ± 2.14	92.99 ± 1.73	93.85 ± 1.69	94.92 ± 1.63
Method		The dense branch was replaced by a normal CNN module.			
AA	89.29 ± 1.99	90.56 ± 1.89	93.35 ± 1.88	94.09 ± 1.74	95.17 ± 1.69
Kappa	89.24 ± 2.06	90.43 ± 2.03	93.08 ± 2.07	93.91 ± 1.78	95.13 ± 1.70
Method		The dilated branch was replaced by a normal CNN module.			
AA	89.43 ± 2.07	90.59 ± 2.36	93.44 ± 2.06	94.18 ± 1.65	95.29 ± 1.45
Kappa	89.36 ± 2.41	90.50 ± 2.13	93.13 ± 2.56	94.04 ± 1.77	95.31 ± 1.46
Method			3D-CNN		
AA	84.63 ± 2.06	86.36 ± 1.82	87.76 ± 1.68	89.36 ± 1.56	91.03 ± 1.54
Kappa	85.36 ± 2.43	86.14 ± 2.03	87.96 ± 1.85	88.76 ± 1.89	90.96 ± 1.63

Table 12. The average accuracy (AA, %) and the Kappa coefficient (%) for IP dataset.

	$L = 5$	$L = 10$	$L = 15$
Method		New Approach	
AA	70.35 ± 1.32	78.58 ± 1.71	82.77 ± 1.88
Kappa	70.15 ± 1.16	77.84 ± 1.54	82.59 ± 1.69
Method	The proposed quadruplet loss was replaced by the siamese loss.		
AA	67.95 ± 2.07	77.75 ± 1.69	80.78 ± 1.68
Kappa	68.12 ± 1.68	77.05 ± 1.74	80.29 ± 2.13
Method	The proposed quadruplet loss was replaced by the triplet loss.		
AA	68.39 ± 1.85	77.96 ± 1.95	81.26 ± 1.64
Kappa	68.45 ± 1.77	77.28 ± 1.86	80.93 ± 1.89
Method	The proposed quadruplet loss was replaced by the original quadruplet loss.		
AA	69.16 ± 1.69	78.08 ± 1.63	81.67 ± 1.65
Kappa	68.65 ± 2.14	77.54 ± 1.91	81.26 ± 1.91
Method	The dense branch was replaced by a normal CNN module.		
AA	68.89 ± 1.64	77.74 ± 1.86	81.43 ± 1.60
Kappa	68.64 ± 1.79	77.47 ± 2.15	80.86 ± 1.88
Method	The dilated branch was replaced by a normal CNN module.		
AA	69.14 ± 1.68	77.96 ± 1.98	81.57 ± 1.58
Kappa	68.73 ± 2.03	77.53 ± 2.36	81.32 ± 1.89
Method		3D-CNN	
AA	64.35 ± 2.58	71.69 ± 1.78	76.64 ± 2.04
Kappa	63.48 ± 2.87	71.26 ± 1.95	76.21 ± 2.36

Table 13. The average accuracy (AA, %) and the Kappa coefficient (%) for UP dataset.

	$L = 5$	$L = 10$	$L = 15$	$L = 20$	$L = 25$
Method			New Approach		
AA	81.15 ± 1.59	86.23 ± 1.04	89.68 ±1.36	93.05 ± 1.21	94.57 ± 1.25
Kappa	80.97 ± 1.41	86.05 ± 1.08	89.49 ± 1.30	92.97 ± 1.13	94.42 ± 1.18
Method		The proposed quadruplet loss was replaced by the siamese loss.			
AA	80.16 ± 2.16	84.36 ± 1.89	87.86 ± 1.69	90.87 ± 1.63	92.86 ± 1.46
Kappa	79.95 ± 2.37	84.16 ± 1.95	87.93 ± 1.86	90.94 ± 1.78	92.78 ± 1.53
Method		The proposed quadruplet loss was replaced by the triplet loss.			
AA	80.39 ± 1.75	84.56 ± 1.78	88.40 ± 1.84	91.56 ± 1.89	92.94 ± 1.63
Kappa	80.06 ± 1.94	84.37 ± 1.85	88.26 ± 2.02	91.36 ± 1.86	92.89 ± 1.74
Method		The proposed quadruplet loss was replaced by the original quadruplet loss.			
AA	80.46 ± 1.86	84.69 ± 1.96	88.46 ± 1.86	91.83 ± 1.76	93.05 ± 1.69
Kappa	80.25 ± 2.03	84.59 ± 2.03	88.37 ± 1.94	91.58 ± 1.88	92.93 ± 1.82
Method		The dense branch was replaced by a normal CNN module.			
AA	80.58 ± 1.89	85.03 ± 1.94	88.59 ± 1.98	91.06 ± 1.89	93.16 ± 1.66
Kappa	80.41 ± 2.36	84.76 ± 2.18	88.44 ± 2.14	92.10 ± 1.83	92.97 ± 1.56
Method		The dilated branch was replaced by a normal CNN module.			
AA	80.68 ± 2.23	85.34 ± 1.94	88.77 ± 1.94	92.16 ± 1.75	93.36 ± 1.79
Kappa	80.49 ± 2.56	85.23 ± 2.00	88.69 ± 2.03	91.95 ± 1.84	93.15 ± 1.84
Method			3D-CNN		
AA	71.26 ± 3.64	80.02 ± 1.69	83.52 ± 3.14	85.53 ± 1.89	89.23 ± 1.36
Kappa	70.89 ± 3.85	79.38 ± 1.87	83.46 ± 3.26	85.06 ± 2.03	88.91 ± 1.41

3.3. Time Consumption

In terms of the overall accuracy of the three datasets, SS-LPSVM, 3D-CNN, DFSL+NN, and DFSL+SVM show the closest accuracy performance with our method. Hence, these methods were selected for the comparative analysis of time consumption. The time consumption of 3D-CNN, DFSL+NN, DFSL+SVM, and the proposed method based on the IP dataset is shown in Table 14. Details regarding the computer configuration and program coding used in analyzing the time consumption are presented in Table 15. Based on the comparative analysis of time consumption, the proposed approach is similar to other classification techniques. SS-LPSVM has been demonstrated that it takes much longer time than DFSL+NN and DFSL+SVM based on the IP dataset [28] (198.30s vs. 11.14s + 0.36s and 11.14s + 2.21s). Hence, it can be inferred that the proposed method shows obvious advantage over SS-LPSVM.

Table 14. The time consumption of the proposed approach and other methods ("+": the time of feature extraction + the time of classification).

Number of Labeled Samples		$L = 5$		
Method	Proposed	DFSL + NN	DFSL + SVM	3D-CNN + NN
Time	10.08 s + 0.30 s	11.09 s + 0.34 s	11.09 s + 2.09 s	13.59 s + 0.38 s
Number of Labeled Samples		$L = 25$		
Method	Proposed	DFSL + NN	DFSL + SVM	3D-CNN + NN
Time	10.08 s + 0.75 s	11.09 s + 0.78 s	11.09 s + 123.48 s	13.59 s + 0.84 s

Table 15. The details about computer configuration and program coding for testing operation time.

Configuration and Program	Details
Processor	Intel (R) Core (TM) i5-9400 @ 2.90 GHz
Memory	Crucial DDR4 2666MHz, 8.00 GB
Graphics	NVIDIA GeForce GTX 1060, 6 GB
CUDA	Version 9.1.0
Programming Language	Python, Version 3.6.10
Deep Learning Platform	Google TensorFlow, Version 1.9.0

4. Conclusions

This study integrates quadruplet loss with deep 3-D CNN with dense and dilated characteristics in proposing a quadruplet deep learning method for few-shot hyperspectral image classification. Verification and comparative analysis were performed using public hyperspectral datasets, and the results suggest the following conclusions:

(1) The proposed approach was found to have higher overall accuracy than existing methods, which suggests that the classification method is state-of-the-art.

(2) An ablation study was conducted replacing each module of the proposed approach (i.e., quadruplet loss, dense branch, and dilated branch) to demonstrate the effectiveness of their contributions. The results show that all modules are effective and necessary in improving classification accuracy, with the proposed quadruplet loss providing the highest contribution.

(3) The time consumption for the different methods was tested under the same operating environment. The analysis shows the proposed methodology has a similar level of time consumption compared to existing methods.

In the future, given the scarcity of training samples in some cases, a sample-synthesis method can be explored for a few-shot hyperspectral image classification.

Author Contributions: C.Z. and J.Y. together proposed the idea and contributed to the experiments, writing, and figures, and contributed equally to this work; Q.Q. contributed to the writing of this paper. All authors have read and agreed to the published version of the manuscript.

Funding: This research was funded by National Natural Science Foundation of China (Grant number 41901291), National key research and development program (Grant number 2018YFC1800102), Open Fund of State Key Laboratory of Coal Resources and Safe Mining (Grant number SKLCRSM19KFA04), the Key Laboratory of Surveying and Mapping Science and Geospatial Information Technology of Ministry of Natural Resources (Grant number 201917), the Key Laboratory for National Geographic Census and Monitoring, National Administration of Surveying, Mapping and Geoinformation (Grant number 2018NGCM07).

Acknowledgments: The authors gratefully acknowledge the National Center for Airborne Laser Mapping (NCALM) for providing the "Houston" dataset, the Space Application Laboratory, Department of Advanced Interdisciplinary Studies, University of Tokyo for providing the "Chikusei" dataset, and Grupo de Inteligencia Computacional (GIC) for providing other datasets. The authors thanks to the professional English editing service from EditX.

Conflicts of Interest: The authors declare no conflict of interest.

References

1. Magendran, T.; Sanjeevi, S. Hyperion image analysis and linear spectral unmixing to evaluate the grades of iron ores in parts of Noamundi, Eastern India. *Int. J. Appl. Earth Obs. Geoinf.* **2014**, *26*, 413–426. [CrossRef]
2. Zhang, C.; Qin, Q.; Chen, L.; Wang, N.; Zhao, S.; Hui, J. Rapid determination of coalbed methane exploration target region utilizing hyperspectral remote sensing. *Int. J. Coal Geol.* **2015**, *150*, 19–34. [CrossRef]
3. Kruse, F. Preliminary results—Hyperspectral mapping of coral reef systems using EO-1 Hyperion. In Proceedings of the 12th JPL Airborne Geoscience Workshop, Buck Island and U.S. Virgin Islands, Buck Island, USVI, USA, 24–28 February 2003.
4. Van der Meer, F.D.; Van der Werff, H.M.A.; Van Ruitenbeek, F.J.A.; Hecker, C.A.; Bakker, W.H.; Noomen, M.F.; van der Meijde, M.; Carranza, E.J.M.; de Smeth, J.B.; Woldai, T. Multi- and hyperspectral geologic remote sensing: A review. *Int. J. Appl. Earth Obs. Geoinf.* **2012**, *14*, 112–128. [CrossRef]

5. Benediktsson, J.A.; Swain, P.H.; Ersoy, O.K. Neural Network Approaches Versus Statistical Methods in Classification of Multisource Remote Sensing Data. In Proceedings of the 12th Canadian Symposium on Remote Sensing Geoscience and Remote Sensing Symposium, Vancouver, BC, Canada, 10–14 July 1989; pp. 489–492.

6. Melgani, F.; Bruzzone, L. Classification of hyperspectral remote sensing images with support vector machines. *IEEE Trans. Geosci. Remote Sens.* **2004**, *42*, 1778–1790. [CrossRef]

7. Gao, L.; Li, J.; Khodadadzadeh, M.; Plaza, A.; Zhang, B.; He, Z.; Yan, H. Subspace-Based Support Vector Machines for Hyperspectral Image Classification. *IEEE Geosci. Remote Sens. Lett.* **2015**, *12*, 349–353.

8. Ham, J.; Chen, Y.C.; Crawford, M.M.; Ghosh, J. Investigation of the random forest framework for classification of hyperspectral data. *IEEE Trans. Geosci. Remote Sens.* **2005**, *43*, 492–501. [CrossRef]

9. Gislason, P.O.; Benediktsson, J.A.; Sveinsson, J.R. Random Forests for land cover classification. *Pattern Recognit. Lett.* **2006**, *27*, 294–300. [CrossRef]

10. Chen, Y.; Lin, Z.; Zhao, X.; Wang, G.; Gu, Y. Deep Learning-Based Classification of Hyperspectral Data. *IEEE J. Sel. Top. Appl. Earth Obs. Remote Sens.* **2014**, *7*, 2094–2107. [CrossRef]

11. Zhao, W.; Guo, Z.; Yue, J.; Zhang, X.; Luo, L. On combining multiscale deep learning features for the classification of hyperspectral remote sensing imagery. *Int. J. Remote Sens.* **2015**, *36*, 3368–3379. [CrossRef]

12. Yue, J.; Mao, S.; Li, M. A deep learning framework for hyperspectral image classification using spatial pyramid pooling. *Remote Sens. Lett.* **2016**, *7*, 875–884. [CrossRef]

13. Singhal, V.; Majumdar, A. Row-Sparse Discriminative Deep Dictionary Learning for Hyperspectral Image Classification. *IEEE J. Sel. Top. Appl. Earth Obs. Remote Sens.* **2018**, *11*, 5019–5028. [CrossRef]

14. Paoletti, M.E.; Haut, J.M.; Plaza, J.; Plaza, A. A new deep convolutional neural network for fast hyperspectral image classification. *ISPRS J. Photogramm. Remote Sens.* **2018**, *145*, 120–147. [CrossRef]

15. Zhao, W.; Du, S. Spectral-Spatial Feature Extraction for Hyperspectral Image Classification: A Dimension Reduction and Deep Learning Approach. *IEEE Trans. Geosci. Remote Sens.* **2016**, *54*, 4544–4554. [CrossRef]

16. Ghamisi, P.; Maggiori, E.; Li, S.; Souza, R.; Tarabalka, Y.; Moser, G.; De Giorgi, A.; Fang, L.; Chen, Y.; Chi, M.; et al. New Frontiers in Spectral-Spatial Hyperspectral Image Classification The latest advances based on mathematical morphology, Markov random fields, segmentation, sparse representation, and deep learning. *IEEE Geosci. Remote Sens. Mag.* **2018**, *6*, 10–43. [CrossRef]

17. Li, J.; Xi, B.; Du, Q.; Song, R.; Li, Y.; Ren, G. Deep Kernel Extreme-Learning Machine for the Spectral-Spatial Classification of Hyperspectral Imagery. *Remote Sens.* **2018**, *10*, 2036. [CrossRef]

18. Song, A.; Choi, J.; Han, Y.; Kim, Y. Change Detection in Hyperspectral Images Using Recurrent 3D Fully Convolutional Networks. *Remote Sens.* **2018**, *10*, 1827. [CrossRef]

19. Hamida, A.B.; Benoit, A.; Lambert, P.; Ben Amar, C. 3-D Deep Learning Approach for Remote Sensing Image Classification. *IEEE Trans. Geosci. Remote Sens.* **2018**, *56*, 4420–4434. [CrossRef]

20. Xu, Y.; Zhang, L.; Du, B.; Zhang, F. Spectral-Spatial Unified Networks for Hyperspectral Image Classification. *IEEE Trans. Geosci. Remote Sens.* **2018**, *56*, 5893–5909. [CrossRef]

21. Shen, H.; Jiang, M.; Li, J.; Yuan, Q.; Wei, Y.; Zhang, L. Spatial-Spectral Fusion by Combining Deep Learning and Variational Model. *IEEE Trans. Geosci. Remote Sens.* **2019**, *57*, 6169–6181. [CrossRef]

22. Choe, J.; Park, S.; Kim, K.; Park, J.H.; Kim, D.; Shim, H. Face Generation for Low-shot Learning using Generative Adversarial Networks. In Proceedings of the IEEE International Conference on Computer Vision, ICCV 2017, Venice, Italy, 22–29 October 2017; pp. 1940–1948.

23. Dong, X.; Zhu, L.; Zhang, D.; Yang, Y.; Wu, F. Fast Parameter Adaptation for Few-shot Image Captioning and Visual Question Answering. In Proceedings of the ACM Multimedia Conference on Multimedia Conference, Seoul, Korea, 22–26 October 2018; pp. 54–62.

24. Shaban, A.; Bansal, S.; Liu, Z.; Essa, I.; Boots, B. One-Shot Learning for Semantic Segmentation. In Proceedings of the British Machine Vision Conference (BMVC), London, UK, 4–7 September 2017; Available online: https://kopernio.com/viewer?doi=arXiv:1709.03410v1&route=6 (accessed on 12 February 2020).

25. Schwartz, E.; Karlinsky, L.; Shtok, J.; Harary, S.; Marder, M.; Kumar, A.; Feris, R.; Giryes, R.; Bronstein, A.M. Delta-encoder: An effective sample synthesis method for few-shot object recognition. In Proceedings of the Annual Conference on Neural Information Processing Systems, Montréal, QC, Canada, 3–8 December 2018.

26. Sung, F.; Yang, Y.; Zhang, L.; Xiang, T.; Torr, P.H.S.; Hospedales, T.M. Learning to Compare: Relation Network for Few-Shot Learning. In Proceedings of the IEEE Conference on Computer Vision and Pattern Recognition, Salt Lake City, UT, USA, 18–22 June 2018; pp. 1199–1208.

27. Koch, G.; Zemel, R.; Salakhutdinov, R. Siamese neural networks for one-shot image recognition. In Proceedings of the 32nd International Conference on Machine Learning (ICML), Lille, France, 6–11 July 2015.

28. Liu, B.; Yu, X.; Yu, A.; Zhang, P.; Wan, G.; Wang, R. Deep Few-Shot Learning for Hyperspectral Image Classification. *IEEE Trans. Geosci. Remote Sens.* **2019**, *57*, 2290–2304. [CrossRef]

29. Xu, S.; Li, J.; Khodadadzadeh, M.; Marinoni, A.; Gamba, P.; Li, B. Abundance-Indicated Subspace for Hyperspectral Classification With Limited Training Samples. *IEEE J. Sel. Top. Appl. Earth Obs. Remote Sens.* **2019**, *12*, 1265–1278. [CrossRef]

30. Chen, W.; Chen, X.; Zhang, J.; Huang, K. Beyond triplet loss: A deep quadruplet network for person re-identification. In Proceedings of the IEEE Conference on Computer Vision and Pattern Recognition, Honolulu, HI, USA, 21–26 July 2017; pp. 1704–1719.

31. Hoffer, E.; Ailon, N. Deep Metric Learning Using Triplet Network. In *Lecture Notes in Computer Science*; Feragen, A., Pelillo, M., Loog, M., Eds.; Springer: New York, NY, USA, 2015; Volume 9370, pp. 84–92.

32. Xiao, Q.; Luo, H.; Zhang, C. Margin Sample Mining Loss: A Deep Learning Based Method for Person Re-Identification. 2017. Available online: https://arxiv.org/pdf/1710.00478.pdf (accessed on 22 October 2019).

33. Huang, G.; Liu, Z.; Maaten, L.V.D.; Weinberger, K.Q. Densely Connected Convolutional Networks. In Proceedings of the IEEE Conference on Computer Vision and Pattern Recognition, Honolulu, HI, USA, 21–26 July 2017; pp. 2261–2269.

34. Yu, F.; Koltun, V.; Funkhouser, T. Dilated Residual Networks. In Proceedings of the IEEE Conference on Computer Vision and Pattern Recognition, Honolulu, HI, USA, 21–26 July 2017; pp. 636–644.

35. Sun, K.; Geng, X.; Chen, J.; Ji, L.; Tang, H.; Zhao, Y.; Xu, M. A robust and efficient band selection method using graph representation for hyperspectral imagery. *Int. J. Remote Sens.* **2016**, *37*, 4874–4889. [CrossRef]

36. Belkin, M.; Niyogi, P.; Sindhwani, V. Manifold regularization: A geometric framework for learning from labeled and unlabeled examples. *J. Mach. Learn. Res.* **2006**, *7*, 2399–2434.

37. Joachims, T. Transductive inference for text classification using Support Vector Machines. In Proceedings of the 16th International Conference on Machine Learning, Bled, Slovenia, 27–30 June 1999; pp. 200–209.

38. Kuo, B.; Huang, C.; Hung, C.; Liu, Y.; Chen, I. Spatial information based support vector machine for hyperspectral image classification. In Proceedings of the IEEE International Symposium on Geoscience and Remote Sensing, Honolulu, Hawaii, USA, 25–30 July 2010; pp. 832–835.

39. Wang, L.; Hao, S.; Wang, Q.; Wang, Y. Semi-supervised classification for hyperspectral imagery based on spatial-spectral Label Propagation. *ISPRS J. Photogra. Remote Sens.* **2014**, *97*, 123–137. [CrossRef]

40. Tan, K.; Hu, J.; Li, J.; Du, P. A novel semi-supervised hyperspectral image classification approach based on spatial neighborhood information and classifier combination. *ISPRS J. Photogra. Remote Sens.* **2015**, *105*, 19–29. [CrossRef]

41. Dópido, I.; Li, J.; Marpu, P.R.; Plaza, A.; Dias, J.M.B.; Benediktsson, J.A. Semisupervised Self-Learning for Hyperspectral Image Classification. *IEEE Trans. Geosci. Remote Sens.* **2013**, *51*, 4032–4044.

42. Wang, Y.A.; Kwok, J.; Ni, L.M.; Yao, Q. Generalizing from a Few Examples: A Survey on Few-Shot Learning. 2019. Available online: https://arxiv.org/pdf/1904.05046.pdf (accessed on 1 February 2020).

Article

Mapping the Topographic Features of Mining-Related Valley Fills Using Mask R-CNN Deep Learning and Digital Elevation Data

Aaron E. Maxwell [1,*], **Pariya Pourmohammadi** [2] **and Joey D. Poyner** [1]

[1] Department of Geology and Geography, West Virginia University, Morgantown, WV 26506, USA; jpoyner@mix.wvu.edu

[2] Davis College of Agriculture, Natural Resources, and Design, Department of Design and Community Development, West Virginia University, Morgantown, WV 26506, USA; papourmohammadi@mix.wvu.edu

[*] Correspondence: Aaron.Maxwell@mail.wvu.edu; Tel.: +1-304-293-2026

Received: 10 December 2019; Accepted: 5 February 2020; Published: 7 February 2020

Abstract: Modern elevation-determining remote sensing technologies such as light-detection and ranging (LiDAR) produce a wealth of topographic information that is increasingly being used in a wide range of disciplines, including archaeology and geomorphology. However, automated methods for mapping topographic features have remained a significant challenge. Deep learning (DL) mask regional-convolutional neural networks (Mask R-CNN), which provides context-based instance mapping, offers the potential to overcome many of the difficulties of previous approaches to topographic mapping. We therefore explore the application of Mask R-CNN to extract valley fill faces (VFFs), which are a product of mountaintop removal (MTR) coal mining in the Appalachian region of the eastern United States. LiDAR-derived slopeshades are provided as the only predictor variable in the model. Model generalization is evaluated by mapping multiple study sites outside the training data region. A range of assessment methods, including precision, recall, and F1 score, all based on VFF counts, as well as area- and a fuzzy area-based user's and producer's accuracy, indicate that the model was successful in mapping VFFs in new geographic regions, using elevation data derived from different LiDAR sensors. Precision, recall, and F1-score values were above 0.85 using VFF counts while user's and producer's accuracy were above 0.75 and 0.85 when using the area- and fuzzy area-based methods, respectively, when averaged across all study areas characterized with LiDAR data. Due to the limited availability of LiDAR data until relatively recently, we also assessed how well the model generalizes to terrain data created using photogrammetric methods that characterize past terrain conditions. Unfortunately, the model was not sufficiently general to allow successful mapping of VFFs using photogrammetrically-derived slopeshades, as all assessment metrics were lower than 0.60; however, this may partially be attributed to the quality of the photogrammetric data. The overall results suggest that the combination of Mask R-CNN and LiDAR has great potential for mapping anthropogenic and natural landscape features. To realize this vision, however, research on the mapping of other topographic features is needed, as well as the development of large topographic training datasets including a variety of features for calibrating and testing new methods.

Keywords: light detection and ranging; LiDAR; deep learning; convolutional neural networks; CNNs; mask regional-convolutional neural networks; mask R-CNN; digital terrain analysis; resource extraction

---

1. Introduction

Light detection and ranging (LiDAR) data provide high spatial resolution, detailed representations of bare earth landscapes, and have been shown to be valuable for mapping features of geomorphic and archeological interest. For example, Joboyedoff et al. [1] suggest that LiDAR is an essential tool for detecting, characterizing, monitoring, and modelling landslides and other forms of mass movement. Chase et al. [2] argue that LiDAR technologies have resulted in a paradigm shift in archeological research, as they allow for the mapping of ancient anthropogenic features and landscapes even under dense canopy cover. For example, LiDAR has recently improved our understanding of ancient Mesoamerican cultures by mapping ancient cities now obscured by dense forest cover, a mapping task that is too labor intensive for field-based survey methods alone [2]. Further, LiDAR data are becoming increasingly available for public download, especially in Europe and North America. For example, the United States has implemented the 3D Elevation Program (3DEP) (https://www.usgs.gov/core-science-systems/ngp/3dep) with a goal of providing LiDAR coverage for the entire country, excluding Alaska [3]. In this spirit, The Earth Archive project has argued for the need for 3D data of the entire Earth surface to create a historic record for future generations, and is currently soliciting donations to support this project [4].

Despite the increasing availability of high spatial resolution digital terrain data, and the wealth of information that can be derived from such data, the extraction of features from these data to support archeological, geomorphic, and landscape change research is in many cases dominated by manual interpretation, as previously noted by [5,6]. With the exception of some notable studies (e.g., [6–9]), generic and automated mapping of topographic features from digital elevation data has proved to be a particularly challenging task. However, deep learning (DL), and in particular mask regional-convolutional neural networks (Mask R-CNN), may make it possible to realize the potential of digital elevation data for automated mapping of topographic features.

Therefore, this study explores the use of Mask R-CNN for mapping valley fill faces (VFFs) resulting from mountaintop removal (MTR) surface coal mining in the Appalachian region of the eastern United States. MTR is a common mining method in this region which results in extensive modifications to the landscape, and therefore mapping VFFs is of significant interest for environmental modelers. From our findings, we comment on the application of this DL method for extracting anthropogenic and natural terrain features from LiDAR-derived data based on distinct topographic and spatial patterns. Since LiDAR data are not commonly available to represent historic terrain conditions due to only recent development of this technology for mapping large spatial extents, we also explore the transferability of the models to older, photogrammetrically-derived elevation data. This study therefore has two objectives:

1. Assess the Mask R-CNN DL algorithm for mapping VFFs using LiDAR-derived digital elevation data.
2. Investigate model performance and generalization by applying the model to LiDAR-derived data in new geographic regions and acquired with differing LiDAR sensors and acquisition parameters, as well as a photogrammetrically-derived digital terrain dataset.

1.1. LiDAR and Digital Terrain Mapping

LiDAR is an active remote sensing method that relies on laser range finding. A laser pulse is emitted by a sensor. When the emitted photons strike an object, a portion of the energy is reflected back to the sensor. Using the two-way travel time of reflected laser pulses detected by the sensor, global position system (GPS) locations, and aircraft orientation and motion from an inertial measurement unit (IMU), horizontal and elevation coordinates of the reflecting surface can be estimated at a high spatial resolution. Further, a single laser pulse can potentially result in multiple returns, allowing for vegetation canopy penetration and the mapping of subcanopy terrain, in contrast to other elevation mapping methods [10].

LiDAR data have been applied to a variety of terrain mapping and analysis tasks. For example, many studies have investigated the mapping of slope failures, such as landslides, using terrain variables derived from LiDAR [1,8,11–13]. Another common application is modeling the likelihood of slope failure occurrence or landslide risk [14–17]. In a 2012 review of the use of LiDAR in landslide investigations, Jaboyedoff et al. [1] suggest that LiDAR is an essential tool for landslide risk management and that there is a need to develop methods to extract useful information from such data. Older, photogrammetrically-derived elevation data have also been used for terrain mapping and analysis tasks and offer a means to characterize historic terrain conditions. For surface mine mapping specifically, Maxwell and Warner [18] found that historic, photogrammetric elevation data were of great value for differentiating grasslands resulting from mine reclamation from other grasslands while DeWitt et al. [19] provided a comparison of different digital elevation data sources for mapping terrain change resulting from surface mining.

Object-based image analysis (OBIA) has been applied to LiDAR data for the mapping of landslides [20] and geomorphic landforms in general [9]. OBIA incorporates segmentation of raster-based data into regions or polygons, based on measures of similarity or homogeneity. These polygons are the spatial unit of analysis and classification [21]. Part of the interest in OBIA for geomorphic mapping is the ability to incorporate spatial context information into the mapping process, facilitated by the data segmentation. Nevertheless, choosing the scale of the segmentation is a major hurdle in OBIA, and indeed some research indicates it is necessary to choose multiple scales [22,23]. In contrast, spatial context information can be included in DL by using convolutional neural networks (CNNs) in a manner that does not require a priori specification of the scale. Thus, applying CNN-based DL to digital terrain data holds great promise.

1.2. Deep Learning

DL algorithms are derived from, and offer an extension to, artificial neural networks (ANNs). Traditional ANNs generally have a small number of hidden layers, whereas DL algorithms have many hidden layers. In contrast to traditional machine learning methods, which are shallow learners, it has been suggested that DL is able to provide a higher level of data abstraction, potentially resulting in improved predictive power, generalization, and transferability [24–30]. Although this results in a model that is much more complex and has many more parameters, it allows for multiple levels of data abstraction to learn complex patterns. Like other supervised machine learning methods, DL requires example training data with associated labels in order to build the model. A measure of error or performance, generally termed loss, is used to guide the algorithm to improve predictions as it iterates through the training data multiple times [24,30].

CNNs extend the deep ANN architecture to incorporate context information into the prediction. CNNs include convolutional layers that learn filters that transform input image values, similar to moving window kernels traditionally used in remote sensing for image edge detection and smoothing. However, in the case of CNNs, the algorithm produces optimal filters to aid in predicting the labels associated with the training images. The addition of this context information has offered substantial advancements in computer vision and scene labeling problems [24,28,30]. In remote sensing applications, CNNs allow for the analysis of spatial context information when applied to high spatial resolution data (for example, [28]), spectral patterns when applied to hyperspectral data (for example, [31]), and temporal patterns when applied to time series products (for example, [32]). Thus, DL with convolution allows for the integration of contextual information in the spatial, spectral, and temporal domains.

Traditional CNNs have primarily been used for scene labeling problems, for example, entire images or image chips categorized by different land cover type. Traditional CNNs do not allow for pixel-level or semantic labeling. However, the introduction of fully convolutional neural networks (FCNs) alleviated this limitation by combining convolution and deconvolution layers with up-sampling, which allows for the final feature map to be produced at the original image resolution with a prediction at each cell location [27,33], similar to traditional remote sensing classification products. Example FCN architectures include SegNet [34] and UNet [35–37].

In this study, we use instance segmentation methods, in which the goal is to distinguish each individual instance of a feature in the scene separately. For example, each tree in a scene can be identified as a separate instance of the tree class. We specifically implement the Mask R-CNN method. This method is an extension of faster R-CNN, which allows for convolution to be applied on regions of the image as opposed to the entire scene. This involves generating convolution feature maps that are then applied to individual subsets of the image, called regions of interest (RoI), defined by the region proposal network (RPN). The process of RoI pooling allows convolution features to be applied to regions of the image of different sizes and rectangular shapes [38,39]. Mask R-CNN extends this framework to allow for polygon masks to be generated within each RoI, essentially performing semantic segmentation within each RoI using FCNs. This requires better alignment between the RoI pooling layers and the RoIs than is provided by faster R-CNN. So, a ROIAlign layer is applied to improve the spatial alignment [39]. Since there are multiple components of the model, multiple loss measures are used to assess performance. Specifically, the total loss is the sum of the loss for the bounding box, classification, and mask predictions [38,39]. For a full discussion of Mask R-CNN, please consult He et al. [39], who introduced this method.

DL methods have shown promise in remote sensing mapping and data processing tasks including scene labeling, pixel-level classification, object-detection, data fusion, and image registration [24]. For example, Microsoft has recently used DL to map 125 million building footprints across the entire US [40]. Kussul et al. [26] explored DL for differentiating crops using a time series of Landsat-8 multispectral and Sentinel-1 synthetic-aperture radar (SAR) data and documented improved overall and class-specific classification performance in comparison to shallow learners, such as random forests (RF). Li et al. [41] used DL and QuickBird satellite imagery to map individual oil palm trees with precision and recall rates greater than 94%.

It should be noted that there are some complexities in implementing these methods and applying them to remotely sensed data, such as the need for a large number of training samples, the difficulty of model optimization and parameterization, and large computational demands [24,30]. Also, the processes of training models and predicting to new data can differ from those used in traditional image classification and machine learning; for example, convolution requires training on and predicting to small rectangular image extents, or image chips, as opposed to individual pixels or image objects. Thus, researchers and analysts must augment workflows and learn new techniques for implementing DL algorithms [24,30].

A review of the literature suggests that the application of DL to LiDAR and digital terrain data is still limited. There has been some research relating to using DL for extracting ground returns from LiDAR point clouds for digital terrain model (DTM) generation (for example, [33,42–44]). Specifically, Hu and Yuan [44] suggest that DL-based algorithms can outperform the current methods that are most commonly used for ground return classification. Others have investigated the classification of features in 3D space represented as point clouds [45–47].

There is a need to investigate mapping anthropogenic and natural terrain features from digital terrain data using DL, as the research on this topic is currently lacking; however, there have been some notable studies. Tier et al. [6] investigated the identification of prehistoric structures from LiDAR-derived raster data. From the LiDAR data, a DTM was interpolated followed by a measure of local relief, which was then provided as input to the ResNet18 CNN algorithm as an RGB image. They reported mixed results, with some areas predicted well and other areas suffering from many false positives. Behrens et al. [48] explored digital soil mapping using DTM raster data and DL and obtained a more accurate output than that produced by RF.

Interestingly, a number of studies attempt to map features that are at least partially characterized by geomorphic and terrain characteristics using spectral data only, without using terrain data. For example, Li et al. [49] mapped craters from image data using faster R-CNN and obtained a mean average precision (mAP) higher than 0.90. As an example of a study that combined spectral and terrain data, Ghorbanzadeh et al. [50] used RapidEye satellite data and measures of plan curvature, topographic slope, and topographic aspect to detect landslides. They noted comparable performance between CNNs and traditional shallow classifiers: ANN, SVM, and RF.

Mask R-CNN has seen limited application in remote sensing at the time of this writing. Zhang et al. [51] assessed the method for mapping artic ice-wedge polygons from high spatial resolution aerial imagery and documented that 95% of individual ice-wedge polygons were correctly delineated and classified, with an overall accuracy of 79%. Zhao et al. [37] found that Mask R-CNN outperformed UNet for pomegranate tree canopy segmentation. Stewart et al. [52] used the method to detect lesions on maize plants from northern leaf blight using unmanned aerial vehicle (UAV) data. Given the small number of studies that have applied this algorithm to remotely sensed data, there is a need for further exploration of this algorithm within the discipline. We found a lack of research associated with mapping terrain features from digital terrain data using DL methods, and no published studies that apply this algorithm to raster-based digital terrain data for mapping geomorphic features. We attribute this to the only recent advancement of DL for semantic and instance segmentation, and lack of available data to train DL models.

1.3. Mountaintop Removal Coal Mining and Valley Fills

In this study we apply Mask R-CNN to detect instances of VFFs from digital terrain data derived from LiDAR. Valley fills are a product of MTR coal mining, which has been practiced in southern West Virginia, eastern Kentucky, and southwestern Virginia in the Appalachian region of the eastern United States for several decades. This surface mining process involves using heavy machinery to extract thin interbedded coal seams. Valley fills are generated from the redistribution of overburden rock material. Since the coal seams are interbedded with other rock types of limited commercial value, a large volume of displaced material is produced. Due to the original steepness of the slopes, it is not possible to reclaim the landscape to the approximate original contour. Therefore, excess overburden material is placed in adjacent valleys, raising the valley elevation and changing the landscape.

The excavation and subsequent reclamation associated with valley fills results in substantial alterations to land cover, soil, and the topography and contour of the landscape [53–62]. Forests are lost and fragmented [63], mountaintop elevations are lowered by tens to hundreds of meters [53,57]; soils are compacted [58,64], and human quality of life and health is affected by exposure to chemicals, dust, and particulates [62]. Because valley fills bury headwater streams [54,61], and the fill material is hydrologically dissimilar to undisturbed land, hydrology is particularly affected. Valley fills tend to increase stream conductivity and alter hydrologic regimes downstream [53,56,58,64]. Wood and Williams [61] documented a decrease in salamander abundance in headwater streams impacted by valley fills in comparison to reference streams. In summary, valley fills profoundly alter the landscape, resulting in a variety of complex effects on the physical environment and its inhabitants, making it of vital importance that these features be monitored and mapped over time to facilitate environmental modeling.

Figure 1 provides examples of valley fills within the study area. Note that these features are generally characterized by steep slopes, a terraced pattern to encourage stability, placement in headwater stream valleys adjacent to mines and reclaimed mines, and drainage ditches to transport water away from the mine site. In short, they have a unique topographic signature and are readily observable in digital terrain data representations, such as hillshades and slopeshades. Due to this unique signature and their potential environmental impacts, we argue that this is a valuable case study in which to assess the use of Mask R-CNN for detecting and mapping topographic features. Here, we are specifically attempting to map the valley fill faces (VFF; i.e., the graded slope that faces the downstream valley not yet filled). Since the true extent of the filled area and excavated areas are not readily observable and grade into one another, the upper extent of each fill is hard to distinguish. Therefore, we focus exclusively on the VFF.

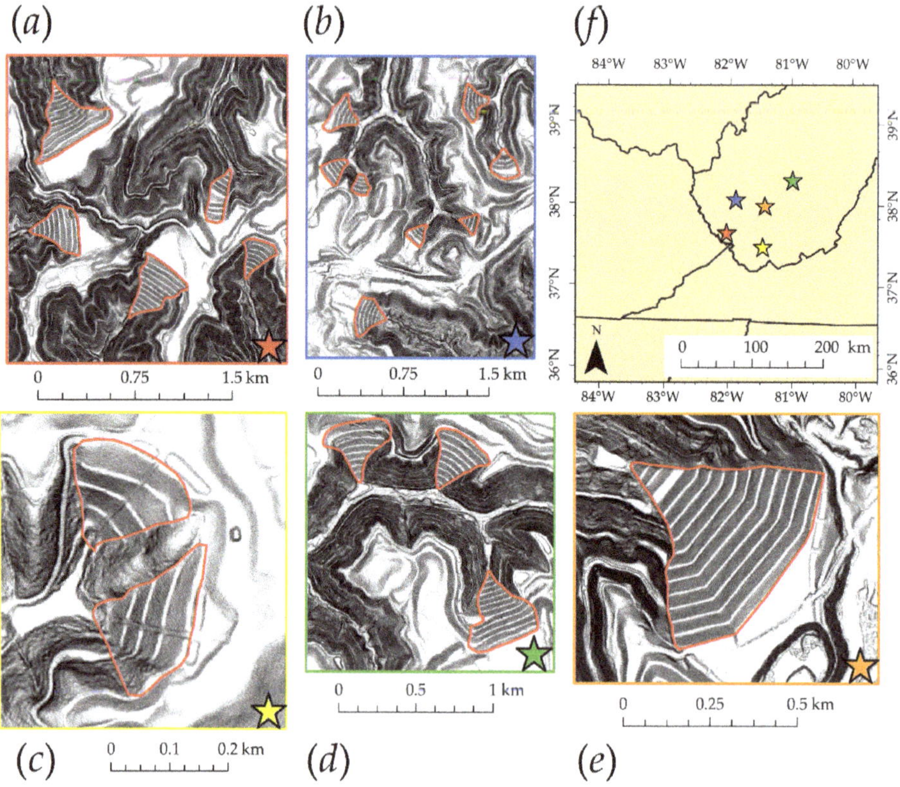

Figure 1. Example of valley fill faces (VFFs) within study area extent. The images are slopeshades, generated from the light-detection and ranging (LiDAR) data. Stars indicate the location of (*a*) through (*e*) in the study areas (*f*).

2. Methods

2.1. Study Area and Training Data Digitizing

The study areas are shown in Figure 2. The training, testing, and validation data partitions are nonoverlapping and defined geographically. The training area (Train) has a size of 9019.6 km^2 and occurs completely within West Virginia. Areas adjacent to the training area, mapped with the same LiDAR sensor and also within West Virginia, were withheld to test model performance during the training process (Test) and to validate the final model once obtained (Val). In order to assess how well the model generalizes to new LiDAR-derived terrain data, we performed additional validations over two areas in Kentucky (KY1 and KY2) and one area in Virginia (VA). Digital terrain data produced using photogrammetric methods were also predicted including a subset of the training area (SAMB1) and the entire validation area (SAMB2); no photogrammetric data were used in the training dataset. In summary, the model was trained over a single large area, tested over an adjacent smaller area, then used to make predictions in a new area collected with the same LiDAR sensor, three additional extents in different states mapped during different LiDAR collections, and two extents of photogrammetric data within West Virginia.

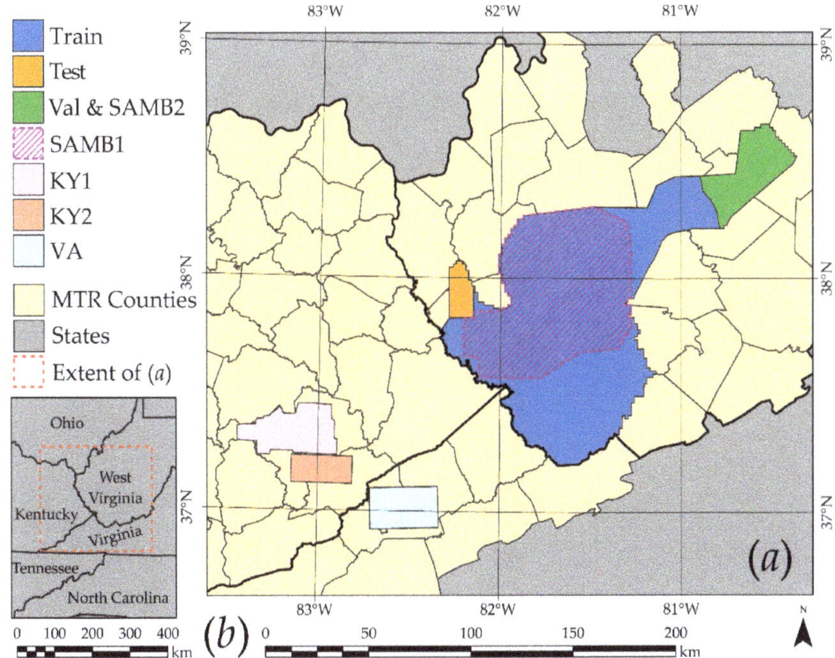

Figure 2. (*a*) Study areas in West Virginia, Kentucky, and Virginia in the Appalachian region of the eastern United States. (*b*) shows the extent of (*a*) in the eastern United States.

The extents were primarily chosen based on the availability of LiDAR data and the abundance of VFFs. In total, 1105 VFFs were provided to train the model, 118 were used to test the model at the end of each epoch in the training process, and 1014 were used to validate the model over different geographic extents (Table 1). Training data were digitized by two analysts by visual interpretation of LiDAR-derived terrain surfaces, specifically hillshades and slopeshades, and additional geospatial data, including aerial imagery. Based on the size distribution of digitized VFFs, a minimal mapping unit (MMU) of 0.2 ha was defined for this study and no VFFs smaller than this size were included.

Table 1. Description of study areas and mapped VFFs.

Study Area	Total Area	Number of Image Chips with Valley Fills	Number of Image Chips to Predict To	Number of Valley Fills
Train	9019.6 km^2	4863	-	1105
Test	279.5 km^2	282	-	118
Val	921.0 km^2	-	3111	182
KY1	773.4 km^2	-	2650	540
KY2	338.9 km^2	-	1138	149
VA	599.3 km^2	-	2093	143
SAMB1	4661.8 km^2	-	17,106	581
SAMB2	921.0 km^2	-	3110	108

2.2. Input Terrain Data and Pre-Processing

Descriptions of the LiDAR data and collections are provided in Table 2. The West Virginia LiDAR data used in this study were obtained as classified point clouds from the West Virginia View/West Virginia GIS Technical Center Elevation and LiDAR Download Tool (http://data.wvgis.wvu.edu/elevation/). LiDAR data for the study sites in Kentucky and Virginia were downloaded from the

3DEP website (https://www.usgs.gov/core-science-systems/ngp/3dep) also as classified point clouds. Data and collection information were obtained from the associated metadata files. Although all data were collected during leaf-off conditions, they differ based on collection dates, sensor used, sensor specifications, and flight specifications. The West Virginia and Kentucky data provide similar point densities at an average of 1 point per square meter (ppsm) while the Virginia data offer a higher point density at 1.7 ppsm. Collectively, the data were collected over nearly seven years. In summary, there are many differences in these datasets to support our goal of assessing transferability of models to new data and geographic extents.

Table 2. Description of LiDAR data.

	LiDAR Dataset		
Specification	**West Virginia**	**Kentucky**	**Virginia**
Collection Dates	4-9-2010 to 12-31-2011	11-8-2011 to 1-19-2013	11-3-2016 to 4-17-2017
Phenology	Leaf-off	Leaf-off	Leaf-off
Sensor	Optech ALTM-3100	Leica ALS70 and Optech Gemini	Riegel 780/680i
Average Post Spacing	1 ppsm	1 ppsm	1.746 ppsm
Flight Height	1524 m AGL	1828 m AGL	1800 m AGL
Approximate Flight Speed	135 knots	116 knots	100 knots
Scanner Pulse Rate	70 kHz	50 kHz	280 kHz
Scan Frequency	35 Hz	30.1 Hz	68 Hz
Maximum Scan Angle	36°	25.6°	60°

AGL = above ground level, ppsm = points per square meter.

The 0.61 m (2 ft) true color stereo imagery used to derive the photogrammetric elevation data used in this study were collected during the spring of 2003 and 2004 during leaf-off conditions as part of a mapping project supported by the West Virginia Statewide Addressing and Mapping Board (WVSAMB). Break lines and elevation mass points were generated using photogrammetric methods at a 3 m interval with a vertical accuracy of ±10 ft. The final 3 m DEM has a tested vertical accuracy of 0.209 m [65].

All LiDAR point clouds were converted to raster grids as DTMs using only the points classified as ground returns and the LAS Dataset to Raster utility in ArcGIS Pro 2 [66]. The average ground return elevation was calculated within each cell, and linear interpolation was used to fill data gaps. Based on data volume and visual assessment of outputs, we gridded all data to a 2 m resolution, as the VFFs and their topographic signature were easily discernable at this spatial resolution. The photogrammetric data were resampled from 3 m to 2 m using cubic convolution to match the LiDAR data.

Many surfaces can be derived from DEMs to characterize and visualize the terrain [67]. We experimented with multiple terrain visualization methods including hillshades, multi-directional hillshades, hypsometrically-tinted hillshades, and slopeshades. Based on visual inspection and initial experimentation with the Mask R-CNN algorithm, the slopeshade was selected to represent the terrain surface because it provides a distinctive and relatively consistent representation of VFFs, unlike hillshades, which vary in appearance based on the local angle of illumination of the solar energy. Since data augmentation, including random rotations and flips of the data, are used to minimize overfitting in this study, as will be discussed below, an illumination invariant terrain representation is preferred.

Slopeshades are generated from a topographic slope surface. A grayscale color ramp is applied to represent steep slopes with darker shades and flat areas with brighter shades [15,68–70]. To produce these surfaces, we first calculated topographic slope in degrees using the Slope Tool from the Spatial

Analyst Extension of ArcGIS Pro 2 [66]. The data were then converted to 8-bit integer data using Equation (1). We used a maximum slope of 90° as opposed to the maximum value in each grid surface so that all study sites could be rescaled consistently.

$$\text{Slopeshade} = (1 - \frac{\text{Slope}}{90}) \times 255 \tag{1}$$

2.3. Image Chip Generation

Since spatial context information is learned using filters, DL methods that include convolution must be trained on rectangular image chips as opposed to single pixels or image objects [24,28]. Image chips were generated using the Export Training Data For Deep Learning Tool in ArcGIS Pro 2 [66]. In order to provide training and testing data for the Mask R-CNN algorithm, the geographic extents were segmented into 512-by-512 pixel image chips. We applied a stride of 256 pixels in the X- and Y-directions for overlap and to produce more training and testing data. Using this method, 4863 training and 282 testing chips were generated that contained at least one instance of the VFF class, as noted in Table 1 above. Training masks were also generated for each image chip using this tool. Background or non-VFF pixels were coded to 0, while each instance of VFFs was coded with a unique value, from 1 to the number of VFFs in the extent, as demonstrated in Figure 3. Instance segmentation methods, such as Mask R-CNN, require that each unique instance be differentiated whereas semantic methods can accept a binary mask to differentiate a class from background pixels [34,35,39]. Once a final model was obtained, it was used to predict VFFs from image chips covering the Val extent and all other study areas. Initial results showed poor predictions near the edge of image chips. To circumvent this issue, we made predictions using larger image chips, with dimensions of 1024-by-1024 pixels and with a stride of 256 pixels, allowing for substantial overlap so that features were not missed and were likely to occur near the center of at least one chip. We then cropped each chip so that only the center 50% was used in the final surface.

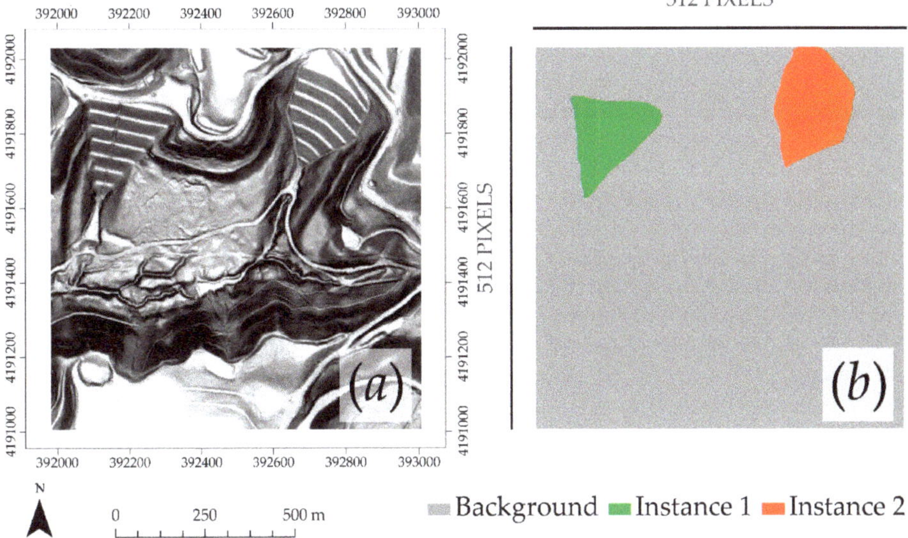

Figure 3. Example of image chip (*a*) and associated training mask with two mapped VFFs (*b*). The coordinate grid in (*a*) represents easting and northing relative to the NAD 83 UTM Zone 17 North projection.

2.4. Mask R-CNN Implementation

We used the Matterport implementation of Mask R-CNN in this study, which is available on GitHub (https://github.com/matterport/Mask_RCNN) This implementation uses Python 3, Keras, and Tensorflow. In order to load our image chips and masks, we had to generate a subclass of the Dataset class provided by the implementation. All experiments were conducted on a workstation equipped with an i9 7900X 3.3 GHz 10 core processor, 128 GB of RAM, and a GeForce RTX 2080 Ti 11 GB graphics card.

To train the model, we used a learning rate of 0.002 to train the head layers for 2 epochs, followed by training all layers at a learning rate of 0.001 for 12 epochs, and lastly training all layers at a learning are of 0.0001 for an additional 10 epochs. An initial experiment was conducted in which learning progressed for 85 epochs. In this experiment, we observed overfitting early and found that 24 epochs were adequate to stabilize the model. The default learning momentum and weight decay values, 0.9 and 0.0001, were used for all epochs. We also maintained the default values for the backbone strides of the feature pyramid network (FPN), RPN anchor scales, RPN anchor scale ratios, and RoI positive ratio. Mask R-CNN includes ROIs for training that do not contain an example of the feature of interest so that it can also learn from negative cases. In this study an ROI positive ratio of 33% was used, or 67% represented negative cases. All loss measures were equally weighted. We use the ResNet101 backbone [71] to learn the convolution filters. For each epoch, 1500 training and 90 validation steps were used with a batch size of 3 image chips, allowing for 4500 training samples and 270 test samples to be used in each epoch. The model that produced the best loss value for predicting to the test samples was selected as the final model.

Initial experimentation with initializing the model from random weights was unsuccessful, perhaps because we did not provide enough training samples to adequately train the complex model [24,30,39]. As a result, we initialized the model using weights learned from the Microsoft Common Objects in Context (MS COCO) dataset [72] (http://cocodataset.org/#home), a process known as transfer learning. Many studies have noted the value of initializing models using pre-trained weights learned from other data and problems, even if the data and classes are different [6,24,30,38,51,73–75]. For example, Tier et al. [6] used pre-trained weights learned from photographs to initialize a model to extract archeological features form digital terrain data. For Mask R-CNN specifically, Zhang et al. [51] initialized their model from the COCO weights for mapping artic ice-wedge polygons from very high spatial resolution aerial imagery. The argument for applying transfer learning is that low-level data abstractions learned from imagery can be valuable when applied even to disparate data [24,30,38,39,74,75]. Since the MS COCO weights were obtained relative to RGB images, the slopeshade data were loaded in as 3-band images by replicating the grayscale values to each band. Although not computationally efficient, this allowed us to make use of transfer learning.

Data augmentation has been shown to minimize overfitting by expanding the number and characteristics of the training samples [24,29,30,50]. Therefore, we implemented random augmentations of the original image chips including rotations at 0°, 90°, 180°, and 270°; left/right flips; up/down flips, brightness and contrast alterations, and blurring and sharpening. We attempted to avoid extreme augmentations of the original data. These augmentations were applied using the imgaug Python library [76]. Example augmentations for the image chip shown in Figure 3 are provided as an example in Figure 4.

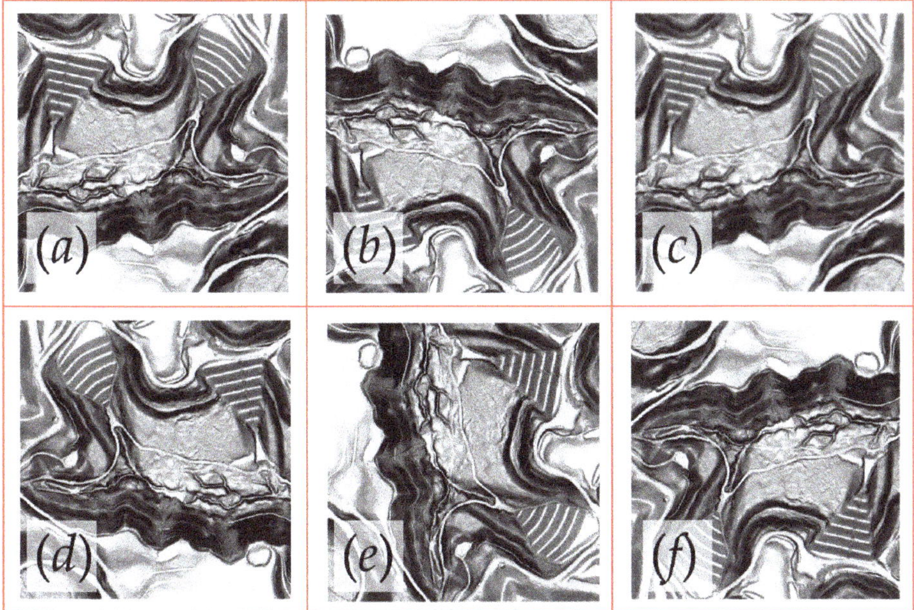

Figure 4. Example image augmentations applied to minimize overfitting. (*a*) Original image chip. (*b*) Vertical flip. (*c*) Gaussian blur. (*d*) Horizontal flip. (*e*) Rotate 90° clockwise. (*f*) Rotate 180° clockwise.

2.5. Prediction and Post-Processing

Once a final model was obtained, it was used to detect features in the Val extent and all other study areas. The final chips were binary surfaces, in which all predicted VFFs were coded to 1 and the background was coded to 0. This process was completed using the Matterport Mask R-CNN code combined with additional Python and R [77] scripts.

Once all image chips within a dataset were processed by the model and cropped, they were merged to a continuous raster surface using the Mosaic to New Raster utility in ArcGIS Pro [66] with the maximum value returned at cells in overlapping area so that all predicted VFF pixels that occurred in the center of the chips were maintained in the final model. The raster grids were then converted to polygon vectors to represent each contiguous area of VFFs as a single feature. Any predicted features smaller than 0.2 ha were removed to satisfy the MMU.

2.6. Accuarcy Assessment

We assessed the Matterport Mask R-CNN model based on mAP at multiple intersection of union (IoU) threshold ranges. IoU is the area of intersection divided by the area of union between the manually digitized and predicted masks as described in Equation (2). mAP represents the interpolated precision at multiple IoU threshold ranges based on the area under the precision-recall curve [78].

$$IoU = \frac{\text{Area of Intersection}}{\text{Area of Union}} \tag{2}$$

Given that the final output of the classification was vector objects occurring over a geographic extent, assessment using overlapping image chips, where the same VFF has the potential to be mapped and evaluated multiple times, will not yield an assessment that approximates the accuracy of the final map product. In this study, our primary interest is the detection of VFFs across the entire dataset as discrete spatial objects. Thus, we focus on true positives (TP), false positives (FP), and false negatives

(FN) [79] in the map. An additional complexity is that the boundaries of VFFs are inherently fuzzy at the fine scale of our map, which has 2 m pixels. Therefore, we assessed the predictions based on a visual comparison with the manually digitized VFFs. If a predicted feature was judged to overwhelming agree with the manually digitized VFF based on shared area and spatial co-occurrence, then it was labelled as a TP. FPs represent areas mapped as VFFs but were in reality not. FNs represent VFFs that were missed.

From the TP, FP, and FN counts of VFFs within each study area extent, we calculated precision, recall, and the F1 score. Precision represents the portion of the predicted VFFs that were VFFs and is equivalent to 1 - commission error. Recall represents the ratio of correctly mapped VFFs relative to the total number of VFFs and is equivalent to 1 - omission error. The F1 score is the harmonic mean of precision and recall. The equations for these metrics are provided below in Equations (3) through (5). We also assessed the TP, FN, and recall for VFFs that were larger than 1 ha in the manually digitized data to explore the performance for only larger features.

$$\text{Precision} \; = \; \frac{\text{TP}}{\text{TP} + \text{FP}} \tag{3}$$

$$\text{Recall} \; = \; \frac{\text{TP}}{\text{TP} + \text{FN}} \tag{4}$$

$$\text{F1 Score} \; = \; 2 \times \frac{\text{Recall} \times \text{Precision}}{\text{Recall} + \text{Precision}} \tag{5}$$

In addition to evaluating the numbers of correctly mapped VFFs, we also assessed the accuracy based on the delineation of VFFs, by focusing on the intersection and union of the derived polygons to estimate area-based producer's and user's accuracy. Using area based measures, larger VFFs have a larger weight in the assessment. Producer's accuracy is similar to recall, while user's accuracy is similar to precision [80–82].

Due to the indeterminate nature of VFF boundaries at the 2 m scale of this project, small differences between the location of the boundary in the reference and map data have little significance. Therefore, we also assessed producer's and user's accuracy for the VFF class using a fuzzy accuracy method modified from Brandtberg et al. [83]. The aim of this approach is to weight the disagreement between predicted and reference polygons, based on distance from the center of the polygon. Thus, a larger weighting is assigned to pixels near the center, and a lower weighting for pixels at the boundary. The fuzzy accuracy was implemented by first creating raster-based Euclidean distance surfaces at a 2 m spatial resolution using ArcGIS Pro [66] for each reference and predicted VFF polygon to represent the straight-line distance of each cell from the feature boundary. We divided each pixel in the VFF feature by the sum of the distances within the polygon, then multiplied the pixel by the area of the polygon, thus achieving our aim of weighting the entire area of the feature higher in the center in comparison to the edges, but keeping the data on a scale that is equivalent to area. We then summed the pixel values in the union and intersecting extents to obtain fuzzy estimates of producer's and user's accuracy for the VFF class from these totals.

3. Results

3.1. Mask R-CNN Model and Visual Assessment

Figure 5 shows loss measures for different components of the Mask R-CNN model as a function of epoch. The lowest overall loss calculated from the withheld test data was obtained after 12 epochs (0.773), so the result from this epoch was chosen for the final model. The graphs suggest some overfitting after 12 epochs; however, the dominant pattern is variability in the test loss measures and no substantial increase in performance. Other authors have noted optimal performance after few epochs, especially when pre-trained weights are used. For example, Zhang et al. [51] note optimal performance after eight epochs when using Mask R-CNN to predict artic ice wedge extents from high spatial resolution aerial imagery, and Stewart al. [52] used only six epochs to map northern leaf blight lesions from UAV data.

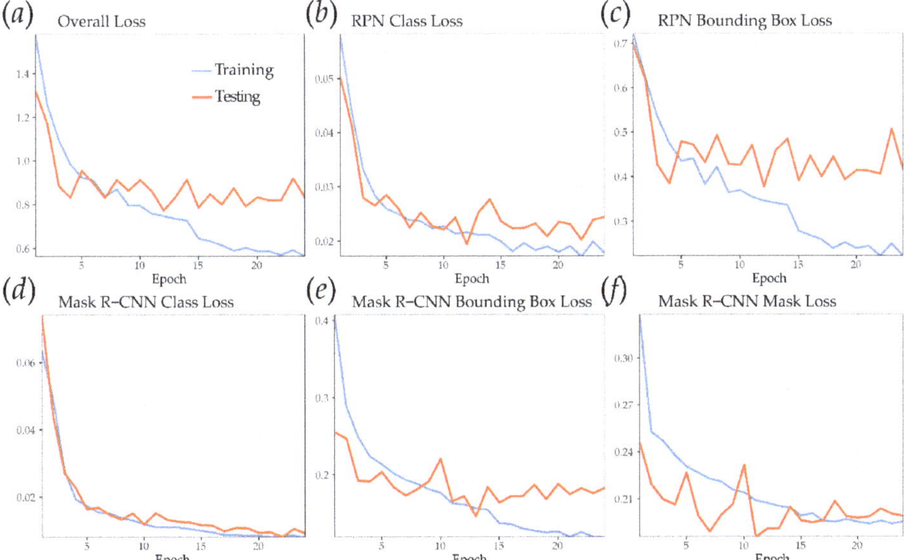

Figure 5. Loss values for training and test data across all epochs. (*b*) region proposal network (RPN) class loss measures how well the Region Proposal Network separates the background from the objects of interest. (*c*) RPN bounding box loss assesses how well the RPN localizes objects. (*d*) Mask R-CNN class loss assesses how well Mask R-CNN recognizes each class of object. (*e*) Mask R-CNN bounding box loss measures how well Mask R-CNN localizes objects. (*f*) Mask R-CNN mask loss measures how well Mask R-CNN segments objects. (*a*) Overall loss is an addition of all other loss measures. For a complete explanation of these metrics see [38,39].

Table 3 shows the mAP results for different IoU threshold ranges calculate from image chips covering the Val area. Performance decreased as the threshold was adjusted to incorporate higher IoU values, as expected. When only IoU thresholds between 0.50 and 0.55 were used, the mAP was 0.596. This generally suggests that with this narrow threshold range, VFFs were detected, but the boundaries did not match well, due to the indeterminate and complex nature of the VFFs.

Table 3. mAP results for validation data using different intersection of union (IoU) ranges.

Start (IoU)	End (IoU)	mAP
0.50	0.95	0.389
0.50	0.90	0.433
0.50	0.85	0.475
0.50	0.80	0.475
0.50	0.75	0.535
0.50	0.70	0.557
0.50	0.65	0.557
0.50	0.60	0.596
0.50	0.55	0.596

Figure 6 shows some predicted VFFs in comparison to those manually digitized in the Val, KY1, KY2, and VA study areas. The figure suggests that VFFs were generally detected with few FPs. Although the overall shapes of the VFFs are similar in the automated and reference (i.e., manually digitized) datasets, the boundaries do not overlay exactly due to their fuzzy nature, as previously discussed.

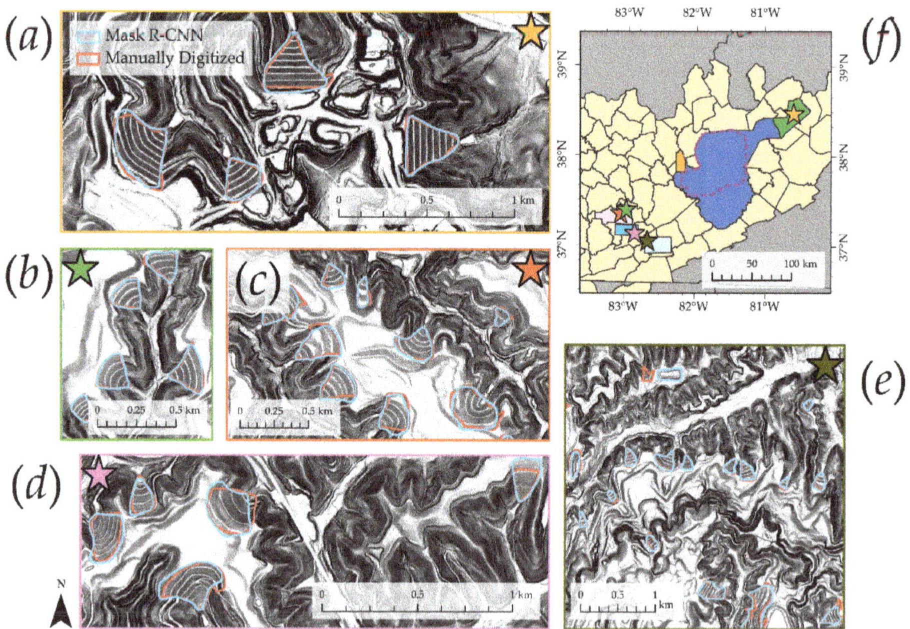

Figure 6. Example of VFF predictions and manually digitized features across multiple study areas. Stars indicate the location of (*a*) through (*e*) in the study areas (*f*).

Figure 7 provides some examples of common FP issues. Some reclaimed mine sites, coal and overburden piles in mine sites, and artificially re-contoured landscapes associated with residential or transportation development, were occasionally misclassified as VFFs (Figure 7*b–d*). Slopes with timber-harvest roads (Figure 7*a*), which result in a pattern similar to terracing, especially within valley-head areas, were sometimes falsely mapped as VFFs. Figure 7*e* shows a misclassified slope that is characterized by deep channeling, which has potentially been confused as the drainage ditches installed on the VFFs. Mapped VFFs that were most often missed (FNs), were those that covered a smaller area and/or lacked the characteristic terracing pattern or drainage ditches. In contrast, VFFs that were larger and had well defined terracing where seldom missed.

3.2. Validation

Table 4 shows the validation based on the manual inspection of TP, FP, and FN VFF counts in all areas that were predicted. For predictions made using the LiDAR-derived data, precision, recall, and F1 score values were all higher than 0.73. The highest precision was obtained for the KY1 area while the highest recall was obtained for the Val area. In general, we documented similar precision, recall, and F1 scores for the LiDAR-derived data, suggesting that the model generalizes well to other geographic regions, collected using different LiDAR sensors. Further, a higher precision was obtained for the KY1 and KY2 test sites than the Val site, which was mapped using the same LiDAR sensor as the data used to develop the model. The VA test site had the lowest precision, recall, and F1 scores of all the LiDAR-derived extents. This could potentially be a result of the higher average post spacing in comparison to the other collections, resulting in a disparate representation of the landscape. Another confounding factor is the characteristics of the VFFs in each area. For example, a visual inspection of the KY1 extent suggests a large number of VFFs that are large and have well defined terracing, which may contribute to comparatively high assessment metrics for this extent. The All LiDAR column represents the pooled results for all LiDAR-derived datasets. Collectively, a precision of 0.878, a recall of 0.858, and a F1 score of 0.868 was obtained when using LiDAR-derived data. For VFFs larger than 1 ha, recall increased for all the study sites, which indicates that the larger fills were generally easier to detect.

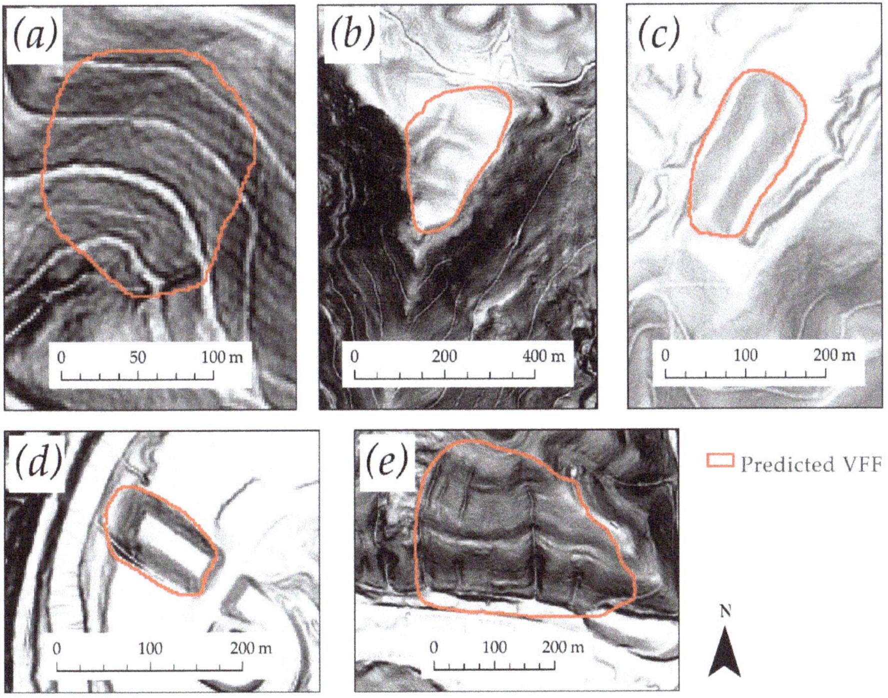

Figure 7. (*a*) through (*e*) Example of false positives.

Table 4. Validation based on manual comparison of numbers of VFFs.

Measure	Val	KY1	KY2	VA	All LiDAR	SAMB1	SAMB2
				Study Area			
No. Mapped VFFs	182	540	149	143	1014	581	108
No. Mask R-CNN VFFs	200	546	143	149	1038	1735	321
TP	170	495	129	117	911	346	39
FP	30	51	14	32	127	1389	282
FN	13	59	37	42	151	239	69
Precision	0.850	0.907	0.902	0.785	0.878	0.199	0.121
Recall	0.929	0.894	0.777	0.736	0.858	0.591	0.361
F1-Score	0.888	0.9	0.835	0.76	0.868	0.278	0.181
No. Mask R-CNN VFFs (>1 ha)	123	463	129	111	826	527	77
TP (>1 ha)	118	418	105	89	730	315	35
FN (>1 ha)	5	48	2	25	80	217	42
Recall (>1 ha)	0.959	0.897	0.809	0.781	0.900	0.592	0.455

Precision, recall, and F1 score were generally low when the model trained on the LiDAR-derived data was used to predict to the photogrammetrically-derived slopeshades in the SAMB1 and SAMB2 extents. All values are lower than 0.6 due in part to the large number of FPs. A visual inspection of the classifications suggests that larger VFFs with well-defined terracing were generally mapped well; however, many VFs were missed, especially smaller features or those without well-defined terracing. Since the photogrammetric methods due not allow for canopy penetration, VFFs that were heavily vegetated with woody vegetation and shrubs generally did not show well defined terracing in the slopeshades even if the pattern was present. Figure 8 provides examples of manually digitized VFFs as represented in the photogrammetric data. Note that the characteristic terracing pattern is not evident for all features. It would be interesting to assess models trained form the photogrammetric data to assess the ability to map historic conditions from such data; however, that is outside the scope of this study, as our goal here is to assess the transferability of the model trained on LiDAR-derived data to disparate datasets.

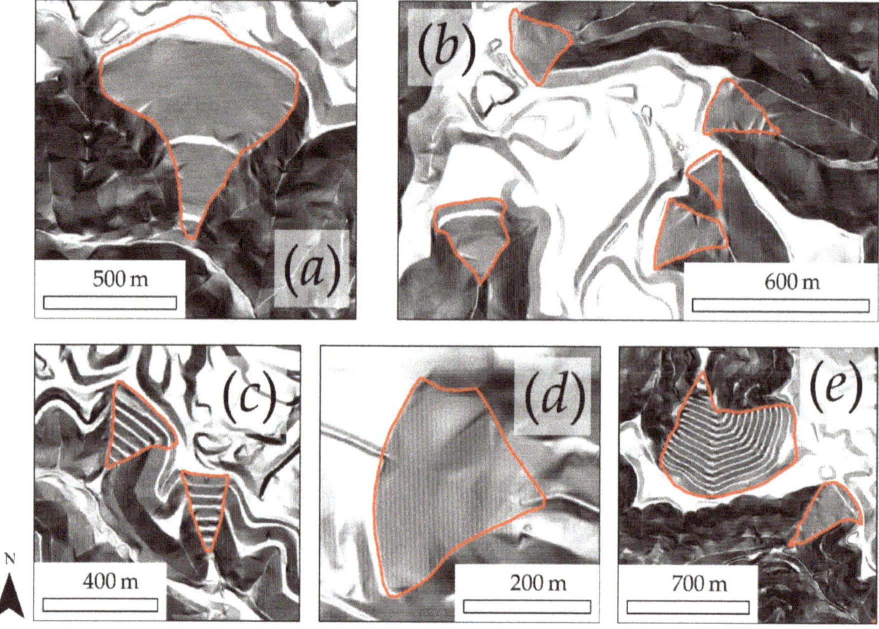

Figure 8. (*a*) Through (*e*) Examples of digitized valley fills as represented in the photogrammetric data.

In summary, these results indicate that the model performed well using LiDAR-derived data; however, photogrammetric data resulted in many FPs and generally poor performance based upon precision and recall. The model was not able to adequately generalize to these very different datasets. It should be noted that it is difficult to note whether or not the poor performance is a result of the inability of the model to generalize to the disparate data or whether it is because the VFFs are not well represented in the photogrammetric datasets.

Validation using area-based user's and producer's accuracy (Table 5) generally yielded lower assessment values than the count-based evaluation (Table 4). For example, all producer's and user's accuracies for the predictions using LiDAR-derived data were lower than the associated count-based recall and precision values, respectively, except for KY2. The photogrammetric data resulted in user's and producer's values between 0.043 and 0.388.

Using the fuzzy, center-weighted method, user's and producer's accuracies increased for all LiDAR-derived results in comparison to the area-based method. This highlights that the center portions of the VFFs were generally mapped well, and the boundaries, which are inherently indeterminate, had lower agreement. Notably, the aggregate All LiDAR measures indicate a fuzzy producer's accuracy of 0.860 and a fuzzy user's accuracy of 0.903.

Table 5. Area- and fuzzy area-based error evaluation.

Measure	Val	KY1	KY2	VA	All LiDAR	SAMB1	SAMB2
				Study Area			
Producer's Accuracy (Area)	0.787	0.797	0.831	0.735	0.793	0.129	0.263
User's Accuracy (Area)	0.841	0.741	0.78	0.603	0.744	0.388	0.043
Fuzzy Producer's Accuracy (Center-Weighted)	0.851	0.866	0.899	0.802	0.860	0.137	0.046
Fuzzy User's Accuracy (Center -Weighted)	0.909	0.944	0.964	0.689	0.903	0.169	0.053

4. Discussion

4.1. Study Findings

Our results, as highlighted by our demonstration of mapping VFFs, suggest that Mask R-CNN can be used to extract terrain features from digital elevation data if the features have a unique topographic signature. Count-based precision, recall, and F1 scores were all above 0.73 across all validation datasets when predicting to LiDAR data. When assessed based on area, producer's and user's accuracies were generally lower than the associated recall and precision values. This highlights the difficulty in mapping and assessing boundaries that are poorly defined or gradational in nature. When center-weighting the areas to reduce the influence of boundaries, we saw increases in user's and producer's accuracies, which suggests that the features were generally successfully mapped, though the boundaries may be disparate. It is notable that this model generally had similar performance when applied to new geographic regions, as long as LiDAR data were used, even if the sensors and acquisition parameters differed. This provides strong evidence that the approach is robust and can be applied to regional mapping. However, the model performed poorly when applied to disparate, photogrammetrically-derived data, suggesting that generalization is limited to comparable data. However, this may partially result from the poor representation of the VFFs in the photogrammetric data, so it is difficult to differentiate the impact of the model and the input data quality. Overall, this case study shows promise in applying instance segmentation to digital terrain data, suggesting that CNN-based deep learning has potential for mapping other topographic classes, for example geomorphological or soils mapping.

Since no prior studies have attempted to map VFFs using automated methods, we are not able to relate our findings to any prior studies that have explored this specific task; however, our findings do reinforce those of Tier et al. [6] and Behrens et al. [48], which note the value of CNNs for extracting features from digital terrain data. More broadly, this study supports prior findings that CNNs in general and Mask R-CNN specifically are of great value for mapping features with a unique spatial, contextual, or textural signature and that may not be spectrally separable from other classes or features [51,52,84,85].

4.2. Limitations and Recommendations

There were some notable limitations in this study. Although we analyzed multiple study sites and datasets, this case study is specific to a single geomorphic feature. Therefore, it would be useful to assess the application of instance segmentation techniques to map and differentiate additional anthropogenic and natural terrain features. For example, these methods could be applied to mapping glacial, fluvial, or eolian landforms. Additionally, there is a need to assess the mapping of landscape change using multitemporal digital terrain data. This could not be pursued in this study due to the lack of pre-mining LiDAR data and our finding that photogrammetric data were not useful for mapping VFFs.

Given the large training dataset requirements of deep learning, there is a need to develop databases of training samples specific to terrain data and features, similar to those that already exist for photographs, such as MS COCO [72] and ImageNet [86]. The remote sensing community should consider investing in the development of large terrain- and image-based datasets to improve our ability to apply deep learning to our data and perform more robust experiments. Trier et al. [6] made a similar argument within the archeological research community. Transfer learning from models trained on terrain data may prove more accurate than the application of weights learned from RGB photographs.

In this project, we specifically relied on slopeshades as a representation of the terrain surface. There is a need to explore additional representations or combinations of representations as input to DL techniques. With LiDAR, it is possible to obtain measures of the height of above ground features, such as trees and buildings, by generating a normalized digital surface model (nDSM). It is also possible to measure the intensity of the returned laser pulse [10]. There is a need to explore these additional measures for mapping terrain features with DL. As noted by Sterenczak et al. [87], the interpolation method used to generate a DTM from LiDAR point cloud data can have an impact on the resulting representation of the terrain surface, so there is also a need to assess how well models perform and transfer to DTMs generated using different interpolation methods.

Due to lengthy computational time and the large number of parameters that can be manipulated, it was not possible to fully optimize the Mask R-CNN method for this task. Training the model for 24 epochs using a single GPU required 12 h; as a result, we could not extensively experiment with the impact of parameter settings as would be possible with traditional, shallow machine learning methods using grid searches combined with k-fold cross validation or bootstrapping. Instead, we had to rely on a limited series of experiments using a small number of model parameters. This issue will need to be addressed in order to support rigorous comparative studies of different algorithms and/or feature spaces. Terrain feature extraction could be performed with semantic segmentation methods, such as SegNet and UNet, so there is a need to compare different CNN methods for mapping terrain features.

It is common for landscape features to have boundaries that are gradational or inherently indeterminate, which adds to the complexity of assessing the quality of the prediction. A review of the remote sensing, computer vision, DL, and machine learning literature suggests a lack of research on this topic. Our method is an extension of an approach proposed by Brandtberg et al. [83] and appears to have potential for widespread use in applications such as mapping wetlands and soils, as well as tree delineation from high resolution images. However, it would be useful to explore this approach more thoroughly, as additional refinements may improve it. For example, measures of distance other than a linear approach may be useful.

5. Conclusions

This exploration of mapping VFFs from digital terrain data suggests that the Mask R-CNN DL instance segmentation method can be applied to map geomorphic and landscape features using LiDAR-derived data. Further, our results suggest that models trained in one area can transfer well to other areas if similar data are available, in this case LiDAR, providing strong evidence of the robustness of the approach. However, the model performed poorly when applied to disparate, photogrammetrically-derive data.

Here we focused on features that have distinctive topographic and geomorphic characteristics, and we suggest that there is a need for further experimentation relating to mapping additional terrain features of variable complexities. Future studies should explore the mapping of additional anthropogenic, fluvial, glacial, and eolian terrain features and landforms. There is also a need to further explore optimization methods for deep learning to foster more rigorous comparisons and develop standardized techniques to assess gradational or uncertain boundaries. As CNN-based semantic and instance segmentation methods mature, there is a need to further explore these techniques for mapping and extracting features from geospatial and remotely sensed data.

The recent developments in CNNs, semantic segmentation, and instance segmentation are providing new opportunities to extract and map digital terrain features. Our hope is that this case study will encourage additional research and data development relating to automated terrain mapping using LiDAR and DL. Further, with the increasing availability of LiDAR data, such methods will likely prove to be of great importance for studying our anthropogenic and natural landscapes and monitoring landscape change.

Author Contributions: Conceptualization, A.E.M.; methodology, A.E.M., and P.P.; validation, A.E.M. and J.D.P.; formal analysis, A.E.M. and J.D.P.; writing—original draft preparation, A.E.M.; writing—review and editing, A.E.M., P.P., and J.D.P. All authors have read and agreed to the published version of the manuscript.

Funding: No funding obtained for this project.

Acknowledgments: We would like to acknowledge the West Virginia GIS Technical Center, West Virginia View, and the United States Geologic Survey (USGS) who provided LiDAR data for this study. We would also like to thank five anonymous reviewers whose suggestions and comments strengthened the work.

Conflicts of Interest: The authors declare no conflict of interest.

References

1. Jaboyedoff, M.; Oppikofer, T.; Abellán, A.; Derron, M.-H.; Loye, A.; Metzger, R.; Pedrazzini, A. Use of LIDAR in landslide investigations: A review. *Nat. Hazards* **2012**, *61*, 5–28. [CrossRef]

2. Chase, A.F.; Chase, D.Z.; Fisher, C.T.; Leisz, S.J.; Weishampel, J.F. Geospatial revolution and remote sensing LiDAR in Mesoamerican archaeology. *Proc. Natl. Acad. Sci. USA* **2012**, *109*, 12916–12921. [CrossRef] [PubMed]

3. Arundel, S.T.; Phillips, L.A.; Lowe, A.J.; Bobinmyer, J.; Mantey, K.S.; Dunn, C.A.; Constance, E.W.; Usery, E.L. Preparing *The National Map* for the 3D Elevation Program–products, process and research. *Cartogr. Geogr. Inf. Sci.* **2015**, *42*, 40–53. [CrossRef]

4. The Earth Archive. Available online: https://www.theeartharchive.com (accessed on 28 October 2019).

5. Passalacqua, P.; Tarolli, P.; Foufoula-Georgiou, E. Testing space-scale methodologies for automatic geomorphic feature extraction from lidar in a complex mountainous landscape. *Water Resour. Res.* **2010**, *46*. [CrossRef]

6. Trier, Ø.D.; Cowley, D.C.; Waldeland, A.U. Using deep neural networks on airborne laser scanning data: Results from a case study of semi-automatic mapping of archaeological topography on Arran, Scotland. *Archaeol. Prospect.* **2019**, *26*, 165–175. [CrossRef]

7. Trier, Ø.D.; Zortea, M.; Tonning, C. Automatic detection of mound structures in airborne laser scanning data. *J. Archaeol. Sci. Rep.* **2015**, *2*, 69–79. [CrossRef]

8. Eeckhaut, M.V.D.; Poesen, J.; Verstraeten, G.; Vanacker, V.; Nyssen, J.; Moeyersons, J.; Beek, L.P.H.; van Vandekerckhove, L. Use of LIDAR-derived images for mapping old landslides under forest. *Earth Surf. Process. Landf.* **2007**, *32*, 754–769. [CrossRef]

9. Verhagen, P.; Drăguţ, L. Object-based landform delineation and classification from DEMs for archaeological predictive mapping. *J. Archaeol. Sci.* **2012**, *39*, 698–703. [CrossRef]

10. Remote Sensing and Image Interpretation, 7th Edition Wiley. Available online: https://www.wiley.com/en-us/Remote+Sensing+and+Image+Interpretation%2C+7th+Edition-p-9781118343289 (accessed on 28 October 2019).

11. Ardizzone, F.; Cardinali, M.; Galli, M.; Guzzetti, F.; Reichenbach, P. Identification and mapping of recent rainfall-induced landslides using elevation data collected by airborne Lidar. *Nat. Hazards Earth Syst. Sci.* **2007**, *7*, 637–650. [CrossRef]

12. Chen, W.; Li, X.; Wang, Y.; Chen, G.; Liu, S. Forested landslide detection using LiDAR data and the random forest algorithm: A case study of the Three Gorges, China. *Remote Sens. Environ.* **2014**, *152*, 291–301. [CrossRef]

13. Van Den Eeckhaut, M.; Kerle, N.; Poesen, J.; Hervás, J. Object-oriented identification of forested landslides with derivatives of single pulse LiDAR data. *Geomorphology* **2012**, *173–174*, 30–42. [CrossRef]

14. Abdulwahid, W.M.; Pradhan, B. Landslide vulnerability and risk assessment for multi-hazard scenarios using airborne laser scanning data (LiDAR). *Landslides* **2017**, *14*, 1057–1076. [CrossRef]

15. Haneberg, W.C.; Cole, W.F.; Kasali, G. High-resolution lidar-based landslide hazard mapping and modeling, UCSF Parnassus Campus, San Francisco, USA. *Bull. Eng. Geol. Environ.* **2009**, *68*, 263–276. [CrossRef]

16. Latif, Z.A.; Aman, S.N.A.; Pradhan, B. Landslide susceptibility mapping using LiDAR derived factors and frequency ratio model: Ulu Klang area, Malaysia. In Proceedings of the 2012 IEEE 8th International Colloquium on Signal Processing and Its Applications, Malacca, Malaysia, 23 March 2012.

17. Youssef, A.M.; Pourghasemi, H.R.; Pourtaghi, Z.S.; Al-Katheeri, M.M. Landslide susceptibility mapping using random forest, boosted regression tree, classification and regression tree, and general linear models and comparison of their performance at Wadi Tayyah Basin, Asir Region, Saudi Arabia. *Landslides* **2016**, *13*, 839–856. [CrossRef]

18. Maxwell, A.E.; Warner, T.A. Differentiating Mine-Reclaimed Grasslands from Spectrally Similar Land Cover using Terrain Variables and Object-Based Machine Learning Classification. *Int. J. Remote Sens.* **2015**, *36*, 4384–4410. [CrossRef]

19. DeWitt, J.D.; Warner, T.A.; Conley, J.F. Comparison of DEMS derived from USGS DLG, SRTM, a Statewide Photogrammetry Program, ASTER GDEM and LiDAR: Implications for Change Detection. *GISci. Remote Sens.* **2015**, *52*, 179–197. [CrossRef]

20. Stumpf, A.; Kerle, N. Object-oriented mapping of landslides using Random Forests. *Remote Sens. Environ.* **2011**, *115*, 2564–2577. [CrossRef]

21. Chen, G.; Weng, Q.; Hay, G.J.; He, Y. Geographic object-based image analysis (GEOBIA): Emerging trends and future opportunities. *Gisci. Remote Sens.* **2018**, *55*, 159–182. [CrossRef]

22. Kim, M.; Madden, M.; Warner, T.A. Forest Type Mapping Using Object-Specific Texture Measures from Multispectral Ikonos Imagery. *Photogramm. Eng. Remote Sens.* **2009**, *75*, 819–829. [CrossRef]

23. Kim, M.; Warner, T.A.; Madden, M.; Atkinson, D.S. Multi-scale GEOBIA with very high spatial resolution digital aerial imagery: Scale, texture and image objects. *Int. J. Remote Sens.* **2011**, *32*, 2825–2850. [CrossRef]

24. Zhu, X.X.; Tuia, D.; Mou, L.; Xia, G.-S.; Zhang, L.; Xu, F.; Fraundorfer, F. Deep Learning in Remote Sensing: A Comprehensive Review and List of Resources. *IEEE Geosci. Remote Sens. Mag.* **2017**, *5*, 8–36. [CrossRef]

25. Ball, J.E.; Anderson, D.T.; Sr, C.S.C. Comprehensive survey of deep learning in remote sensing: Theories, tools, and challenges for the community. *JARS* **2017**, *11*, 042609. [CrossRef]

26. Kussul, N.; Lavreniuk, M.; Skakun, S.; Shelestov, A. Deep Learning Classification of Land Cover and Crop Types Using Remote Sensing Data. *IEEE Geosci. Remote Sens. Lett.* **2017**, *14*, 778–782. [CrossRef]

27. Maggiori, E.; Tarabalka, Y.; Charpiat, G.; Alliez, P. Fully convolutional neural networks for remote sensing image classification. In Proceedings of the 2016 IEEE International Geoscience and Remote Sensing Symposium (IGARSS), Beijing, China, 10–15 July 2016.

28. Maggiori, E.; Tarabalka, Y.; Charpiat, G.; Alliez, P. Convolutional Neural Networks for Large-Scale Remote-Sensing Image Classification. *IEEE Trans. Geosci. Remote Sens.* **2017**, *55*, 645–657. [CrossRef]

29. Yu, X.; Wu, X.; Luo, C.; Ren, P. Deep learning in remote sensing scene classification: A data augmentation enhanced convolutional neural network framework. *Gisci. Remote Sens.* **2017**, *54*, 741–758. [CrossRef]

30. Zhang, L.; Zhang, L.; Du, B. Deep Learning for Remote Sensing Data: A Technical Tutorial on the State of the Art. *IEEE Geosci. Remote Sens. Mag.* **2016**, *4*, 22–40. [CrossRef]

31. Hu, W.; Huang, Y.; Wei, L.; Zhang, F.; Li, H. Deep Convolutional Neural Networks for Hyperspectral Image Classification. *J. Sens.* **2015**, *2015*, 258619. [CrossRef]

32. Ji, S.; Zhang, C.; Xu, A.; Shi, Y.; Duan, Y. 3D Convolutional Neural Networks for Crop Classification with Multi-Temporal Remote Sensing Images. *Remote Sens.* **2018**, *10*, 75. [CrossRef]

33. Rizaldy, A.; Persello, C.; Gevaert, C.M.; Oude Elberink, S.J. Fully Convolutional Networks for Ground Classification from LiDAR Point Clouds. *ISPRS Ann. Photogramm. Remote Sens. Spat. Inf. Sci.* **2018**, *IV–2*, 231–238. [CrossRef]

34. Badrinarayanan, V.; Kendall, A.; Cipolla, R. SegNet: A Deep Convolutional Encoder-Decoder Architecture for Image Segmentation. *IEEE Trans. Pattern Anal. Mach. Intell.* **2017**, *39*, 2481–2495. [CrossRef]

35. Christ, P.F.; Elshaer, M.E.A.; Ettlinger, F.; Tatavarty, S.; Bickel, M.; Bilic, P.; Rempfler, M.; Armbruster, M.; Hofmann, F.; D'Anastasi, M.; et al. Automatic Liver and Lesion Segmentation in CT Using Cascaded Fully Convolutional Neural Networks and 3D Conditional Random Fields. Available online: https://arxiv.org/abs/1610.02177 (accessed on 5 February 2020).

36. Li, X.; Chen, H.; Qi, X.; Dou, Q.; Fu, C.-W.; Heng, P.-A. H-DenseUNet: Hybrid Densely Connected UNet for Liver and Tumor Segmentation from CT Volumes. *IEEE Trans. Med. Imaging* **2018**, *37*, 2663–2674. [CrossRef] [PubMed]

37. Zhao, T.; Yang, Y.; Niu, H.; Wang, D.; Chen, Y. Comparing U-Net convolutional network with mask R-CNN in the performances of pomegranate tree canopy segmentation. In Proceedings of the Multispectral, Hyperspectral, and Ultraspectral Remote Sensing Technology, Techniques and Applications VII. *Int. Soc. Opt. Photonics* **2018**, *10780*, 107801J.

38. Ren, S.; He, K.; Girshick, R.; Sun, J. Faster R-CNN: Towards Real-Time Object Detection with Region Proposal Networks. In *Advances in Neural Information Processing Systems 28*; Cortes, C., Lawrence, N.D., Lee, D.D., Sugiyama, M., Garnett, R., Eds.; Curran Associates, Inc.: Red Hook, NY, USA, 2015.

39. He, K.; Gkioxari, G.; Dollar, P.; Girshick, R. Mask R-CNN. Available online: https://arxiv.org/abs/1703.06870 (accessed on 5 February 2020).

40. Microsoft/USBuildingFootprints; Microsoft. Available online: https://github.com/microsoft/USBuildingFootprints (accessed on 5 February 2020).

41. Li, W.; Fu, H.; Yu, L.; Cracknell, A. Deep Learning Based Oil Palm Tree Detection and Counting for High-Resolution Remote Sensing Images. *Remote Sens.* **2017**, *9*, 22. [CrossRef]

42. Janssens-Coron, E.; Guilbert, E. Ground Point Filtering from Airborne LiDAR Point Clouds using Deep Learning: A Preliminary Study. *Int. Arch. Photogramm. Remote Sens. Spat. Inf. Sci.* **2019**, *XLII-2/W13*, 1559–1565. [CrossRef]

43. Hackel, T.; Savinov, N.; Ladicky, L.; Wegner, J.D.; Schindler, K.; Pollefeys, M. Semantic3D.net: A new Large-scale Point Cloud Classification Benchmark. Available online: https://www.semanticscholar.org/paper/Semantic3D.net%3A-A-new-Large-scale-Point-Cloud-Hackel-Savinov/5d9b36e296e6f61177c2f1739a6ca8c553303c09 (accessed on 5 February 2020).

44. Hu, X.; Yuan, Y. Deep-Learning-Based Classification for DTM Extraction from ALS Point Cloud. *Remote Sens.* **2016**, *8*, 730. [CrossRef]

45. Zou, X.; Cheng, M.; Wang, C.; Xia, Y.; Li, J. Tree Classification in Complex Forest Point Clouds Based on Deep Learning. *IEEE Geosci. Remote Sens. Lett.* **2017**, *14*, 2360–2364. [CrossRef]

46. Guan, H.; Yu, Y.; Ji, Z.; Li, J.; Zhang, Q. Deep learning-based tree classification using mobile LiDAR data. *Remote Sens. Lett.* **2015**, *6*, 864–873. [CrossRef]

47. Liu, F.; Li, S.; Zhang, L.; Zhou, C.; Ye, R.; Wang, Y.; Lu, J. 3DCNN-DQN-RNN: A Deep Reinforcement Learning Framework for Semantic Parsing of Large-Scale 3D Point Clouds. In Proceedings of the 2017 IEEE International Conference on Computer Vision (ICCV), Venice, Italy, 22–29 October 2017.

48. Behrens, T. Multi-scale digital soil mapping with deep learning. *Sci. Rep.* **2018**, *8*, 1–9. [CrossRef]

49. Li, W.; Zhou, B.; Hsu, C.-Y.; Li, Y.; Ren, F. Recognizing terrain features on terrestrial surface using a deep learning model: An example with crater detection. In Proceedings of the 1st Workshop on Artificial Intelligence and Deep Learning for Geographic Knowledge Discovery-GeoAI 17, Los Angeles, CA, USA, 7–10 November 2017.

50. Ghorbanzadeh, O.; Blaschke, T.; Gholamnia, K.; Meena, S.R.; Tiede, D.; Aryal, J. Evaluation of Different Machine Learning Methods and Deep-Learning Convolutional Neural Networks for Landslide Detection. *Remote Sens.* **2019**, *11*, 196. [CrossRef]

51. Zhang, W.; Witharana, C.; Liljedahl, A.K.; Kanevskiy, M. Deep Convolutional Neural Networks for Automated Characterization of Arctic Ice-Wedge Polygons in Very High Spatial Resolution Aerial Imagery. *Remote Sens.* **2018**, *10*, 1487. [CrossRef]

52. Stewart, E.L.; Wiesner-Hanks, T.; Kaczmar, N.; DeChant, C.; Wu, H.; Lipson, H.; Nelson, R.J.; Gore, M.A. Quantitative Phenotyping of Northern Leaf Blight in UAV Images Using Deep Learning. *Remote Sens.* **2019**, *11*, 2209. [CrossRef]

53. Bernhardt, E.S.; Palmer, M.A. The environmental costs of mountaintop mining valley fill operations for aquatic ecosystems of the Central Appalachians. *Ann. N. Y. Acad. Sci.* **2011**, *1223*, 39–57. [CrossRef] [PubMed]

54. Griffith, M.B.; Norton, S.B.; Alexander, L.C.; Pollard, A.I.; LeDuc, S.D. The effects of mountaintop mines and valley fills on the physicochemical quality of stream ecosystems in the central Appalachians: A review. *Sci. Total Environ.* **2012**, *417–418*, 1–12. [CrossRef] [PubMed]

55. Fritz, K.M.; Fulton, S.; Johnson, B.R.; Barton, C.D.; Jack, J.D.; Word, D.A.; Burke, R.A. Structural and functional characteristics of natural and constructed channels draining a reclaimed mountaintop removal and valley fill coal mine. *J. N. Am. Benthol. Soc.* **2010**, *29*, 673–689. [CrossRef]

56. Hartman, K.J.; Kaller, M.D.; Howell, J.W.; Sweka, J.A. How much do valley fills influence headwater streams? *Hydrobiologia* **2005**, *532*, 91–102. [CrossRef]

57. Maxwell, A.E.; Strager, M.P. Assessing landform alterations induced by mountaintop mining. *Nat. Sci.* **2013**, *5*, 229–237. [CrossRef]

58. Miller, A.J.; Zégre, N.P. Mountaintop Removal Mining and Catchment Hydrology. *Water* **2014**, *6*, 472. [CrossRef]

59. Ross, M.R.V.; McGlynn, B.L.; Bernhardt, E.S. Deep Impact: Effects of Mountaintop Mining on Surface Topography, Bedrock Structure, and Downstream Waters. *Environ. Sci. Technol.* **2016**, *50*, 2064–2074. [CrossRef]

60. Wickham, J.; Wood, P.B.; Nicholson, M.C.; Jenkins, W.; Druckenbrod, D.; Suter, G.W.; Strager, M.P.; Mazzarella, C.; Galloway, W.; Amos, J. The Overlooked Terrestrial Impacts of Mountaintop Mining. *BioScience* **2013**, *63*, 335–348. [CrossRef]

61. Wood, P.B.; Williams, J.M. Impact of Valley Fills on Streamside Salamanders in Southern West Virginia. *J. Herpetol.* **2013**, *47*, 119–125. [CrossRef]

62. Zullig, K.J.; Hendryx, M. Health-Related Quality of Life among Central Appalachian Residents in Mountaintop Mining Counties. *Am. J. Public Health* **2011**, *101*, 848–853. [CrossRef] [PubMed]

63. Wickham, J.D.; Riitters, K.H.; Wade, T.G.; Coan, M.; Homer, C. The effect of Appalachian mountaintop mining on interior forest. *Landsc. Ecol.* **2007**, *22*, 179–187. [CrossRef]

64. Miller, A.J.; Zégre, N. Landscape-Scale Disturbance: Insights into the Complexity of Catchment Hydrology in the Mountaintop Removal Mining Region of the Eastern United States. *Land* **2016**, *5*, 22. [CrossRef]

65. WVGISTC. Resources. Available online: http://www.wvgis.wvu.edu/resources/resources.php?page=dataProductDevelopment/SAMBElevation (accessed on 29 October 2019).

66. ArcGIS Pro 2.2, ESRI, 2018. Available online: https://www.esri.com/arcgis-blog/products/arcgis-pro/uncategorized/arcgis-pro-2-2-now-available/ (accessed on 5 February 2020).

67. Li, Z.; Zhu, Q.; Gold, C.M. *Digital Terrain Modeling-Principles and Methodology*; CRC: Boca Raton, FL, USA, 2004.

68. Reed, M. *How Will Anthropogenic Valley Fills in Appalachian Headwaters Erode*; West Virginia University Libraries: Morgantown, WV, USA, 2018.

69. Gold, R.D.; Stephenson, W.J.; Odum, J.K.; Briggs, R.W.; Crone, A.J.; Angster, S.J. Concealed Quaternary strike-slip fault resolved with airborne lidar and seismic reflection: The Grizzly Valley fault system, northern Walker Lane, California. *J. Geophys. Res. Solid Earth* **2013**, *118*, 3753–3766. [CrossRef]

70. Kweon, I.S.; Kanade, T. Extracting Topographic Terrain Features from Elevation Maps. *CVGIP. Image Underst.* **1994**, *59*, 171–182. [CrossRef]

71. He, K.; Zhang, X.; Ren, S.; Sun, J. Deep Residual Learning for Image Recognition. Available online: https://www.cv-foundation.org/openaccess/content_cvpr_2016/papers/He_Deep_Residual_Learning_CVPR_2016_paper.pdf (accessed on 5 February 2020).

72. Lin, T.-Y.; Maire, M.; Belongie, S.; Bourdev, L.; Girshick, R.; Hays, J.; Perona, P.; Ramanan, D.; Zitnick, C.L.; Dollár, P. Microsoft COCO: Common Objects in Context. Available online: https://arxiv.org/abs/1405.0312 (accessed on 5 February 2020).

73. Hu, F.; Xia, G.-S.; Hu, J.; Zhang, L. Transferring Deep Convolutional Neural Networks for the Scene Classification of High-Resolution Remote Sensing Imagery. *Remote Sens.* **2015**, *7*, 14680–14707. [CrossRef]

74. Kawaguchi, K.; Kaelbling, L.P.; Bengio, Y. Generalization in Deep Learning. Available online: https://arxiv.org/abs/1710.05468 (accessed on 5 February 2020).

75. Penatti, O.A.B.; Nogueira, K. Do Deep Features Generalize from Everyday Objects to Remote Sensing and Aerial Scenes Domains. In Proceedings of the 2015 IEEE Conference on Computer Vision and Pattern Recognition Workshops (CVPRW), Boston, MA, USA, 7–12 June 2015.

76. Imgaug—Imgaug 0.3.0 Documentation. Available online: https://imgaug.readthedocs.io/en/latest/ (accessed on 30 October 2019).

77. R Core Team. *R: A Language and Environment for Statistical Computing*; R Foundation for Statistical Computing: Vienna, Austria, 2018.

78. Henderson, P.; Ferrari, V. End-to-End Training of Object Class Detectors for Mean Average Precision. In *Computer Vision–ACCV 2016*; Lai, S.-H., Lepetit, V., Nishino, K., Sato, Y., Eds.; Springer International Publishing: Berlin/Heidelberg, Germany, 2017.

79. Goutte, C.; Gaussier, E. A Probabilistic Interpretation of Precision, Recall and F-Score, with Implication for Evaluation. In *Proceedings of the Advances in Information Retrieval*; Losada, D.E., Fernández-Luna, J.M., Eds.; Springer: Berlin/Heidelberg, Germany, 2005.

80. Stehman, S.V. Comparison of Systematic and Random Sampling for Estimating the Accuracy of Maps Generated from Remotely Sensed Data. *Photogramm. Eng.* **1992**, *58*, 1343–1350.

81. Stehman, S.V. Estimating area from an accuracy assessment error matrix. *Remote Sens. Environ.* **2013**, *132*, 202–211. [CrossRef]

82. Stehman, S.V. Thematic map accuracy assessment from the perspective of finite population sampling. *Int. J. Remote Sens.* **1995**, *16*, 589–593. [CrossRef]

83. Brandtberg, T.; Warner, T.A.; Landenberger, R.E.; McGraw, J.B. Detection and analysis of individual leaf-off tree crowns in small footprint, high sampling density lidar data from the eastern deciduous forest in North America. *Remote Sens. Environ.* **2003**, *85*, 290–303. [CrossRef]

84. Zhang, Y.; You, Y.; Wang, R.; Liu, F.; Liu, J. Nearshore vessel detection based on Scene-mask R-CNN in remote sensing image. In Proceedings of the 2018 International Conference on Network Infrastructure and Digital Content (IC-NIDC), Guiyang, China, 22–24 August 2018.

85. You, Y.; Cao, J.; Zhang, Y.; Liu, F.; Zhou, W. Nearshore Ship Detection on High-Resolution Remote Sensing Image via Scene-Mask R-CNN. *IEEE Access* **2019**, *7*, 128431–128444. [CrossRef]

86. ImageNet. Available online: http://www.image-net.org/ (accessed on 4 November 2019).

87. Stereńczak, K.; Ciesielski, M.; Balazy, R.; Zawiła-Niedźwiecki, T. Comparison of various algorithms for DTM interpolation from LIDAR data in dense mountain forests. *Eur. J. Remote Sens.* **2016**, *49*, 599–621. [CrossRef]

Article

Improved Winter Wheat Spatial Distribution Extraction from High-Resolution Remote Sensing Imagery Using Semantic Features and Statistical Analysis

Feng Li [1,2,†], Chengming Zhang [3,4,*,†], Wenwen Zhang [3,†], Zhigang Xu [5], Shouyi Wang [3], Genyun Sun [6] and Zhenjie Wang [6]

1   School of Geosciences, China University of Petroleum (East China), Qingdao 266580, China; B18010057@s.upc.edu.cn
2   Shandong Provincial Climate Center, NO.12 Wuying Mountain Road, Jinan 250001, China
3   College of Information Science and Engineering, Shandong Agricultural University, 61 Daizong Road, Taian 271000, China; 2018120639@sdau.edu.cn (W.Z.); 2017110610@sdau.edu.cn (S.W.)
4   Shandong Technology and Engineering Center for Digital Agriculture, 61 Daizong Road, Taian 271000, China
5   School of Computer Science, Hubei University of Technology, 28 Nanli Road, Wuhan 430068, China; bocog@hotmail.com
6   College of Ocean and Space Information, China University of Petroleum (East China), Qingdao 266580, China; sungenyun@upc.edu.cn (G.S.); sdwzj@upc.edu.cn (Z.W.)
*   Correspondence: chming@sdau.edu.cn; Tel.: +86-139-5382-3659
†   These authors are co-first authors as they contributed equally to this work.

Received: 23 December 2019; Accepted: 4 February 2020; Published: 6 February 2020

Abstract: Improving the accuracy of edge pixel classification is an important aspect of using convolutional neural networks (CNNs) to extract winter wheat spatial distribution information from remote sensing imagery. In this study, we established a method using prior knowledge obtained from statistical analysis to refine CNN classification results, named post-processing CNN (PP-CNN). First, we used an improved RefineNet model to roughly segment remote sensing imagery in order to obtain the initial winter wheat area and the category probability vector for each pixel. Second, we used manual labels as references and performed statistical analysis on the class probability vectors to determine the filtering conditions and select the pixels that required optimization. Third, based on the prior knowledge that winter wheat pixels were internally similar in color, texture, and other aspects, but different from other neighboring land-use types, the filtered pixels were post-processed to improve the classification accuracy. We used 63 Gaofen-2 images obtained from 2017 to 2019 of a representative Chinese winter wheat region (Feicheng, Shandong Province) to create the dataset and employed RefineNet and SegNet as standard CNN and conditional random field (CRF) as post-process methods, respectively, to conduct comparison experiments. PP-CNN's accuracy (94.4%), precision (93.9%), and recall (94.4%) were clearly superior, demonstrating its advantages for the improved refinement of edge areas during image classification.

Keywords: convolutional neural network; semantic features; statistical features; Gaofen-2 imagery; winter wheat; post-processing; spatial distribution; Feicheng; China

---

1. Introduction

Determining the accurate spatial distribution of winter wheat is of great significance for agricultural production management, crop yield estimation, and national food security [1,2]. Remote sensing imagery has become the main source of such data characterizing this information. Image segmentation technology is now widely used to produce pixel-by-pixel classification results that can extract a wide range of spatial distribution information [3,4]. The specific pixel feature extraction method and the classifier both have decisive impacts on the accuracy of the classification results [5].

Effective features can improve the accuracy of the classification result. The fundamental goal of feature extraction methods is to clearly differentiate the feature value of a given object type from that of other types [6,7]. Based on statistical analysis, an effective feature extraction method can be obtained. For example, spectral indexes, which have been widely used in the classification of middle- and low-resolution remote sensing imagery, are obtained by statistical analysis of the spectral information of the pixels [8]. Commonly used methods include various vegetation indexes [9,10], the Automated Water Extraction Index (AWEI) [11], the Normalized Difference Built-up Index (NDBI) [12], and the Remote Sensing Ecological Index (RSEI) [13]. The Enhanced Vegetation Index (EVI) [8], Normalized Different Vegetation Index (NDVI) [10], and other indexes derived from NDVI are effective at extracting vegetation information and have been widely used for extracting crop spatial distributions from low-resolution remote sensing imagery. Some researchers have taken advantage of the high temporal resolution of middle- and low-spatial resolution remote sensing imagery to obtain the spectral index characteristics of a time series before extracting crop information with good results [14–16]. When applying statistical analysis technology to high-resolution remote sensing images, it is necessary to fully consider the impact of increasingly detailed pixel information on the extraction results [6,8,10].

When classifying high spatial resolution remote sensing imagery, information for both the target pixel and adjacent pixels must be considered [17,18]. Texture features are commonly used to express information related to adjacent pixels [19]; these can be extracted by methods including the gray level of co-occurrence matrix (GLCM) [20], Gabor filters [21], Markov random fields [22], and wavelet transforms [23]. As texture features can accurately express the spatial correlation between pixels, combining these with spectral features can effectively improve the classification accuracy of high-resolution remote sensing imagery [24]. The combination of traditional texture feature extraction methods can obtain more effective features [23,25].

The development of machine learning has allowed researchers to use machine learning abilities to improve pixel feature extraction. However, early machine learning methods such as neural networks [26,27], support vector machines [28,29], decision trees [30,31], and random forests [32,33] still use pixel spectral information as input. Although these methods can be effective at obtaining features, these remain single-pixel features, without utilizing the spatial relationships between adjacent pixels.

The development of convolutional neural networks (CNNs) has greatly improved feature extraction. CNNs use trained convolution kernels to form a feature extractor and then generate a feature vector for each pixel in the input image block [34,35]. Unlike other feature extraction methods, CNNs can simultaneously extract the features of a given pixel and the spatial correlation features between adjacent pixels [36,37]. Classic CNNs include fully convolutional networks (FCNs) [38], SegNet [39], DeepLab [40], RefineNet [41], and U-Net [42]. FCNs and SegNet only use high-level semantic features to generate the feature vectors of pixels, yielding very rough object edges [38,39]. DeepLab uses CRFs to post-process the segmentation results outputted by CNN; this significantly improves the quality of the results [40]. RefineNet and U-Net use low-level fine features and high-level rough features to generate pixel-level feature vectors. This strategy is conducive to the expression of multi-depth information [41,42].

RefineNet and most other classic CNNs use two-dimensional convolution. The two-dimensional convolution method is suitable for processing images with a small number of channels, such as camera images and optical remote sensing images [43,44]. Improved classic CNNs have been widely applied to remote sensing image segmentation [45] as well as target identification [46–48], monitoring [49–51], and other fields. For example, CNNs have been successfully used to extract spatial distribution information for various crops, including wheat [52], rice [53], and corn [54]. Two-dimensional convolution methods are unsuitable for processing images with many channels, such as hyperspectral remote sensing images [55]. Aiming to preserve the spectral and spatial features of hyperspectral remote sensing images, researchers use three-dimensional convolution to extract spectral–spatial information [55,56]. Because three-dimensional convolution can fully utilize the abundant spectral and spatial information of hyperspectral imagery, three-dimensional convolution has achieved remarkable success in the classification of hyperspectral images.

When remote sensing images are segmented by CNNs, the intended results can be obtained only by using appropriate feature extraction methods and classification methods according to the characteristics of the images [57,58]. CNN and traditional feature extraction methods have different advantages, and CNN cannot completely replace traditional feature extraction methods. The fusion of different feature extraction methods can improve the accuracy of the segmentation results [59].

When CNNs are used for pixel classification, the accuracy is high in the inner area but low in the edge area, resulting in rough edges [60,61]. Because the rough edges are caused by the differences in feature values between pixels of the same type, it is necessary to introduce appropriate post-processing methods to improve the accuracy of edge pixel classification [62–64]. The fully connected CRF comprehensively uses the pixel spatial distance information and the semantic information generated by the CNN to effectively improve the edge accuracy of segmentation, but the amount of data required for model calculation is too large. Researchers used recurrent neural networks [62] and convolution [63] to improve the calculation efficiency. Reference [65] comprehensively used the pixel spatial distance information and category information as constraints for network training to improve the accuracy of image segmentation results.

Object-level information is an information category commonly used in post-processing methods; it includes object shape information [65] and position information [65,66]. Using object-level information to post-process the CNN segmentation results can improve the fineness of the edges. Multiresolution segmentation algorithms [67] and patch-based learning [65,68] have been used to successfully generate image object information. Classifiers are equally important; using more powerful classifiers such as decision trees, the results obtained are better than those obtained by simple linear classifiers [69]. Methods for extracting more knowledge and more suitable post-processing methods still require further research.

In order to obtain fine winter wheat spatial distribution information from high spatial resolution remote sensing imagery using CNNs, we proposed a post-process CNN (PP-CNN) that uses prior knowledge of the similarity in color and texture between the inner and edge pixels of the target type and their differences from other types to post-process CNN segmentation results and effectively improve the accuracy of edge pixel classification (and thus overall classification). The main contributions of this work are as follows.

- PP-CNN uses confidence to evaluate the reliability of the pixel-by-pixel classification results obtained using CNN and clarifies the calculation method of confidence.
- PP-CNN proposes a new hierarchical classification strategy. Features generated by standard CNN from the large receipt fields are used for the first-level classifier; features generated from the small receipt fields are used for the second-level classifier. As this hierarchical classification strategy combines the advantage of the large receipt field and the small receipt field, it thus achieves the goal of obtaining fine edges.

2. Study Area and Data

2.1. Study Area

Feicheng is a county-level city covering 1277 km^2 in central-western Shandong Province, China (35°53′ to 36°19′N, 116°28′ to 116°59′E; Figure 1). This is an important Chinese production area for commodity grains such as winter wheat (the main local crop). The area has a warm temperate continental sub-humid monsoon climate with four distinct seasons; the average annual precipitation is 645.7 mm, the average annual temperature is 13.6 °C, and the average annual sunshine duration is

2281.3 h. Feicheng's variable terrain includes mountains along its northern border and central hills, separated by several plains and rivers; its landscape and climate are representative of many Chinese regions, making it an appropriate study area for our purposes.

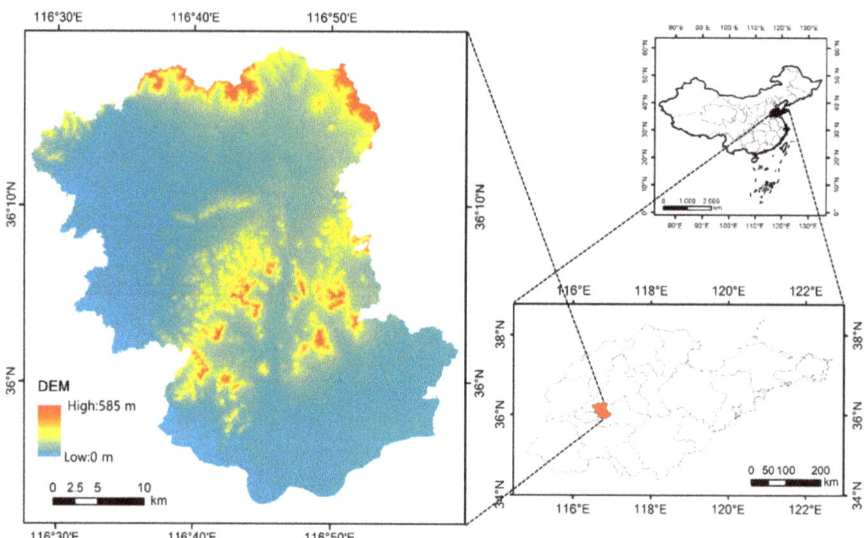

Figure 1. Location and terrain of Feicheng in Shandong Province, China.

2.2. Remote Sensing Imagery

We collected 63 Gaofen-2 (GF-2) remote sensing images as experimental data: 19 from 2017, 23 from 2018, and 21 from 2019. Each image contained multi-spectral bands (blue, green, red, and near-infrared) with 4-m resolution and panchromatic bands with 1-m resolution. After mosaicking, the images from each year covered Feicheng completely. As winter wheat has distinct characteristics during winter, all images were chosen during that time to improve the accuracy of the extraction results.

We used Environment for Visualizing Images (ENVI) software to conduct four pre-processing steps for all images. First, multi-spectral and panchromatic orthographic correction was completed using measured ground control points and the rational polynomial coefficient (RPC) model, based on 30-m resolution DEM data from the Shuttle Radar Topography Mission (https://earthexplorer.usgs.gov/).

Second, radiometric calibration involved calibrating the multi-spectral data from the original digital number (DN) value to the equivalent radiance by:

$$l = DN * g + b,\tag{1}$$

where l is the equivalent radiance obtained after conversion, DN is the DN value of the pixel, g is the calibration coefficient, and b is the calibration offset; both g and b were published by the China Resource Satellite Application Center (http://www.cresda.com/cn/).

Third, atmospheric correction used the Fast Line-of-sight Atmospheric Analysis of Hypercubes (FLAASH) module in ENVI. Given the acquisition season, latitude, and land cover, the Sub-Arctic Summer model was adopted for the atmospheric model and the Rural model was adopted for the aerosol model; the initial visibility was 40 km.

Fourth, fusion processing of the panchromatic and multi-spectral data used the Gram-Schmidt Pan Sharpening module in ENVI. After fusion, the spatial resolution of the resulting image was 1 m with the red, blue, green, and near-infrared bands; each image was 7300 × 6900 pixels.

2.3. Ground Survey Data

The main land-use types in the study area during winter include winter wheat, buildings, roads, woodland, water bodies, agricultural buildings, unplanted farmland, and other. In the GF-2 imagery, buildings, roads, water bodies, agricultural buildings, unplanted farmland, and other all have obvious color and texture features that can be easily distinguished visually. However, winter wheat and woodland (especially some evergreen trees) are more similar in color and texture. To address this, we conducted ground investigations throughout the study area from December 2017 to January 2019, obtaining 119 samples (83 winter wheat and 36 woodland) for which the coordinates, type, and other information were recorded, along with photos (Figure 2).

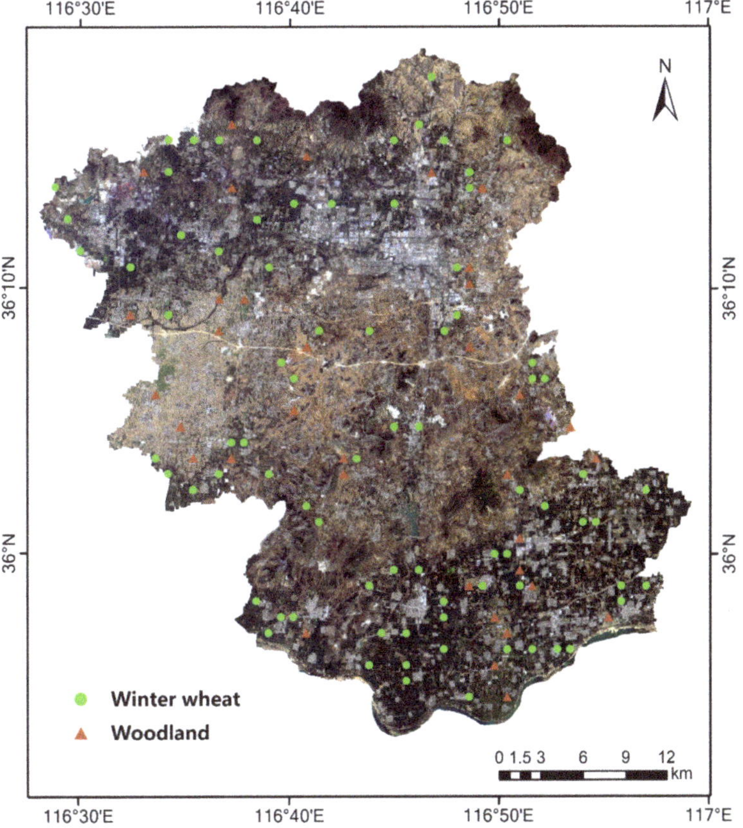

Figure 2. Distribution of ground sampling points used to distinguish winter wheat from woodland within the study area.

2.4. Labeled Image Dataset

We selected 317 non-overlapping 960×720-pixel sub-regions within the fused image (Section 2.2), then labeled each manually. After labeling was completed, each sub-region corresponded to a label file, forming an image–label pair (Figure 3). These files were single-band files in which the number of pixel rows and columns was consistent with the corresponding image. Each labeled pixel was given a category number: winter wheat (1), buildings (2), roads (3), water bodies (4), agricultural buildings (5), unplanted farmland (6), woodland (7), and other (8).

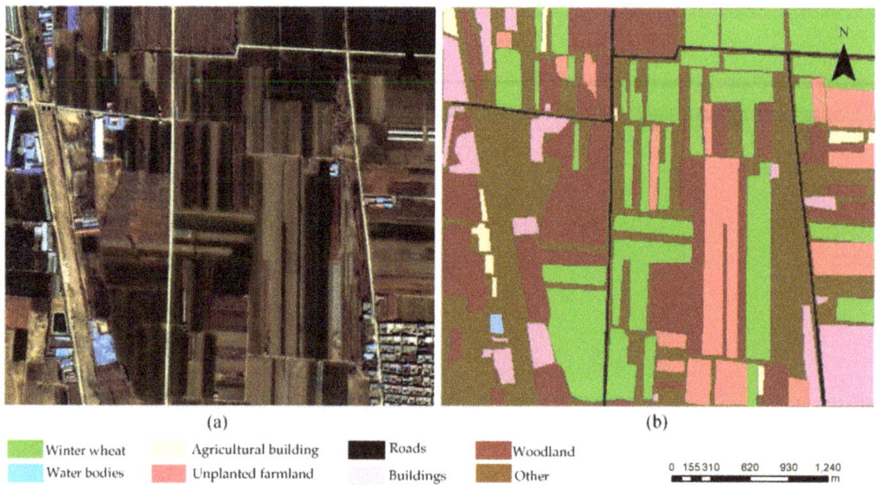

Figure 3. Example of image–label pair: (**a**) original Gaofen-2 image and (**b**) labeled image by pixel.

3. Method

Our method consisted of three steps. First, the improved RefineNet generated the initial segmentation and outputted a category probability vector for each pixel (Section 3.1). Second, these initial segmentations were statistically analyzed using manual labels as a reference to determine the confidence threshold (Section 3.2). Third, all pixels below the confidence threshold were post-processed to generate their final category label (Section 3.3).

3.1. Initial Segmentation by CNN

In the common CNN structure, the feature extractor comprises multiple overlapping convolutional layers, each of which was followed by pooling, batch normalization, and activation layers (Figure 4). The convolution layer contained several convolution kernels, most of which were 3×3. The pooling layer aggregated the features, which was beneficial for screening out features with good discrimination. The batch normalization layer was used to normalize the feature values. The activation layer adopted a nonlinear function. According to Hornik [70], the use of an activation layer facilitates better expressions of the correlation features between similar pixels and better optimization of features.

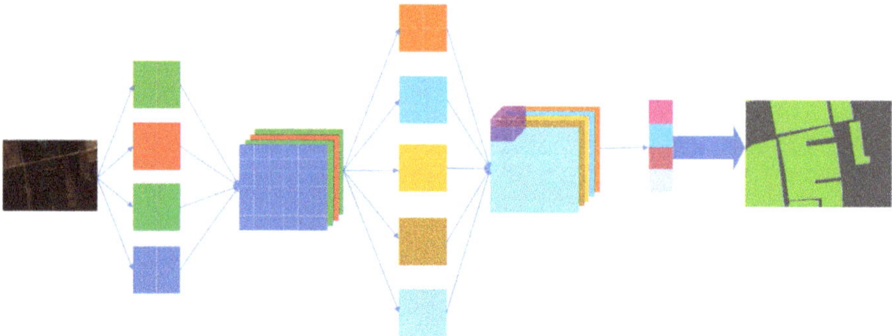

Figure 4. Basic structure of convolutional neural networks (CNNs) used for image segmentation.

The feature vector generator is generally composed of deconvolution layers, which can generate feature vectors of equal length for each pixel. These generated feature vectors are used as the inputs for the classifier to determine the pixel category. Therefore, the deconvolution performance directly determines the model performance. At present, most CNNs used for image segmentation have similar feature extractor structures; they are mainly distinguished by their feature vector generators. For example, FCN uses the interpolation method as a feature vector generator, while SegNet uses the deconvolution kernel. More recent CNNs generate pixel-level feature vectors using trained deconvolution kernels.

Unlike other CNNs, RefineNet [42] uses a new "multipath" structure to fuse fine low-level features and rough high-level features, effectively improving the distinguishability of features and greatly improving the accuracy of segmentation results. The RefineNet feature vector generator consists of four levels. Each level uses the results of both the higher-level semantic feature deconvolution and the feature extractor at the same level as the input. This multi-level feature fusion strategy improves the distinguishability of features.

Considering the superior performance of the RefineNet model, we chose this as the initial segmentation model in our study. Similar to other CNNs, RefineNet also employs the Softmax model as a classifier.

We used a modified Softmax model as a classifier. The modified SoftMax model also takes a pixel-level feature vector as the input, and calculates the probability of classifying the pixel into each category. The category corresponding to the maximum probability value was assigned as the category of the pixel. The probabilities were organized into a category probability vector. The output included the category probability vector and initial category for each pixel.

3.2. Statistics for Initial Classification Results

Statistical analysis showed that the most pixels which had been correctly classified were located inside the winter wheat planting area, and the most pixels which had been incorrectly classified were located at the edge of this area. Statistical analysis also showed that the difference between the maximum probability value and the second-highest probability value was generally large in the category probability vectors of pixels that had been correctly classified, but that it was generally small or nearly equivalent in the category probability vectors of pixels that had been incorrectly classified.

We proposed the confidence level (CL) as an indicator for the credibility of the CNN segmentation results. The CL of a category probability vector was calculated as:

$$CL = p_i - p_j,\qquad(2)$$

where p is a category probability vector, p_i is the maximum value in p, and p_j is the second-highest value in p.

Our analysis showed that the classification result for a pixel could be considered credible if the CL of this pixel was higher than the minimum confidence threshold (minCL); otherwise, it was considered non-credible. Those pixels with CL values lower than minCL required post-processing. In our study, based on the statistical analysis of the training results, 0.21 was selected as minCL.

3.3. Low-Confidence Pixel Post-Processing

3.3.1. Feature Selection

Based on the prior knowledge that the inner pixels and edge pixels in winter wheat planting areas have very similar colors and textures, and the near-infrared (NIR) band is sensitive to crops, we created a feature vector for each pixel using the red, blue, green, and near-infrared bands along with NDVI, uniformity (UNI), contrast (CON), entropy (ENT), and inverse difference (INV). NDVI was calculated following Wang et al. [10],

$$NDVI = \frac{NIR - Red}{NIR + Red}, \tag{3}$$

UNI, CON, ENT, and INV were extracted using the methods proposed by Yang and Yang, based on GLCM [23],

$$UNI = \sum_{i=1}^{q} \sum_{j=1}^{q} (g(i,j))^2, \tag{4}$$

$$CON = \sum_{n=0}^{q-1} n^2 \left\{ \sum_{i=1}^{q} \sum_{j=1}^{q} g(i,j) \right\} where \; |i - j| = n, \tag{5}$$

$$ENT = -\sum_{i=1}^{q} \sum_{j=1}^{q} (g(i,j) \log\{g(i,j)\}, \tag{6}$$

$$INV = \sum_{i=1}^{q} \sum_{j=1}^{q} \frac{g(i,j)}{1 + (i-j)^2}, \tag{7}$$

In (4)–(7), q is the gray level quantization and $g(i,j)$ is the element of GLCM. The feature vector v of each pixel had nine elements, structured as:

$$v = (red, green, blue, NIR, NDVI, UNI, CON, ENT, INV) \tag{8}$$

3.3.2. Vector Distance Calculation Method

We used the improved Euclidean distance to calculate the vector distance of the two feature vectors. The standard Euclidean distance is defined as:

$$d(x,y) = \sqrt{\sum_{i=1}^{b} (x_i - y_i)^2}, \tag{9}$$

where x and y are the feature vectors to be compared, x_i and y_i are the feature components, and b is the length of the feature vector. Smaller distances between the two feature vectors correspond to greater similarity. In the standard Euclidean distance, all elements are considered to have equal weight, without considering the influence of the aggregation degree of elements on the distance.

Statistically, among the features of the samples of the same category, a higher concentration of the value of a certain feature corresponds to stronger distinguishability of this feature and greater weight that should be assigned to this feature. Similarly, greater dispersion in the value of a certain feature corresponds to weaker distinguishability and smaller assigned weight of this feature.

Based on prior knowledge, we introduced the reciprocal of the feature value distance as the weight factor to improve the Euclidean distance, thus better reflecting the influence of feature value aggregation on the vector distance. This weight factor was calculated as:

$$w_i = \frac{1}{|max_i - min_i|}, \tag{10}$$

where i is the position number of the component in the feature vector, w_i is the weight of the component, max_i is the maximum value of the ith components of all feature vectors, and min_i is the minimum value of the ith components of all feature vectors. On this basis, the vector distance calculation formula was:

$$d(x,y) = \sqrt{\sum_{i=1}^{n} w_i (x_i - y_i)^2}, \tag{11}$$

where x and y are the feature vectors to be compared, x_i and y_i are the feature components, w_i is the weight of component i, and n is the component number of the feature vector.

3.3.3. Vector Distance Threshold Determination

- Firstly, each complete crop planting area in the training image was set as a statistical unit. The vector distance d between each pixel and other pixels was calculated individually, and the maximum vector distance d_i of the unit was recorded, where i was the number of the statistical unit.
- Secondly, the vector distance threshold (vdt) was obtained by:

$$vdt = \max_{1 \le i \ll n} d_i, \tag{12}$$

where n is the number of statistical units.

3.3.4. Low-confidence Pixel Classification

We used the following steps to optimize the results of winter wheat planting areas outputted by the improved RefineNet model:

- NDVI for each pixel was calculated;
- UNI, CON, ENT, and INV for each pixel was calculated;
- CL was calculated pixel by pixel;
- Winter wheat pixels with continuous position and CL > minCL were divided into a separate group;
- For each group, the adjacent pixels for which CL < minCL were processed individually. For a certain adjacent pixel p, we calculated the vector distances between p and each pixel in the adjacent group and then chose the minimum value as the minimum distance $mind$. If $mind < vdt$, p was re-classified as a winter wheat pixel.

3.4. Experimental Setup

We conducted a comparative experiment on a graphics workstation with a 12-GB internal graphics card and a Linux Ubuntu 16.04 operating system. TensorFlow 1.10 software was used to write the statistical analysis and post-processing code in the Python language. Using a RefineNet model from the GitHub platform, we modified the output of the SoftMax model used by RefineNet. We used this for initial segmentation and used the output as basic data for statistical analysis.

We selected the SegNet and unmodified RefineNet models as standard CNN and CRF as the post-process method for comparison with PP-CNN (Table 1). SegNet works like RefineNet, except it uses only high-level semantic features to generate feature vectors for each pixel.

Table 1. Models used in the comparative experiment.

Name	Description
PP-CNN	The proposed method
SegNet	Classifier using only high-level semantic features
SegNet-CRF	SegNet was used as the initial segmentation model, CRF was used as the post-processing method
PP-SegNet	As in PP-CNN, SegNet was used as the initial segmentation model
RefineNet	Linear model was adopted for feature fusion
RefineNet-CRF	Classic RefineNet was used as the initial segmentation model, CRF was used as the post-processing method

By comparing the results from SegNet and RefineNet, we hoped to verify that the strategy of generating features with RefineNet was better than that of generating features with SegNet. By comparing the results of SegNet-CRF, RefineNet, and RefineNet-CRF with PP-CNN, we hoped to show that post-processing could effectively improve the accuracy of segmentation results. By comparing the results of SegNet with PP-SegNet, we hoped to show that the proposed post-processing method had strong adaptability.

We applied data augmentation techniques onto the training dataset, including horizontal flip, color adjustment, and vertical flip steps. The color adjustment factors used included brightness, hue, saturation, and contrast. Each image in the training dataset was processed 10 times. All images created by the data augmentation techniques were only used in training the CNNs.

We used cross-validation techniques in the comparative experiments. Each CNN model was trained over four rounds; in each round, 87 images were selected as test images and the other images were used as training images. Each image was used at least once as the test image (Table 2).

Table 2. Percent of every category sample used in experiments.

Category	Percent of Total Samples
Winter wheat	39.00%
Agricultural buildings	0.10%
Woodland	9.01%
Buildings	19.01%
Roads	0.81%
Water bodies	0.90%
Unplanted farmland	24.12%
Other	7.05%

Table 3 shows the hyper-parameter setup we used to train our model. In the comparison experiments, the hyper-parameters were also applied to the comparison model.

Table 3. The hyper-parameter setup.

Hyper-Parameter	Value
mini-batch size	32
learning rate	0.0001
momentum	0.9
epochs	20000

4. Results

We randomly selected ten test images from the test data set and assessed their segmentation results using the SegNet, SegNet-CRF, PP-SegNet, RefineNet, RefineNet-CRF, and PP-CNN models (Figure 5).

The six methods had very similar performances within the winter wheat planting areas, with virtually no misclassifications. However, differences were obvious at the edges of these areas. PP-CNN and PP-SegNet misclassified only very small numbers of discrete pixels, while SegNet had the most errors in a more continuous pattern, with errors being more common at corners than at edges. RefineNet had significantly fewer errors than the SegNet model, with most located near corners and few in continuous patterns.

Comparing SegNet-CRF and PP-SegNet, RefineNet, and PP-CNN, respectively, it can be seen that, on the premise that the initial segmentation results are the same, the results obtained by post-processing using the proposed method are better than those obtained by using CRF. Considering that CRF has very good performance in processing camera images, this may be because the resolution of remote sensing images is lower than that of camera images, which reduces the performance of CRF. It shows that the appropriate post-processing method should be selected according to the image characteristics.

Whether using CRF or the method proposed in this paper, the accuracy of the results after post-processing is improved, which also shows the importance of post-processing methods when CNN is applied to image segmentation.

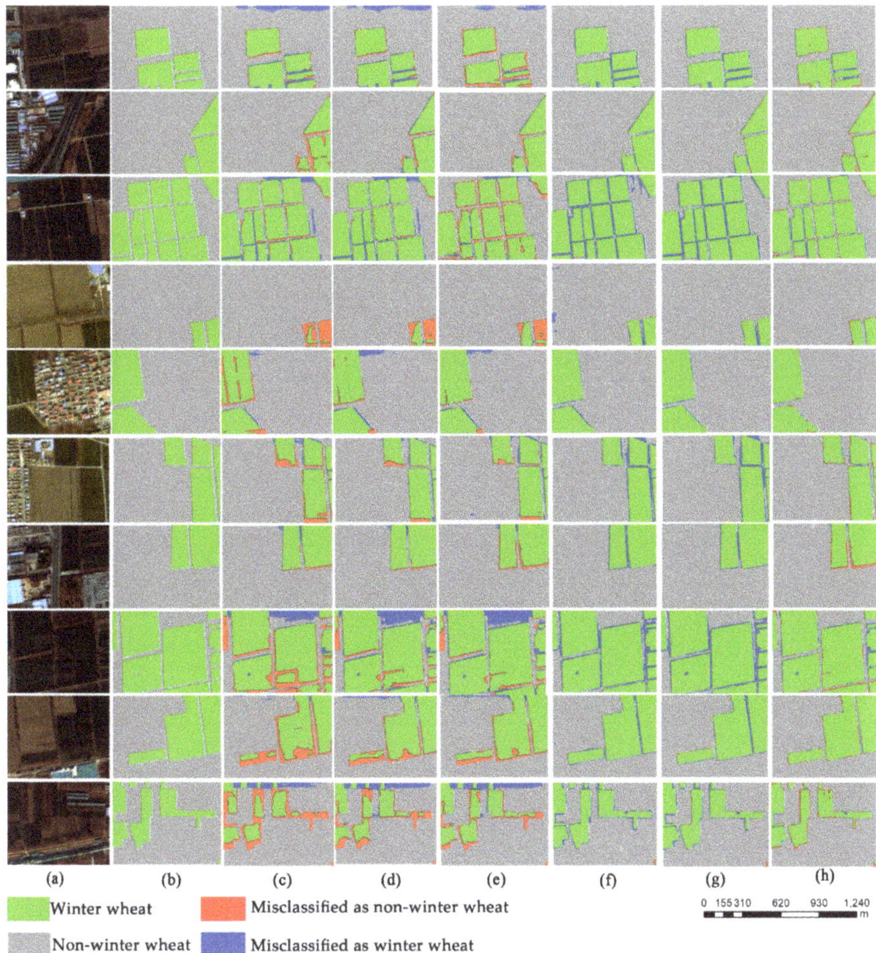

Figure 5. Comparison of segmentation results for GF-2 satellite imagery for six test images: (**a**) original image; (**b**) manually labeled image; (**c**) SegNet; (**d**) SegNet-CRF (conditional random field); (**e**) PP-SegNet; (**f**) RefineNet; (**g**) RefineNet-CRF; (**h**) PP-CNN.

We then produced a confusion matrix for the segmentation results for all four methods (Table 4), where each column represents the classification result obtained from the segmentation results and each row represents the actual category defined by manual classification. PP-CNN was clearly superior, with classification errors accounting for only 5.6%, lower than the 13.7% for SegNet, 9.8 for SegNet-CRF, 6.2% for PP-SegNet, 7.2% for RefineNet, and 5.9% for RefineNet-CRF.

We used the accuracy, precision, recall, and Kappa coefficient to evaluate the performance of the four models [45] (Table 5). The average accuracy of PP-CNN was 13.7% higher than SegNet, 7.2% higher than RefineNet, and 6.2% higher than PP-SegNet.

Table 6 shows the average time required for each method to complete the testing of one image. The proposed post-processing method requires an approximate increase of 2% in time and improves the accuracy by 7.2%. The time consumed by CRF is higher than that consumed by the proposed method because the CRF must calculate the distances between all pixel–pixel pairs for a single image, while the proposed method must calculate the distances for only a small number of pixel–pixel pairs.

Table 4. Confusion matrix for winter wheat classification.

Approach	Predicted	Winter Wheat	Non-Winter Wheat
SegNet	Winter wheat	29.6%	9.4%
	Non-winter wheat	9.9%	51.1%
SegNet-CRF	Winter wheat	31.9%	7.1%
	Non-winter wheat	8.3%	52.7%
PP-SegNet	Winter wheat	33.1%	5.9%
	Non-winter wheat	5.9%	55.1%
RefineNet	Winter wheat	32.5%	6.5%
	Non-winter wheat	6.3%	54.7%
RefineNet-CRF	Winter wheat	35.3%	3.7%
	Non-winter wheat	7.8%	53.2%
PP-CNN	Winter wheat	36.9%	2.1%
	Non-winter wheat	3.5%	57.5%

Table 5. Statistical comparison of model performance.

Index	SegNet	SegNet-CRF	PP-SegNet	RefineNet	RefineNet-CRF	PP-CNN
Accuracy	80.7%	84.6%	88.2%	87.2%	88.5%	94.4%
Precision	79.7%	83.7%	87.6%	86.6%	87.7%	93.9%
Recall	79.8%	84.1%	87.6%	86.5%	88.9%	94.4%
Kappa	0.663	0.722	0.779	0.763	0.786	0.889

Table 6. Statistical comparison of model performance.

Index	SegNet	SegNet-CRF	PP-SegNet	RefineNet	RefineNet-CRF	PP-CNN
Time [ms]	295	375	301	297	361	302

*ms: millisecond

5. Discussion

5.1. Advantages of PP-CNN

When an image is segmented pixel-wise by a CNN, the accuracy of the results is determined by the feature extractor, feature generator, and classifier. The first two use trained feature extraction rules to process remote sensing images and obtain feature vectors for each pixel, while the third uses trained classification rules to process the acquired feature vectors and determine the pixel category. Therefore, both sets of rules aim to express the main common features of similar objects. In remote sensing images, the number of inner pixels for most objects is much larger than the number of edge pixels, such that the trained rules tend to reflect the inner features, making classification errors more likely at the edge of the object.

In order to further illustrate the influence of pixel position on feature extraction, we defined the pixel blocks used to calculate feature values as three types: internal (type A), in which the pixel blocks used are all composed of the same kind of pixel; edge (type B), in which the pixel blocks used contain ~50% of other types of pixel; and corner (type C), in which the pixel blocks used contain 75% or more of other categories of pixel (Figure 6). Considering that CNNs use the same convolution kernel for feature extraction, it is clear that when the channel values of other categories of pixels in the calculated pixel blocks are different from the category of interest, the feature values of pixels in types A, B, and C will be quite different. Especially for type C pixels, if the difference between the pixel value and the neighboring pixel value is large, the calculated feature value may be closer to the feature value range of the neighboring category. This makes it difficult to effectively solve the problem of higher error occurrence in edge pixel segmentation simply by using a CNN.

Figure 6. Examples of the effect of pixel position on the extracted features; pixel boxes (red) centered on edge areas contain 50% or more non-winter wheat pixels.

Statistical analysis showed that, although crop planting areas may have clear differences between inner and edge pixels in high-level semantic features, these remain quite similar in low-level features (such as color or texture). Considering the high accuracy of inner pixel classification in our extraction results, PP-CNN clearly integrated the advantages of CNNs and statistical features, thus significantly improving the accuracy of the extraction results.

5.2. Influence of Maximum Vector Distance Threshold on PP-CNN Segmentation Results

The PP-CNN method uses color, texture, and other features to compose feature vectors and combines statistical analysis techniques to post-process the results of the CNN model, thereby providing improved spatial distribution data for winter wheat. When performing post-processing, we first calculated the vector distance between low-confidence pixels and nearby crop pixels with high confidence. We then compared the obtained vector distance with the vector distance threshold obtained by statistical analysis to determine whether low-confidence pixels could be classified as winter wheat. We took the maximum vector distance calculated by all statistical units as the vector distance threshold.

To compare the impact of vector distance thresholds on model performance, we used the minimum vector distance (method I), the average of all vector distances (method II), and the maximum vector distance (method III) as the vector distance threshold, respectively, with the results shown in Table 7.

Table 7. Comparison of PP-CNN model performance for minimum (I), average (II), and maximum (III) vector distances.

Index	I	II	III
Precision	96.1%	94.8%	93.9%
Recall	90.1%	92.5%	94.4%

Method III had the lowest precision but the highest recall rate, because using the maximum distance as the threshold means that similar pixels of other categories are classified as winter what pixels, thus reducing the accuracy. However, method I ensures maximum winter pixel extraction. Therefore, when PP-CNN is applied in the real world, researchers should choose among the three methods according to the extraction target and research goals.

5.3. Influence of Feature Strategy on Classification Results

We further compared SegNet and RefineNET by analyzing the impact of feature extraction strategies on the classification results. We selected a group of semantic features from the last layer of the SegNet and RefineNet models having the greatest difference. We divided these features into three groups of pixels: winter wheat edge, winter wheat inner, and non-winter wheat (Figure 7). Here, "inner" meant that when extracting the pixel features, only the winter wheat pixels included in the pixel block participated in the feature calculation; "edge" meant that pixels mixed with other categories participated. The feature results extracted by RefineNet were more concentrated by type and better discriminated between type; in comparison, the SegNet results were far less coherent. The feature fusion strategy adopted by the RefineNet model was clearly more conducive to improving the accuracy of the results than SegNet's strategy of only using high-level semantic features.

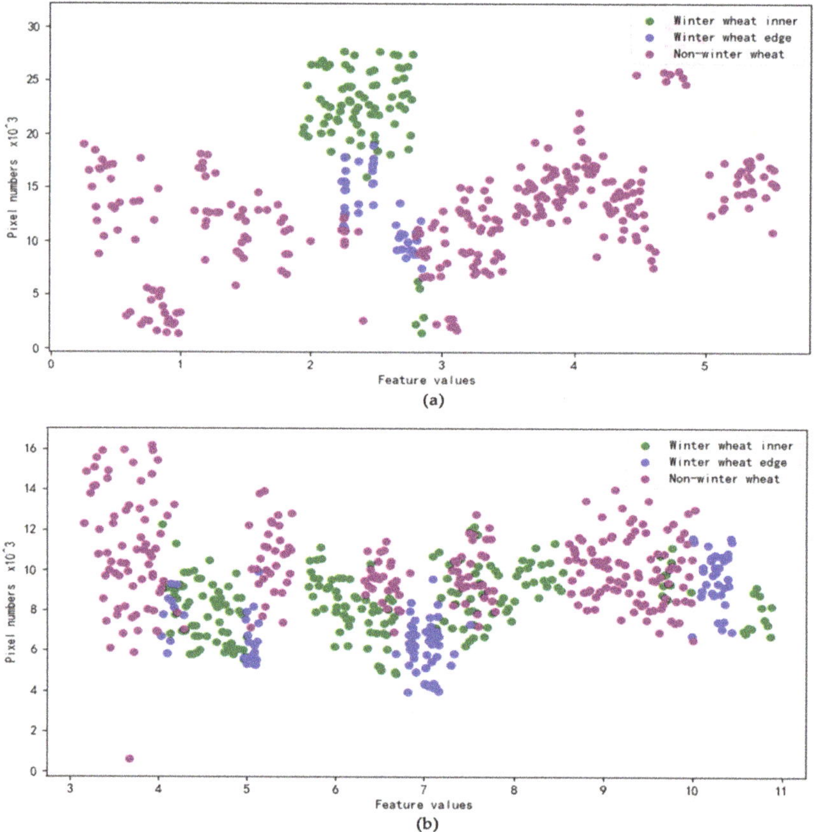

Figure 7. Statistical comparison of extracted features for (**a**) RefineNet and (**b**) SegNet.

6. Conclusions

Using CNNs to extract crop spatial distribution information from satellite remote sensing imagery has become increasingly common. However, the use of CNNs alone usually results in very rough edge areas, with a corresponding negative influence on overall accuracy. We used prior knowledge and statistical analysis to optimize winter wheat CNN extraction results, especially with regard to edge areas.

We analyzed the root cause of increased errors in CNN edge pixel classification, then used the category probability vector output to calculate the results' credibility, dividing these into high-credibility and low-credibility pixels for subsequent processing. We then optimized the accuracy of the latter's classification by analyzing the characteristics of planting area pixels using prior knowledge of the segmentation results. This new extraction strategy effectively improved the accuracy of crop extraction results.

Although the PP-CNN post-processing method proposed here was mainly established for crop extraction, it could be applied to the extraction of water, forest, grassland, and other land-use types with small internal pixel differences. However, for land-use types with larger internal differences, such as residential land, other post-treatment feature organization methods must be developed. The main disadvantage of our approach is the need for more manually classified images; future research should test the use of semi-supervised classification to reduce this dependence.

Author Contributions: Conceptualization, C.Z. and F.L.; methodology, C.Z.; software, F.L. and S.W.; validation, W.Z. and F.L.; formal analysis, C.Z., F.L., and W.Z.; investigation, W.Z. and S.W.; resources, F.L.; data curation, Z.X.; writing—original draft preparation, C.Z., F.L., and W.Z.; writing—review and editing, C.Z., F.L., W.Z., G.S., and Z.W.; visualization, Z.X., S.W., and W.Z.; supervision, C.Z.; project administration, C.Z.; funding acquisition, C.Z. All authors have read and agreed to the published version of the manuscript.

Funding: This research was funded by the Science Foundation of Shandong, grant numbers ZR2017MD018; the Key Research and Development Program of Ningxia, Grant numbers 2019BEH03008; the National Key R and D Program of China, grant number 2017YFA0603004; the Open Research Project of the Key Laboratory for Meteorological Disaster Monitoring, Early Warning and Risk Management of Characteristic Agriculture in Arid Regions, Grant numbers CAMF-201701 and CAMF-201803; the arid meteorological science research fund project by the Key Open Laboratory of Arid Climate Change and Disaster Reduction of CMA, Grant numbers IAM201801. The APC was funded by ZR2017MD018.

Conflicts of Interest: The authors declare no conflict of interest.

References

1. Atzberger, C. Advances in remote sensing of agriculture: Context description, existing operational monitoring systems and major information needs. *Remote Sens.* **2013**, *5*, 949–981. [CrossRef]
2. Zhang, J.; Feng, L.; Yao, F. Improved maize cultivated area estimation over a large scale combining MODIS–EVI time series data and crop phenological information. *ISPRS J. Photogramm. Remote Sens.* **2014**, *94*, 102–113. [CrossRef]
3. Mhangara, P.; Odindi, J. Potential of texture-based classification in urban landscapes using multispectral aerial photos. *S. Afr. J. Sci.* **2013**, *109*, 1–8. [CrossRef]
4. Wang, F.; Kerekes, J.P.; Xu, Z.Y.; Wang, Y.D. Residential roof condition assessment system using deep learning. *J. Appl. Remote Sens.* **2018**, *12*, 016040. [CrossRef]
5. Jiang, T.; Liu, X.N.; Wu, L. Method for mapping rice fields in complex landscape areas based on pre-trained convolutional neural network from HJ-1 A/B data. *ISPRS Int. J. Geo Inf.* **2018**, *7*, 418. [CrossRef]
6. El-naggar, A.M. Determination of optimum segmentation parameter values for extracting building from remote sensing images. *Alex. Eng. J.* **2018**, *57*, 3089–3097. [CrossRef]
7. Zhang, B.; Liu, Y.Y.; Zhang, Z.Y.; Shen, Y.L. Land use and land cover classification for rural residential areas in China using soft-probability cascading of multifeatures. *J. Appl. Remote Sens.* **2017**, *11*, 045010. [CrossRef]
8. Younes, N.; Joyce, K.E.; Northfield, T.D.; Maier, S.W. The effects of water depth on estimating Fractional Vegetation Cover in mangrove forests. *Int. J. Appl. Earth Obs. Geoinf.* **2019**, *83*, 101924. [CrossRef]

9. Blaschke, T.; Feizizadeh, B.; Hölbling, D. Object-based image analysis and digital terrain analysis for locating landslides in the Urmia Lake Basin, Iran. *IEEE J. Select. Top. Appl. Earth Obs. Remote Sens.* **2014**, *7*, 4806–4817. [CrossRef]

10. Wang, L.; Chang, Q.; Yang, J.; Zhang, X.H.; Li, F. Estimation of paddy rice leaf area index using machine learning methods based on hyperspectral data from multi-year experiments. *PLoS ONE* **2018**, *13*, e0207624. [CrossRef]

11. Feyisa, G.L.; Meilby, H.; Fensholt, R.; Proud, S.R. Automated Water Extraction Index: A new technique for surface water mapping using Landsat imagery. *Remote Sens. Environ.* **2014**, *140*, 23–35. [CrossRef]

12. Bhatti, S.S.; Tripathi, N.K. Built-up area extraction using Landsat 8 OLI imagery. *GISci. Remote Sens.* **2014**, *51*, 445–467. [CrossRef]

13. Xu, H.Q. A remote sensing index for assessment of regional ecological changes. *China Environ. Sci.* **2013**, *33*, 889–897. [CrossRef]

14. Wang, W.J.; Zhang, X.; Zhao, Y.D.; Wang, S.D. Cotton extraction method of integrated multi-features based on multi-temporal Landsat 8 images. *J. Remote Sens.* **2017**, *21*, 115–124. [CrossRef]

15. Kussul, N.; Lavreniuk, M.; Skakun, S.; Shelestov, A. Deep learning classification of land cover and crop types using remote sensing data. *IEEE Geosci. Remote Sens. Lett.* **2017**, *14*, 778–782. [CrossRef]

16. Beyer, F.; Jarmer, T.; Siegmann, B. Identification of agricultural crop types in northern Israel using multitemporal RapidEye data. *Photogramm. Fernerkund. Geoinf.* **2015**, *2015*, 21–32. [CrossRef]

17. Warner, T.A.; Steinmaus, K. Spatial classification of orchards and vineyards with high spatial resolution panchromatic imagery. *Photogramm. Eng. Remote Sens.* **2005**, *71*, 179–187. [CrossRef]

18. Li, L.; Liang, J.; Weng, M.; Zhu, H. A multiple-feature reuse network to extract buildings from remote sensing imagery. *Remote Sens.* **2018**, *10*, 1350. [CrossRef]

19. Reis, S.; Taşdemir, K. Identification of hazelnut fields using spectral and Gabor textural features. *ISPRS J. Photogramm. Remote Sens.* **2011**, *66*, 652–661. [CrossRef]

20. Moya, L.; Zakeri, H.; Yamazaki, F.; Liu, W.; Mas, E.; Koshimura, S. 3D gray level co-occurrence matrix and its application to identifying collapsed buildings. *ISPRS J. Photogramm. Remote Sens.* **2019**, *149*, 14–28. [CrossRef]

21. Chen, J.; Deng, M.; Xiao, P.F.; Yang, M.H.; Mei, X.M. Rough set theory based object-oriented classification of high resolution remotely sensed imagery. *J. Remote Sens.* **2010**, *14*, 1139–1155. [CrossRef]

22. Zhao, Y.D.; Zhang, L.P.; Li, P.X. Universal Markov random fields and its application in multispectral textured image classification. *J. Remote Sens.* **2006**, *10*, 123–129. [CrossRef]

23. Yang, P.; Yang, G. Feature extraction using dual-tree complex wavelet transform and gray level co-occurrence matrix. *Neurocomputing* **2016**, *197*, 212–220. [CrossRef]

24. Mao, L.; Zhang, G.M. Complex cue visual attention model for harbor detection in high-resolution remote sensing images. *J. Remote Sens.* **2017**, *21*, 300–309. [CrossRef]

25. Liu, P.H.; Liu, X.P.; Liu, M.X.; Shi, Q.; Yang, J.X.; Xu, X.C.; Zhang, Y.Y. Building footprint extraction from high-resolution images via spatial residual inception convolutional neural network. *Remote Sens.* **2019**, *11*, 830. [CrossRef]

26. Kim, S.; Son, W.J.; Kim, S.H. Double weight-based SAR and infrared sensor fusion for automatic ground target recognition with deep learning. *Remote Sens.* **2018**, *10*, 72. [CrossRef]

27. Gao, J.; Wang, K.; Tian, X.Y.; Chen, J. A BP-NN Based Cloud Detection Method For FY-4 Remote Sensing images. *J. Infrared Millim. Waves* **2018**, *37*, 477–485. [CrossRef]

28. Li, X.; Lyu, X.; Tong, Y.; Li, S.; Liu, D. An object-based river extraction method via Optimized Transductive Support Vector Machine for multi-spectral remote-sensing images. *IEEE Access* **2019**, *7*, 46165–46175. [CrossRef]

29. He, T.; Sun, Y.J.; Xu, J.D.; Wang, X.J.; Hu, C.R. Enhanced land use/cover classification using support vector machines and fuzzy k-means clustering algorithms. *J. Appl. Remote Sens.* **2014**, *8*, 083636. [CrossRef]

30. Zhang, K.W.; Hu, B.X. Individual urban tree species classification using very high spatial resolution airborne multi-spectral imagery using longitudinal profiles. *Remote Sens.* **2012**, *4*, 1741–1757. [CrossRef]

31. Sang, X.; Guo, Q.Z.; Wu, X.X.; Fu, Y.; Xie, T.Y.; He, C.W.; Zang, J.L. Intensity and stationarity analysis of land use change based on CART algorithm. *Nat. Sci. Rep.* **2019**, *9*, 12279. [CrossRef] [PubMed]

32. Santos Pereira, L.F.; Barbon, S.; Valous, N.A.; Barbin, D.F. Predicting the ripening of papaya fruit with digital imaging and random forests. *Comput. Electron. Agric.* **2018**, *145*, 76–82. [CrossRef]

33. Wang, N.; Li, Q.Z.; Du, X.; Zhang, Y.; Zhao, L.C.; Wang, H.Y. Identification of main crops based on the univariate feature selection in Subei. *J. Remote Sens.* **2017**, *21*, 519–530. [CrossRef]
34. Krizhevsky, A.; Sutskever, I.; Hinton, G. ImageNet classification with deep convolutional neural networks. *Commun. ACM* **2017**, *60*, 84–90. [CrossRef]
35. Szegedy, C.; Liu, W.; Jia, Y.Q.; Sermanet, P.; Reed, S.; Anguelov, D.; Erhan, D.; Vanhoucke, V.; Rabinovich, A. Going deeper with convolutions. In Proceedings of the 2015 IEEE Conference on Computer Vision and Pattern Recognition, Boston, MA, USA, 7–12 June 2015. [CrossRef]
36. Simonyan, K.; Zisserman, A. Very deep convolutional networks for large-scale image recognition. *arXiv* **2014**, arXiv:1409.1556.
37. He, K.M.; Zhang, X.Y.; Ren, S.Q.; Sun, J. Deep residual learning for image recognition. In Proceedings of the 2016 IEEE Conference on Computer Vision and Pattern Recognition, Las Vegas, NV, USA, 27–30 June 2016. [CrossRef]
38. Long, J.; Shelhamer, E.; Darrell, T. Fully convolutional networks for semantic segmentation. In Proceedings of the 2015 IEEE Conference on Computer Vision and Pattern Recognition, Boston, MA, USA, 7–12 June 2015. [CrossRef]
39. Badrinarayanan, V.; Kendall, A.; Cipolla, R. SegNet: A deep convolutional encoder-decoder architecture for image segmentation. *IEEE Trans. Pattern Anal. Mach. Intell.* **2017**, *39*, 2481–2495. [CrossRef]
40. Chen, L.C.; Papandreou, G.; Kokkinos, I.; Murphy, K.; Yuille, A.L. DeepLab: Semantic image segmentation with deep convolutional nets, Atrous convolution, and fully connected CRFs. *IEEE Trans. Pattern Anal. Mach. Intell.* **2018**, *40*, 834–848. [CrossRef]
41. Lin, G.S.; Milan, A.; Shen, C.H.; Reid, I. RefineNet: Multi-path refinement networks for high-resolution semantic segmentation. In Proceedings of the 2017 IEEE Conference on Computer Vision and Pattern Recognition, Honolulu, HI, USA, 21–26 July 2017. [CrossRef]
42. Ronneberger, O.; Fischer, P.; Brox, T. U-Net: Convolutional networks for biomedical image segmentation. In *Medical Image Computing and Computer-Assisted Intervention—MICCAI 2015*; Lecture Notes in Computer Science; Navab, N., Hornegger, J., Wells, W., Frangi, A., Eds.; Springer: Berlin, Germany, 2015; Volume 9351. [CrossRef]
43. Cui, W.; Wang, F.; He, X.; Zhang, D.Y.; Xu, X.X.; Yao, M.; Wang, Z.W. Multi-scale semantic segmentation and spatial relationship recognition of remote sensing images based on an attention model. *Remote Sens.* **2019**, *11*, 1044. [CrossRef]
44. Fu, G.; Liu, C.J.; Zhou, R.; Sun, T.; Zhang, Q.J. Classification for high resolution remote sensing imagery using a fully convolutional network. *Remote Sens.* **2017**, *9*, 498. [CrossRef]
45. Lu, J.Y.; Wang, Y.Z.; Zhu, Y.Q.; Ji, X.H.; Xing, Y.T.; Li, W.; Zomaya, A.Y. P_segnet and NP_segnet: New neural network architectures for cloud recognition of remote sensing images. *IEEE Access* **2019**, *7*, 87323–87333. [CrossRef]
46. Shustanov, A.; Yakimov, P. CNN design for real-time traffic sign recognition. *Procedia Eng.* **2017**, *201*, 718–725. [CrossRef]
47. Dai, X.B.; Duan, Y.X.; Hu, J.P.; Liu, S.C.; Hu, C.Q.; He, Y.Z.; Chen, D.P.; Luo, C.L.; Meng, J.Q. Near infrared nighttime road pedestrians recognition based on convolutional neural network. *Infrared Phys. Technol.* **2019**, *97*, 25–32. [CrossRef]
48. Wang, D.D.; He, D.J. Recognition of apple targets before fruits thinning by robot based on R-FCN deep convolution neural network. *Trans. Chin. Soc. Agric. Eng.* **2019**, *35*, 156–163. [CrossRef]
49. Ferentinos, K.P. Deep learning models for plant disease detection and diagnosis. *Comput. Electron. Agric.* **2018**, *145*, 311–318. [CrossRef]
50. Cheng, X.; Zhang, Y.H.; Chen, Y.Q.; Wu, Y.Z.; Yue, Y. Pest identification via deep residual learning in complex background. *Comput. Electron. Agric.* **2017**, *141*, 351–356. [CrossRef]
51. Liu, F.; Shen, T.; Ma, X.; Zhang, J. Ship recognition based on multi-band deep neural network. *Opt. Precis. Eng.* **2017**, *25*, 166–173. [CrossRef]
52. Chen, Y.; Zhang, C.M.; Wang, S.Y.; Li, J.P.; Li, F.; Yang, X.X.; Wang, Y.Y.; Yin, L.K. Extracting crop spatial distribution from Gaofen 2 imagery using a convolutional neural network. *Appl. Sci.* **2019**, *9*, 2917. [CrossRef]
53. Xie, B.; Zhang, H.K.; Xue, J. Deep convolutional neural network for mapping smallholder agriculture using high spatial resolution satellite image. *Sensors* **2019**, *19*, 2398. [CrossRef]

54. Yang, W.; Yang, C.; Hao, Z.Y.; Xie, C.Q.; Li, M.Z. Diagnosis of plant cold damage based on hyperspectral imaging and convolutional neural network. *IEEE Access* **2019**, *7*, 118239–118248. [CrossRef]
55. Li, Y.; Zhang, H.; Shen, Q. Spectral–spatial classification of hyperspectral imagery with 3D convolutional neural network. *Remote Sens.* **2017**, *9*, 67. [CrossRef]
56. Sellami, A.; Farah, M.; Farah, I.R.; Solaiman, B. Hyperspectral imagery classification based on semi-supervised 3-D deep neural network and adaptive band selection. *Expert Syst. Appl.* **2019**, *129*, 246–259. [CrossRef]
57. Alonzo, M.; Andersen, H.E.; Morton, D.C.; Cook, B.D. Quantifying boreal forest structure and composition using UAV structure from motion. *Forests* **2018**, *9*, 119. [CrossRef]
58. Zhang, C.; Pan, X.; Li, H.; Gardiner, A.; Sargent, I.; Hare, J.; Atkinson, P.M. A hybrid MLP-CNN classifier for very fine resolution remotely sensed image classification. *ISPRS J. Photogramm. Remote Sens.* **2018**, *140*, 133–144. [CrossRef]
59. Jozdani, S.E.; Johnson, B.A.; Chen, D. Comparing deep neural networks, ensemble classifiers, and support vector machine algorithms for object-based urban land use/land cover classification. *Remote Sens.* **2019**, *11*, 1713. [CrossRef]
60. Carranza-García, M.; García-Gutiérrez, J.; Riquelme, J.C. A framework for evaluating land use and land cover classification using convolutional neural networks. *Remote Sens.* **2019**, *11*, 274. [CrossRef]
61. Zhang, C.M.; Han, Y.J.; Li, F.; Gao, S.; Song, D.J.; Zhao, H.; Fan, K.Q.; Zhang, Y.N. A new CNN-Bayesian model for extracting improved winter wheat spatial distribution from GF-2 imagery. *Remote Sens.* **2019**, *11*, 619. [CrossRef]
62. Zheng, S.; Jayasumana, S.; Romera-Paredes, B.; Vineet, V.; Su, Z.; Du, D.; Huang, C.; Torr, P.H. Conditional random fields as recurrent neural networks. In Proceedings of the 2015 IEEE International Conference on Computer Vision (ICCV), Santiago, Chile, 7–13 December 2015; pp. 1529–1537.
63. Teichmann, M.T.T.; Cipolla, R. Convolutional CRFs for Semantic Segmentation. *arXiv* **2018**, arXiv:1805.04777.
64. Audebert, N.; Boulch, A.; Saux, B.E.; Lefèvre, S. Distance transform regression for spatially-aware deep semantic segmentation. *Comput. Vis. Image Underst.* **2019**, *189*, 102809. [CrossRef]
65. Fu, T.; Ma, L.; Li, M.; Johnson, B.A. Using convolutional neural network to identify irregular segmentation objects from very high-resolution remote sensing imagery. *J. Appl. Remote Sens.* **2018**, *12*, 025010. [CrossRef]
66. Mboga, N.; Georganos, S.; Grippa, T.; Lennert, M.; Vanhuysse, S.; Wolff, E. Fully Convolutional Networks and Geographic Object-Based Image Analysis for the Classification of VHR Imagery. *Remote Sens.* **2019**, *11*, 597. [CrossRef]
67. Zhao, W.; Du, S.; Emery, W.J. Object-based convolutional neural network for high-resolution imagery classification. *IEEE J. Sel. Top. Appl. Earth Obs. Remote Sens.* **2017**, *10*, 3386–3396. [CrossRef]
68. Papadomanolaki, M.; Vakalopoulou, M.; Karantzalos, K. A Novel Object-Based Deep Learning Framework for Semantic Segmentation of Very High-Resolution Remote Sensing Data: Comparison with Convolutional and Fully Convolutional Networks. *Remote Sens.* **2019**, *11*, 684. [CrossRef]
69. Mi, L.; Chen, Z. Superpixel-enhanced deep neural forest for remote sensing image semantic segmentation. *ISPRS J. Photogramm. Remote Sens.* **2020**, *159*, 140–152. [CrossRef]
70. Hornik, K. Approximation capabilities of multilayer feedforward networks. *Neural Netw.* **1991**, *4*, 251–257. [CrossRef]

 *remote sensing*

Article

Comparative Research on Deep Learning Approaches for Airplane Detection from Very High-Resolution Satellite Images

Ugur Alganci [1],*, Mehmet Soydas [2] and Elif Sertel [1]

1   Geomatics Engineering Department, Istanbul Technical University, ITU Ayazaga Campus, Civil Engineering Faculty, Sariyer, 34469 Istanbul, Turkey; sertele@itu.edu.tr
2   Institute of Informatics, Satellite Communication and Remote Sensing Program, Istanbul Technical University, ITU Ayazaga Campus, Institute of Informatics Building, Sariyer, 34469 Istanbul, Turkey; mehmets@cscrs.itu.edu.tr
*   Correspondence: alganci@itu.edu.tr; Tel.: +90-212-285-3810

Received: 18 December 2019; Accepted: 30 January 2020; Published: 1 February 2020

Abstract: Object detection from satellite images has been a challenging problem for many years. With the development of effective deep learning algorithms and advancement in hardware systems, higher accuracies have been achieved in the detection of various objects from very high-resolution (VHR) satellite images. This article provides a comparative evaluation of the state-of-the-art convolutional neural network (CNN)-based object detection models, which are Faster R-CNN, Single Shot Multi-box Detector (SSD), and You Look Only Once-v3 (YOLO-v3), to cope with the limited number of labeled data and to automatically detect airplanes in VHR satellite images. Data augmentation with rotation, rescaling, and cropping was applied on the test images to artificially increase the number of training data from satellite images. Moreover, a non-maximum suppression algorithm (NMS) was introduced at the end of the SSD and YOLO-v3 flows to get rid of the multiple detection occurrences near each detected object in the overlapping areas. The trained networks were applied to five independent VHR test images that cover airports and their surroundings to evaluate their performance objectively. Accuracy assessment results of the test regions proved that Faster R-CNN architecture provided the highest accuracy according to the F1 scores, average precision (AP) metrics, and visual inspection of the results. The YOLO-v3 ranked as second, with a slightly lower performance but providing a balanced trade-off between accuracy and speed. The SSD provided the lowest detection performance, but it was better in object localization. The results were also evaluated in terms of the object size and detection accuracy manner, which proved that large- and medium-sized airplanes were detected with higher accuracy.

Keywords: convolutional neural networks (CNNs); end-to-end detection; transfer learning; remote sensing; single shot multi-box detector (SSD); You Look Only Once-v3 (YOLO-v3); Faster RCNN

---

1. Introduction

Object detection from satellite imagery has considerable importance in areas, such as defense and military applications, urban studies, airport surveillance, vessel traffic monitoring, and transportation infrastructure determination. Remote sensing images obtained from satellite sensors are much complex than computer vision images since these images are obtained from high altitudes, including interference from the atmosphere, viewpoint variation, background clutter, and illumination differences [1]. Moreover, satellite images cover larger areas (at least 10kmx10km for one image frame) and represent the complex landscape of the Earth's surface (different land categories) with two-dimensional images with less spatial details compared to digital photographs obtained from cameras. As a result, the data size and areal coverage of satellite images are also bigger compared to natural images. In object detection studies with satellite imagery, the visual interpretation approach that benefits from experts' knowledge for the identification of different objects/targets is still widely used. The accuracy of this approach is dependent on the level of expertise and the approach is time consuming due to the manual process [2].

Several studies have been conducted on the automatic identification of different targets, such as buildings, aircraft, ships, etc., to reduce human-induced errors and save time and effort [1,3,4]. However, the complexity of the background; differences in data acquisition geometry, topography, and illumination conditions; and the diversity of objects make automatic detection challenging for satellite images. The object detection task can be considered as a combination of two fundamental tasks, which are the classification of the objects and determination of their location on the images. Studies conducted so far have focused on improving these two tasks separately or together [1,5].

In the early studies, the majority of the studies were conducted with unsupervised methods using different attributes. For example, the scale-invariant feature transform (SIFT) key points and the graph theorem were used for building detection from panchromatic images [6]. Alternatively, a wavelet transform was utilized in ship detection from synthetic aperture radar (SAR) images [7]. However, such unsupervised methods generally provided efficient results for simple structure types, and the results were successful for a limited variety of objects. Later studies focused on supervised learning methods so that objects with different constructions could be identified with high performance from more complex scenes [8,9]. The main reason behind the more successful results with supervised learning is that the learning process during the training phase is performed with previously manually labeled samples. Before the use of convolutional neural network (CNN) structures became widespread, different supervised learning methods were utilized with handcrafted features. In previous research, a spatial sparse coding bag-of-words (BOW) model was developed for aircraft recognition through the SVM classifier and the results were better than the traditional BOW model [10]. Gabor filters with SVM were used to detect aircraft, and achieved a 91% detection rate (DR) with a 7.5% false alarm rate (FAR) [11]. A deformation model representing the relation of the roots and parts of the objects by utilizing an extracted histogram of oriented gradient (HOG) features at different scales of images was developed and trained in a discriminatory manner as a framework for object recognition using a mixed model [12]. In another study, a probabilistic latent semantic analysis model (pLSA) and a K-Nearest Neighbor (k-NN) classifier with bag-of-visual-words (BoVW) was used for landslide detection [13]. In a more recent research, visually saliency and sparse coding methods were combined to efficiently and simultaneously recognize the multi-layered targets from optical satellite images [14].

In summary, the location of objects on the image is generally determined by scanning the entire image with a sliding window approach and a classifier, which may be selected from the abovementioned methods, which does the recognition task. The classifiers trained with these methods have a low size of parameters. Therefore, scanning the entire image with small strides allows an acceptable pace at object detection.

In 2012, following the remarkable success of AlexNet's [15] at the ImageNet Large-Scale Visual Recognition Challenge [16], the CNN architectures, which are also known as deep learning methods, have begun to be used in different image processing problems. After the evolution of AlexNet, which can be accepted as a milestone for deep learning, deeper architectures, such as visual geometry group (VGG) [17], GoogleNet [18], which came up with inception modules, and residual network (ResNet) [19], were developed and the error rate in the competition decreased gradually. Along with these advancements, researchers started to use CNN structures in object classification with satellite images [20–24]. Although the remote sensing images have less spatial details and complex background, these methods can achieve highly accurate results near the visual interpretation performance.

In the challenges for object detection from natural images, competitors have tended to use state-of-the-art deep learning architectures, such as PASCAL VOC (Pattern Analysis, Statistical Modelling and Computational Learning Visual Object Classes) and COCO (Common Objects in

Context) as a base network with a large amount of labeled data to beat the previous results. They applied different approaches, fine-tuning the base networks and performing some modifications; not only for increasing the accuracy of the classification part of the object detection task but also for improving the localization performance.

For the classification stage of object detection, the success of deep architectures is promising, but as they include a large number of parameters, direct use of the sliding window method, which has a high computational cost, is being abandoned. New architectures, such as the R-CNN (regions with CNN features), SPP-NET (Spatial Pyramid Pooling), Fast R-CNN, and Faster R-CNN, have emerged to overcome the computational cost disadvantage of the sliding window approach. These architectures use CNN networks as a base network for classification and solve the problem of localization by creating object candidates from the image [25–28]. With these structures, high performance and high speed could be achieved in real-time applications, such as object detection from a video stream. Additionally, object proposal approaches have become more widely used in remote sensing applications, with improvements in speed and performance [29–33]. Detection by producing an object proposal achieved successful results, but there is a trade-off between the detection performance and processing speed according to the number of proposals produced. The number and the accuracy of the object candidates could directly affect the precision of the trained model or reduce the detection speed [34].

In recent years, You Only Look Once (YOLO) [35] and Single Shot MultiBox Detector (SSD) [36] networks, which convert the classification and localization steps of the object detection task into a regression problem, can perform object detection tasks with a single neural network structure. These new methods have also overwhelmed the object proposal techniques in major competitions, such as PASCAL VOC (Pattern Analysis, Statistical Modelling and Computational Learning Visual Object Classes) [37] and COCO (Common Objects in Context) [38], where objects are detected from natural images. However, few studies have implemented these techniques on remotely sensed images. This is mainly due to an imbalanced dataset, where there are a large number of labeled natural images for detection tasks but less for remote sensing images. In addition, unlike the natural images, some of the objects that should be detected in satellite images are represented with few numbers of pixels due to the limitations of the sensor's spatial resolution. Moreover, the presence of a multi-perspective data set is important to obtain highly accurate detection results. Although it is not a difficult task to create multi-perspective data sets using natural images, this could be challenging with satellite sensors. This challenge could be partially overcome by using satellite images obtained with different incidence angles to account for perspective differences in the training phase. Lastly, the atmospheric conditions and sun angle should be considered for satellite images as they affect the spectral response of the objects. Cheng et al. proposed the creation of a large-sized dataset named NWPU-RESISC45 to overcome the lack of training samples that are derived from satellite images. Their network consists of 31,500 image chips related to 45 land classes and they reported an obvious improvement in scene classification by implementing this dataset on pre-trained networks [22]. Radovic et al. worked on the detection of aircraft from unmanned aerial vehicle (UAV) imageries with YOLO and achieved a 99.6% precision rate [39]. Nie et al. used SSD to detect the various sizes of ships inshore and offshore areas using a transfer-learned SSD model with 87.9% average precision and outperformed the Faster R-CNN model, which provided an 81.2% average precision [40]. Wang et al. tried two sizes of detectors (SSD300 and SSD512) with SAR images for the same purpose and achieved 92.11% and 91.89% precisions, respectively [41].

The main objective of this research was to develop a framework with a comparative evaluation of the state-of-the-art CNN-based object detection models, which are Faster R-CNN, SSD, and YOLO, to increase the speed and accuracy of the detection of aircraft objects. To increase the detection accuracy, VHR satellite images obtained from different incidence angles and atmospheric conditions were introduced into the evaluation. As mentioned above, the trained data availability for the satellite images is limited, which is an important drawback in CNN-based architectures. Thus, this research proposes the use of a pre-trained network as a base, and improves the training data by comparatively less number of samples obtained from satellite images. In addition, default bounding boxes are generated with six different aspect ratios at every feature map layer to detect objects more accurately and faster. The detection models were trained with a labeled dataset produced from satellite images with different acquisition characteristics and by the use of the transfer learning approach. The training processes were performed repeatedly with different optimization methods and hyper-parameters. Although the accuracy is very important, it must be taken into account that the framework needs to process very large-scale satellite images quickly. Thus, a detection flow was developed to use trained models in the simultaneous detection of multiple objects from satellite images with large coverage. This research aimed to significantly contribute to the CNN-based object detection field by:

- Improving the performance of state-of-the-art object detectors on the satellite image domain by improving the learning with a patched and augmented "A Large-scale Dataset for Object DeTection in Aerial Images (DOTA)" satellite dataset (transfer learning) and hyperparameter tuning.

- Providing a detection flow that includes the slide-and-detect approach and non-Maximum suppression algorithm, to enable fast and accurate detection on large-scale satellite images.

- Providing a comparative evaluation of object detection models across different object sizes and different IOUs and preform an independent evaluation with full-sized (large-scale) Pleiades satellite images that have different resolution specs than the training dataset to investigate the transferability.

2. Data and Methods

In this section, information about the used satellite images and data augmentation process are given initially. Next, a detailed description of the evaluated network architectures is provided. Lastly, the steps and parameterization of the training process are explained.

2.1. Data and Augmentation

The DOTA dataset was used for training and testing purposes. It is an open-source dataset for object detection purposes from remote sensing images. The dataset includes satellite image patches obtained from the Google Earth© platform, and Jilin 1 (JL-1) and Gaofen 2 (GF-2) satellites. It contains 15 object categories as airplane, ship, storage tank, baseball diamond, tennis court, basketball court, ground track field, harbor, bridge, large vehicle, small vehicle, helicopter, roundabout, soccer ball field, and swimming pool. The image sizes are in the range of 800×800 to 4000×4000. In this study, airplane detection was aimed for; therefore, 1631 images that contained 5209 commercial airplane objects were selected from the dataset. The images were split to the size of 1024×1024 patches to train Faster R-CNN and 608×608 for training SSD and YOLO-v3 detectors. The spatial resolution of the images varies in range 0.11 to 2 m and they contain various orientations, aspect ratios, and pixel sizes of the objects. In addition, the images vary according to the altitude, nadir-angles of the satellites, and the illumination conditions. The selected images were separated as 90% for training and the rest for testing. The DOTA training and test sets also include different samples in terms of airplane dimensions, background complexity, and illuminance conditions. Some image patches have some cropped objects, and some examples are black and white panchromatic images. These variations in the DOTA dataset enable the trained object detection architectures to achieve a similar performance in different image conditions (Figure 1).

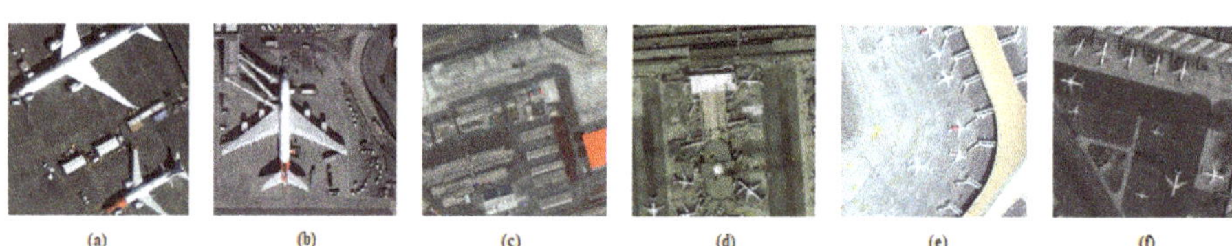

Figure 1. Patches from the DOTA test set; (**a**) cropped, (**b**) very big, (**c**) very small, (**d**) complex background, (**e**) illuminance effect, and (**f**) panchromatic samples.

Moreover, independent testing was performed with five image scenes obtained from very high-resolution pan-sharpened Pleiades 1A&1B satellite images with a 0.5-m spatial resolution and four spectral channels. In this research, Red/Green/Blue (RGB) bands of the Pleiades images were used. Satellite images were acquired in different atmospheric conditions but mostly at cloudless days and at different times in daylight. Images were collected between 2015 and 2017 in different seasons except for the winter. The images contain the Istanbul Ataturk, Istanbul Sabiha Gokcen, Izmir Adnan Menderes, Ankara Esenboga, and Antalya airport districts. They cover about a 53 km^2 area and contain 280 commercial airplanes. Properties of the Pleiades VHR images are provided in Table 1.

Table 1. Properties of Pleiades satellite images used for model construction and independent testing.

Image Area	Acquisition Date	Incidence Angle (°)	Sun Elevation Angle (°)	Surface Area (km^2)
Ataturk Airport	4/13/2017	21.16	54.43	15.03
Esenboga Airport	9/27/2015	19.4	46.42	14.97
Sabiha Gokcen Airport	4/29/2017	29.77	62.28	17.93
Antalya Airport	5/3/2017	16.88	64.61	23.42
Izmir Airport	3/28/2017	21.19	52.31	8.3

When the bounding box area distributions of aircraft samples were investigated for the DOTA training, DOTA test, and Pleiades image datasets, it was revealed that the DOTA train set includes almost the same distribution as the DOTA test set, with areas between 0 and 15,000 pixels, while it differs slightly from the samples in the large-scale Pleiades image data set. There is no object sample over 20,000 pixels in the large-scale test set and the areas of the samples are mostly between 3000 and 6000 pixels (Figure 2).

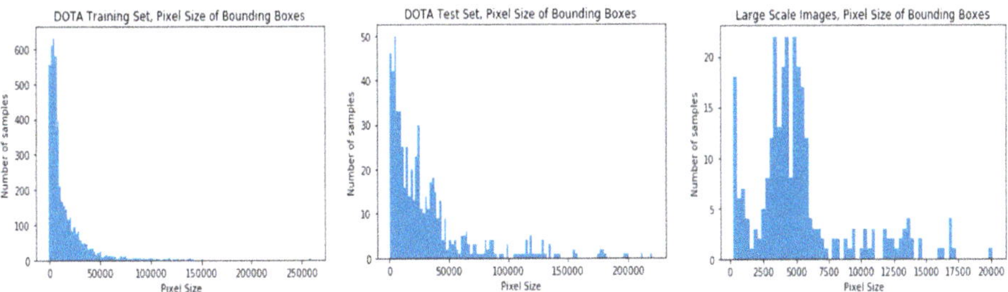

Figure 2. Number of the pixel size of the airplane bounding boxes in the dataset.

In deep architectures, a large number of labeled data is significant. Thus, the data augmentation has vital importance to cope with a lack of labeled data and to have robustness in the training step. Horizontal rotation and random cropping were applied as augmentation techniques. Besides, the image chips were scaled in HSV (hue-saturation-value) to imitate atmospheric and lighting conditions (Figure 3).

2.2. SSD Network Framework

In this sub-section, the general architecture of the SSD framework is presented initially. After, the default bounding box and negative sample generation procedures are explained. Next, the loss function and detection flow steps are presented.

(a)   (b)   (c)

Figure 3. Augmented data: (**a**) original image chip; (**b**) Rotated and randomly scaled in HSV; (**c**) Rotated, randomly cropped, and scaled in HSV.

2.2.1. General Architecture

The SSD is an object detector in the form of a single convolutional neural network. The SSD architecture works with the corporation of extracted feature maps and generated bounding boxes, which are called default bounding boxes. The network simply performs the loss calculation by comparing the offsets of the default bounding boxes and predicted classes with the ground truth values of the training samples at every iteration by the use of different filters. After that, it updates all the parameters according to that calculated loss value with the back propagation algorithm. In this way, it tries to learn best filter structures that can detect the features of the objects and generalize the training samples to reduce the loss value, thus attaining high accuracy at the evaluation phase [36].

In the SSD method, a state-of-the-art CNN architecture was used as a base network for feature extraction with additional convolution layers, which produce smaller feature maps to detect the objects with different scales. Also, SSD allows more aspect ratios for generating default bounding boxes. In this way, SSD boxes can wrap around the objects in a tighter and more accurately. Lastly, the SSD network used in this research has a smaller input size, which positively affects the detection speed compared to YOLO architectures (Figure 4). Besides, YOLO has just two fully connected layers instead of additional convolution layers. These modifications are the main differences of SSD from the YOLO and they help to obtain a higher precision rate and faster detection [36].

In the original SSD research, the VGG-16 model was used as a base network. In this research, the InceptionV2 model was used to reach a higher precision and faster detection as it has a deeper structure than the VGG models. In addition, it uses fewer parameters than VGG models thanks to the inception modules that are composed of multiple connected convolution layers [42]. As an example, GoogleNet, which is one of the first networks with inception modules, employed only 5 million parameters, which represented a 12x reduction compared to AlexNet and it gives slightly more accurate results than VGG. Furthermore, VGGNet has 3x more parameters than AlexNet [18].

2.2.2. Default Bounding Boxes and Negative Sample Generation

In the initial phase of training, it is necessary to find out which default bounding box matches well with the bounding boxes of the ground truth samples. The default generated bounding boxes vary with the location and aspect ratio, and a scale process is applied by matching each ground truth box to a default box with the best jaccard overlapping value, which should be higher than 0.5 threshold. This condition facilitates the learning process and allows the network to predict high scores for multiple overlapping default boxes, rather than selecting only those that have the maximum overlap.

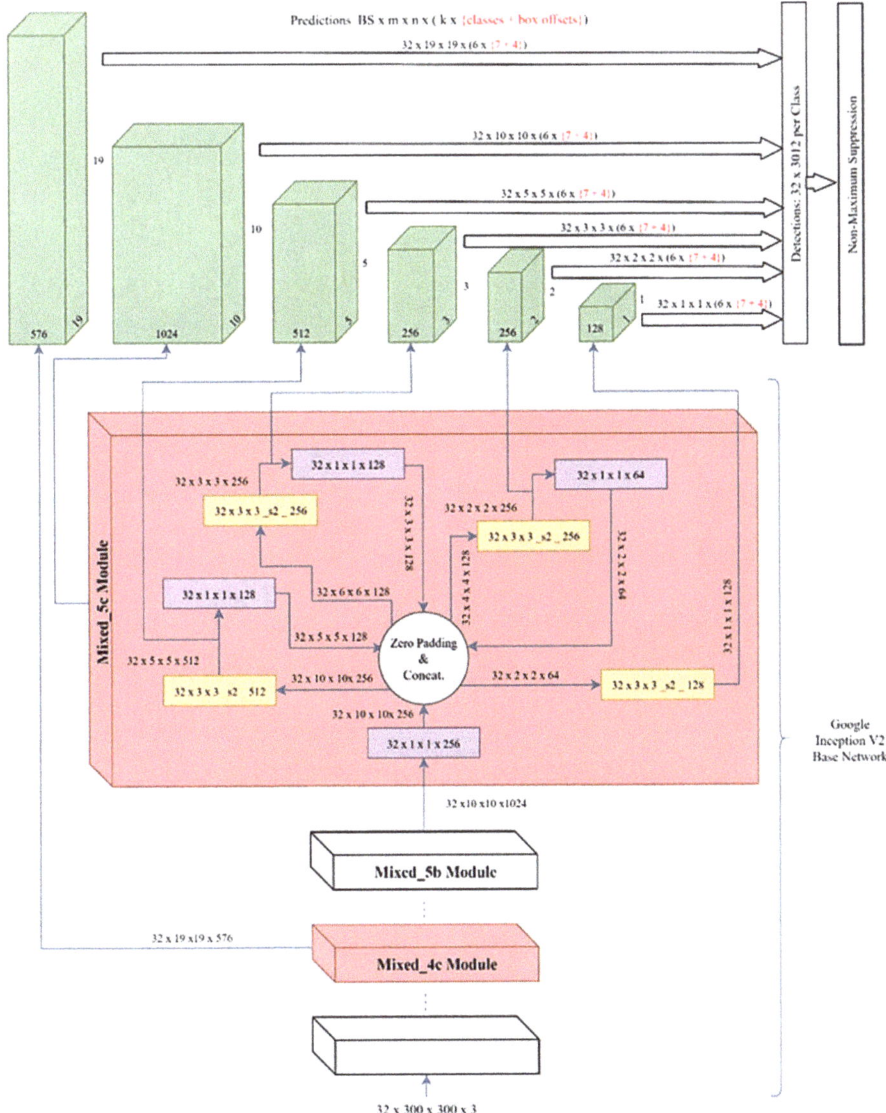

Figure 4. SSD architecture that uses Inception V2 as a base network with 32 as the batch size at training.

To handle different object scales, SSD utilizes feature maps that were extracted from several different layers in a single network. For this aim, a fixed number of default bounding boxes should be produced at different scales and aspect ratios in each region of the extracted feature maps. Six levels of aspect ratios were set supposing $a_r \in \{1, 2, 3, 1/2, 1/3\}$ and s_k is the scale of the k-th square feature map for generating default boxes. The sixth one is generated for the aspect ratio of 1 with the scale of $s'_k = \sqrt{s_k s_k + 1}$. Therefore, the width ($w_k^a = s_k \sqrt{a_r}$) and height ($h_k^a = s_k \sqrt{a_r}$) can be computed for each default box. Figure 3 illustrates how the generated default bounding boxes on a 5 × 5-feature map are represented on the input image and overlap with the possible objects (Figure 5). For this research,

150 bounding boxes were generated. At the same time, each of them represents the predictions in the evaluation step.

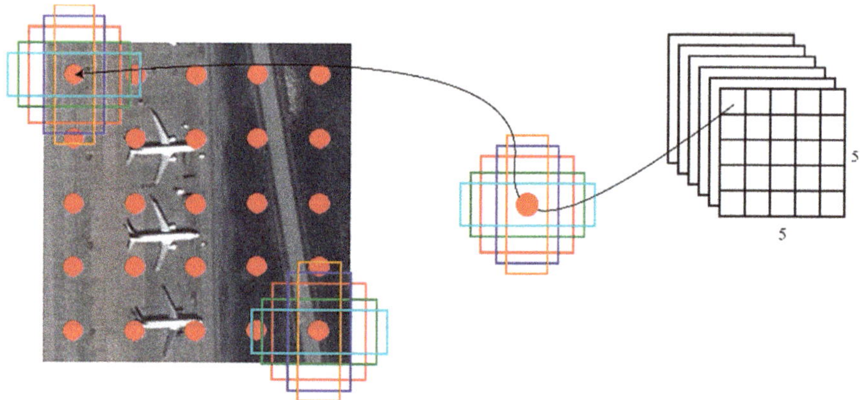

Figure 5. Illustration of the 5×5 feature map and generated default boxes with six aspect ratios.

After the matching phase, which is performed at the beginning of the training, most of the default boxes are set as negatives. Instead of using all the negative examples to protect the balance with the positive examples, the confidence loss for each default box was calculated and three of them with the highest scores were selected, so the ratio between the negatives and positives is not more than 3:1. This ratio is found to provide faster optimization and training with higher accuracy [36].

2.2.3. Loss Function

The loss (objective) value was calculated as a combination of the confidence of the predicted class scores and the accuracy of the location. The total loss value (localization loss + confidence loss) given in Equation (1) is an indication of the pairing of the *i*-th default box with *j*-th ground truth box of class p, such that $x_{ij}^p = \{1, 0\}$:

$$L(x, c, l, g) = \frac{1}{N} \left(L_{conf}(x, c) + \alpha L_{loc}(x, l, g) \right),\tag{1}$$

where N corresponds to the number of matching default boxes. If there is no match ($N = 0$), the total loss is determined as zero directly. The α value is the balance of two types of losses, and it is equal to 1 during the cross-validation phase. The localization loss is calculated as the Smooth L1 loss between the offsets of the predicted box (l) and the ground truth box (g). If the center location of the boxes denoted as cx, cy, the default boxes d, width w, and height as h:

$$L_{loc}(x, l, g) = \sum_{i \in Pos}^{N} \sum_{m \in \{cx, cy, w, h\}} x_{ij}^k smooth_{L1}(l_i^m - \hat{g}_j^m),\tag{2}$$

in which:

$$smooth_{L1}(x) = \begin{cases} 0.5x^2 & if \; |x| < 1 \\ |x| - 0.5 & otherwise, \end{cases}$$

$$\hat{g}_j^{cx} = (g_j^{cx} - d_i^{cx})/d_i^w \quad \hat{g}_j^{cy} = (g_j^{cy} - d_i^{cy})/d_i^h$$

$$\hat{g}_j^w = \log \frac{g_j^w}{d_i^w} \quad \hat{g}_j^h = \log \frac{g_j^h}{d_i^h}.$$

Additionally, the confidence loss (c) was calculated as a softmax loss of the predicted class relative to other classes:

$$L_{conf}(x,c) = -\sum_{i \in Pos}^{N} x_{ij}^{p} \log \hat{c}_{i}^{p} - \sum_{i \in Neg}^{N} \log(\hat{c}_{i}^{0}) \tag{3}$$

$$\hat{c}_{i}^{p} = \frac{\exp(c_{i}^{p})}{\sum_{p} \exp(c_{i}^{p})}. \tag{4}$$

The above-mentioned equations are detailed in Liu et al.'s article [37].

2.2.4. Detection Flow

While the usual sliding window technique slides the whole image at a fixed sliding step, it cannot ensure that the windows cover the objects exactly. Moreover, small sliding steps result in huge computation costs and larger window sizes, thus decreasing the accuracy. As shown in Figure 6, a detection flow was created with the sliding window approach and an optimized sliding step to achieve higher accuracy and faster detection [43]. As an example schema, when the sliding was performed with a 300-pixel size over a 500 × 500 pixel image patch, the objects at the edges of the window could not be detected or the bounding box offsets of them would be incorrect. To tackle this problem, an overlapping area between two windows was determined as 100 pixels, which covers the object size for this research. In the sliding process for an image with a certain overlap, k × l windows were obtained to detect by the object detector for the horizontal and vertical directions, respectively:

$$k = \left\lceil \frac{height - overlap}{ssd\ height - overlap} \right\rceil, \tag{5}$$

$$l = \left\lceil \frac{width - overlap}{ssd\ width - overlap} \right\rceil. \tag{6}$$

Figure 6. Process of the proposed detection flow of a 500 × 500 image with 100 pixels overlapping; the colored parts in the middle represents overlapping areas.

After the sliding and detection step, the non-maximum suppression (NMS) algorithm [44] (Appendix A) was used to eliminate multiple detection occurrences over an object in the overlapping regions and a score threshold was also applied to decrease the number of false detections (Figure 7).

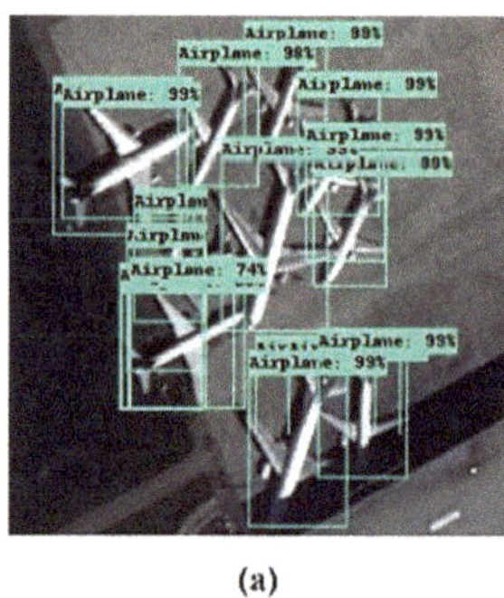
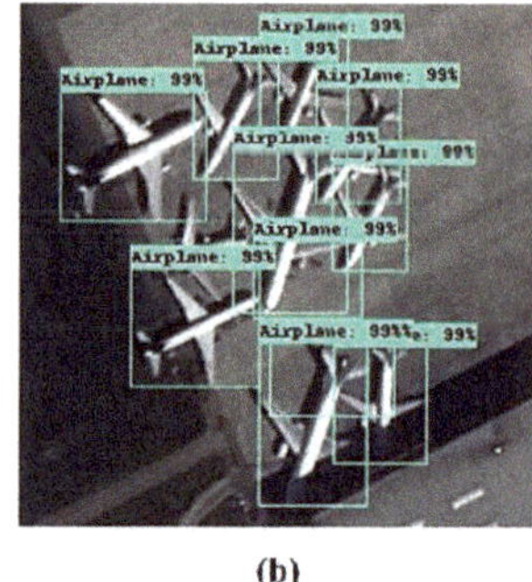

(a)   (b)

Figure 7. Detection results of occluded objects (**a**) Without the NMS algorithm, (**b**) With the NMS algorithm.

2.3. You Look Only Once (YOLO) v3 Network Framework

Yolo-v3 is grounded upon the custom CNN architecture, which is called DarkNet-53 [45]. The initial Yolo v1 architecture was inspired by GoogleNet, and it performs downsampling of the image and produces final predictions from a tensor. This tensor is obtained in a similar way as in the ROI pooling layer of the Faster R-CNN network. The next-generation Yolo v2 architecture uses a 30-layer architecture, which consists of 19 layers from Darknet-19 and an additional 11 layers adopted for object detection purposes. This new architecture provides more accurate and faster object detection results, but it often struggles with the detection of small objects in the region of interest. Moreover, it does not benefit from the advantages of the residual blocks or upsampling operations while Yolo v3 does.

Yolo v3 consists of a fully convolutional architecture, which uses a variant of Darknet, which has 53 layers trained with the Imagenet classification dataset. For the object detection tasks, an additional 53 layers were added onto it, and the improved architecture trained with the Pascal VOC dataset. With this structural design, the Yolo v3 outperformed most of the detection algorithms, while it is still fast enough for the real-time applications. With the help of the residual connections and upsampling, the architecture can perform detections at three different scales from the specific layers of the structure [45]. This makes the architecture more efficient at the detection of smaller objects but results in slower processing than the previous versions due to the complexity of the framework (Figure 8).

The shape of the detection kernel is $1 \times 1 \times (B \times (5 + C))$. In the v3 network, 3 pieces of an anchor are used for detection for each scale. Here, B is the number of the anchors on the feature map, 5 is for the 4 bounding box offsets, and one for object confidence. C is the number of categories. In the current research, the Yolo v3 network was used and the class was the only airplane, so the detection kernel shape was designed as $1 \times 1 \times (3 \times (5 + 1))$ for each scale. The first detection process was performed from the 82nd layer, as the first 81 layers downsampled the input image by the size of 32 strides. If the input image has a size of 608×608 pixels, that will be output as a feature map of 18×18 pixels in that layer. This corresponds to $18 \times 18 \times 18$ detection features being obtained from this layer. After the first detection operation, the feature map was upsampled by a factor of 2. This upsampled feature map is was with the feature map arising from the 61st layer. Then, a few 1×1 convolution operations were performed to fuse features and reduce the depth dimension. After that, the second detection is performed from the 94th layer, which returns a detection feature map of $36 \times 36 \times 18$. The same procedure was performed for the third scale at the 106th layer, which yields a feature map of the $72 \times 72 \times 18$ size. This means it produced 20,412 predicted boxes for each image. As in the SSD network, the final predictions were proposed after the NMS algorithm was applied.

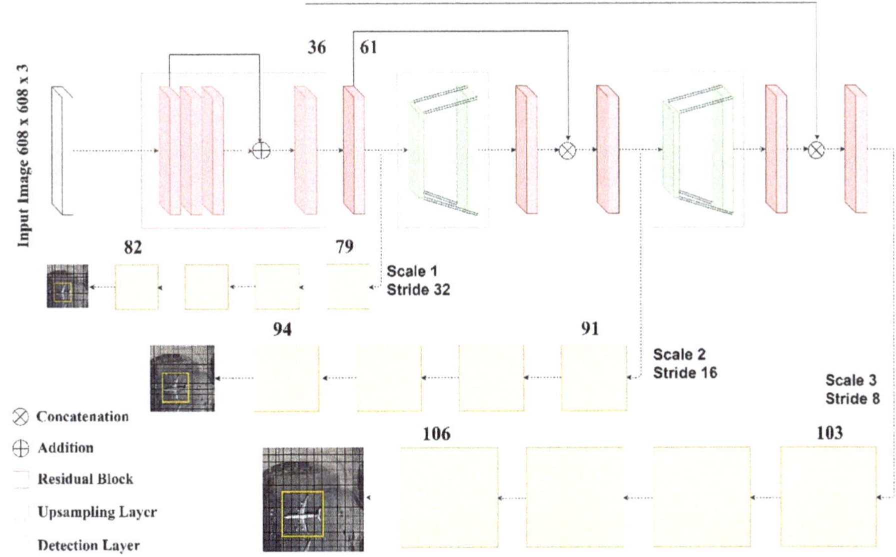

Figure 8. Yolo v3 architecture schema.

2.4. Faster R-CNN Network Framework

In this sub-section, the general architecture of the faster R-CNN framework is presented initially. After, the loss function and residual blocks are explained in detail.

2.4.1. General Architecture

Faster R-CNN is one of the most used object detection networks, which achieves accurate and quick results with CNN structures. It was initially used for nearly real-time applications, such as video indexing tasks, due to these capabilities. Faster R-CNN has developed progressively over time. The first version of it, the R-CNN, uses a selective search algorithm that utilizes a hierarchical grouping method to produce object proposals. It produces 2000 objects as the rectangular boxes, and they are passed to a pre-trained CNN model. Then, the feature maps of them are extracted from the CNN model to pass them to an SVM for classification [25].

In 2015, Girshick R. et al. [27] came up again with the Fast R-CNN, which moves the R-CNN solution one step forward. The main advantage of Fast R-CNN over the R-CNN is gained by producing the object proposals from the feature map of the CNN, instead of getting them from the complete input image. In this way, there is no need to apply the CNN process 2000 times to extract feature maps. In the next step, the region of interest (ROI) pooling is applied to ensure a standard and pre-defined output size is obtained. Finally, the future maps are classified with a softmax classifier and bounding box localizations are performed with linear regression.

In the Faster R-CNN, the selective search method is replaced by a region proposal network (RPN). This network aims to learn the proposal od an object from the feature maps. The RPN is the first stage of this object detection method. The feature maps extracted from a CNN are passed to the RPN for proposing the regions. For each location of the feature maps, k anchor boxes are used to generate region proposals. The anchor box number k is defined as 9 considering the 3 different scales and 3 aspect ratios in the original research [36]. With a size of W × H feature map, there are W × H × k anchor boxes in total, which are comprised of the negative (not object) and positive (object) samples. This means that there are many negative anchor boxes for an image, and to prevent bias occurring due to this imbalance, the negative and positive samples are chosen randomly by a 1:1 ratio (128 negative and

128 positives) as a mini-batch. The RPN learns to generate the region proposals at the training phase by utilizing these anchor boxes by comparing the ground truth boxes of the objects. The bounding box classification layer (cls) of the RPN outputs 2 × k scores whether there is an object or not for k boxes. A regression layer is used to predict the 4 × k coordinates (center coordinates of box, width, and height) of k boxes. After generation of the region proposals, the ROI pooling operation is performed as in the Fast R-CNN at the second stage of the network. Again, as in Fast R-CNN, a ROI feature vector is obtained from fully connected layers and this vector is classified by softmax to determine which category it belongs. A box regressor is applied to it to adapt the bounding box of that object. In the current research, the Faster R-CNN was used with a residual neural network (ResNet) that was comprised of 101 residual layers. This network won the COCO 2015 challenge by utilizing the ResNet-101, instead of VGG-16 in Faster R-CNN. Moreover, one additional scale parameter was added for generating the anchor boxes to detect smaller airplanes (4 scales, 3 aspect ratios, k = 12).

2.4.2. Loss Function

The loss function of the RPN network for an image was defined as:

$$L(\{p_i\}, \{t_i\}) = \frac{1}{N_{cls}} \sum_i L_{cls}(p_i, p_i^*) + \lambda \frac{1}{N_{reg}} \sum_i p_i^* L_{reg}(t_i, t_i^*),$$ (7)

where i is the index of an anchor, p_i is the prediction probability of anchor i being an object, and p_i^* is the ground truth label and it is 1 if the anchor is an object; otherwise, it is 0. L_{cls} and L_{reg} represent the classification loss, respectively, which is a log loss over two classes (object or not object) and the regression loss is the smooth L_1 function used for the t_i and t_i^* parameters. t_i is a vector representation of the predicted bounding box, and t_i^* is a ground truth bounding box associated with a positive anchor. Lastly, the parameter λ is used for balancing the loss function terms, and N_{cls} and N_{reg} are the normalization parameters of the classification and regression losses according to the mini-batch size and anchor locations.

2.4.3. Residual Blocks

When the CNN networks are designed with a deeper structure, degradation problems can occur. As the architecture becomes deeper, the layers of the higher level can act simply as an identity function. The output of them, which are the feature maps, becomes more similar to the input data. This phenomenon causes saturation in the accuracy, which is followed by fast degradation. To solve this problem, the residual blocks can be used. Instead of learning from a direct mapping of x → y with a function H(x), the residual blocks can be used to modify the function as H(x) = F(x) + x, where F(x) and x represent the stacked non-linear layers and identity function, respectively.

2.5. Training

In this work, all the experiments were performed with the Tensorflow and Keras open-source deep learning framework, which was developed by the Google research team [46]. The transfer learning technique was applied by using the pre-trained network with the COCO dataset. Additionally, fine-tuning of the parameters and extending the training set with the sample collection were performed to improve the performance as much as possible.

Through the transfer learning approach, the training was started with the implementation of the pre-trained parameters to include the useful information gathered from a previously trained network with different data used for another problem in the computer vision area. Although the COCO dataset contains natural images, the pre-trained model of the networks, which was utilized from COCO, can be used for the current research as well, because features, such as the edge, corner, shape, and color, can be implemented, which form the basis of all of the vision tasks. After starting the network with the parameters of the pre-trained model, it was fed with training examples from the produced DOTA image chips.

For Faster R-CNN, 1024 × 1024 sized image patches were used to train the model. For the RPN stage, the bounding box scales were defined as 0.25, 0.5, 1.0, and 2.0 with 0.5, 1.0, and 2.0 aspect ratios, which ensured that the network generated 12 anchor boxes for each location of the feature maps. The batch size was defined as 1 to prevent memory allocation errors. For the first attempt of the training, the process continued until 400,000 iterations, which took 72 h. The learning rate was started at 0.003 and was reduced to half of it in each further 75,000 step. The training loss did not decrease more, thus a new training process was initialized, with the learning rate corresponding to a tenth of the previous value, and the process continued for 900,000 iterations by reducing the learning rate to a quarter for each 50,000 step after the 150,000th iteration.

For the SSD network, 608 × 608-sized image patches were used for training. The sizes and aspect ratios of the default bounding boxes of each feature map layers remained the same as the original SSD research [36]. The RMSProp optimization method was used for gradient calculations with a 0.001 learning rate and 0.9 decay factor for each 25,000 iteration. The batch size was defined as 16 and the training process was continued till the 200,000th step, which took 60 h. The first attempt at the SSD training provided unsatisfactory results similar to Faster R-CNN. Therefore, a new training process initialized with a 0.0004 learning rate value and the same decay factor for each 50,000th iteration along with 450,000 iterations.

The Yolo-v3 architecture was trained with the Adam optimizer by a learning rate of 5×10^{-5} with a decay factor of 0.1 for every 3 epochs, with which the validation loss did not decrease. We used 9 anchor boxes with different sizes, 3 for each stage of the network, as in the original paper. Before the training, the bounding boxes of the entire data were clustered according to their sizes with the k-means clustering algorithm to find 9 optimum anchor box sizes. In the next step, bounding boxes were sorted from smallest to largest. For the validation purpose, 10% of the training data was split for monitoring the validation loss during the training process. The batch size was defined as 8 and the whole training was continued for 80 epochs. One epoch means the feed forward and back propagation processes are completed for the whole training dataset. Training of the Yolo-v3 took about 36 h.

3. Results and Discussion

In this section, the evaluation metrics used in this research are introduced in the first place. Secondly, the comparative results of each network according to COCO metrics across different datasets are presented and discussed. Next, the overall performance of the networks is discussed with respect to the precision, recall, and F1 scores. Lastly, a visual evaluation of the results is provided.

3.1. Evaluation Metrics

In the object detection tasks, two widely used performance metrics are the average precision (AP) and F1 score. At the training process, a detector compares the predicted bounding boxes with the ground truth bounding boxes according to the intersection over union (IOU) at each iteration to update its parameters. Generally, a 0.5 IOU ratio for each prediction at the training stage is aimed for. This means that if the network predicts an object with a bounding box that overlaps with the ground truth box by at least 50%, it is considered as a true prediction. When the localization is a matter for a computer vision task, this ratio could be set higher. In this research, the value remained as 0.5, and it was expected to detect the objects at least with this ratio at the evaluation phase. Therefore, this ratio was used for calculating the performance metrics.

The F1 score evaluation metric is used to understand the success rate by calculating the precision and recall rate. The precision is the ratio of the actual matches of all objects that are detected as matches and the recall is the ratio of the number of objects that are detected correctly to the number of all ground truth samples. Neither the recall rate nor precision rate is individually enough to measure the performance of the framework; therefore, the harmonic mean of them, which is the F1 score, was also calculated. By defining the true positive (TP) as truly detected objects, the false negative (FN) as non-detected objects, and the false positive (FP) as falsely detected objects, the precision, recall and F1 score was calculated as:

$$Precision = \frac{TP}{TP + FP},\tag{8}$$

$$Recall = \frac{TP}{TP + FN},\tag{9}$$

$$F1\ score = 2 * \frac{Precision * Recall}{Precision + Recall}.\tag{10}$$

Additionally, 12 different metrics were used to measure the characteristics and performance of the object detection algorithms with the COCO metric API (Table 2). Unless defined otherwise, the average precision (AP) and average recall (AR) weree calculated by averaging over 10 different IOU ranging from 0.5 to 0.95 with 0.05 intervals. Besides, the values where IOU is 0.5 and 0.75 were calculated for AP. AP is the average precision calculation according to all categories and IOU values. In this research, there was only one detection category, which is airplane. AR is the maximum number of detections per image, averaged over categories and IoUs. These calculations were also checked by interpreting the bounding box areas. According to COCO, objects with a size smaller than 32^2 pixels are defined as small, between 32^2 and 96^2 as medium, and more than 96^2 pixels as large. The metric calculations were performed according to all scale levels and for separate scales [47].

Table 2. COCO performance metrics with calculation rules.

Calculated for	Metric Name
AP for [IoU = 0.50:0.95 | area = all | maxDets = 100]	1. Metric
AP for [IoU = 0.50 | area = all | maxDets = 100]	2. Metric
AP for [IoU = 0.75 | area = all | maxDets = 100]	3. Metric
AP for [IoU = 0.50:0.95 | area = small | maxDets = 100]	4. Metric
AP for [IoU = 0.50:0.95 | area = medium | maxDets = 100]	5. Metric
AP for [IoU = 0.50:0.95 | area = large | maxDets = 100]	6. Metric
AR for [IoU = 0.50:0.95 | area = all | maxDets = 1]	7. Metric
AR for [IoU = 0.50:0.95 | area = all | maxDets = 10]	8. Metric
AR for [IoU = 0.50:0.95 | area = all | maxDets = 100]	9. Metric
AR for [IoU = 0.50:0.95 | area = small | maxDets = 100]	10. Metric
AR for [IoU = 0.50:0.95 | area = medium | maxDets = 100]	11. Metric
AR for [IoU = 0.50:0.95 | area = large | maxDets = 100]	12. Metric

3.2. Evaluation with COCO API

The DOTA dataset was randomly divided into two as a training and test with 90% and 10% ratios, respectively. However, there is a difference in the distribution of object scales for the training and test groups. Moreover, for the independent large-scale image set produced from Pleiades satellite images, most of the objects are in the medium range (Table 3).

Table 3. The ratios (%) of data sets according to the object scale.

Data Set/Object Scale	Small	Medium	Large
DOTA Training Set	0.06	0.52	0.42
DOTA Test Set	0.03	0.28	0.69
Large Scale Image Set	0.10	0.76	0.14

To evaluate the converge rates of the models on the training data, the performance metrics were also calculated for the DOTA training set, in addition to the test data. The performances of all trained models were examined with the COCO metric API, except for the first training attempts of SSD and Faster R-CNN as their learning rate was low (Table 4). According to the COCO metrics, the Faster

R-CNN model provided the best results when considering the mean of the precision for different IoU values. Yolo-v3 provided promising results for 0.5 IoU and above, while Faster R-CNN is better if 0.75 IOU and above is desired. For metrics 4, 5, and 6, Faster R-CNN provided the best AP result for different IOUs in small, medium, and large objects for the DOTA test set. However, in the large-scale image set, the Yolo-v3 model provided better results for small and medium objects. The reason for these results is that the architectures have different structures to learn different attributes from the training data. The seventh, eighth, and ninth metrics provide information about the recall rates for all object sizes according to the detection number per image. Similarly, the Faster R-CNN provided better results according to these metrics. When the AR results were investigated according to metrics 10, 11, and 12, it was revealed that the recall rates of Yolo-v3 are worse than the SSD for large-scale image sets. In addition, the SSD is also ahead of the Faster R-CNN for small and medium aircrafts.

Table 4. Performance of DOTA training, DOTA test, and large-scale image set according to COCO metrics.

	DOTA Training Set			DOTA Test Set			Large Scale Image Set		
	Yolo-v3	SSD (2nd)	F RCNN (2nd)	Yolo-v3	SSD (2nd)	F RCNN (2nd)	Yolo-v3	SSD (2nd)	F RCNN (2nd)
Metric 1	0.428	0.411	0.481	0.391	0.371	0.451	0.148	0.151	0.172
Metric 2	0.806	0.711	0.757	0.76	0.618	0.717	0.434	0.431	0.364
Metric 3	0.397	0.445	0.573	0.345	0.415	0.513	0.079	0.078	0.136
Metric 4	0.044	0.020	0.119	0.077	0.040	0.088	0.055	0.026	0.046
Metric 5	0.417	0.399	0.463	0.358	0.290	0.394	0.179	0.188	0.169
Metric 6	0.491	0.475	0.540	0.414	0.422	0.485	0.060	0.146	0.300
Metric 7	0.187	0.188	0.210	0.252	0.273	0.302	0.006	0.006	0.009
Metric 8	0.482	0.458	0.515	0.458	0.431	0.501	0.069	0.056	0.073
Metric 9	0.500	0.475	0.536	0.458	0.432	0.501	0.242	0.276	0.265
Metric 10	0.064	0.067	0.167	0.075	0.087	0.138	0.052	0.078	0.074
Metric 11	0.484	0.462	0.520	0.407	0.355	0.427	0.275	0.295	0.262
Metric 12	0.568	0.536	0.596	0.487	0.471	0.539	0.198	0.305	0.408

The fact that the DOTA training and test performances are similar for the three architectures indicates that the models can successfully learn the object characteristics from the DOTA dataset. However, when these results were compared with the results from the large-scale Pleiades image set, there is a big performance gap. The main reasons behind the performance gap are that the dimensions of the aircrafts inside the large-scale image set are distributed differently than the DOTA dataset and large-scale image sets contain different types of aircraft (Figure 9).

With the COCO metric API, precision–recall curves were plotted according to the object size, and the differences between these curves provides valuable insights about the detection efficiencies of models. As presented in Figures A1–A3, precision–recall (PR) curves were plotted for small-, medium-, large-scale objects and for all object sizes across the three models. The evaluations were performed for the DOTA test set and large-scale image set separately. The orange area out of the curves represents the false negative (FN) portion of the evaluated data set. In other words, it is the PR after all errors are removed. The purple area presents the falsely detected objects, which are the backgrounds in the dataset (BG). The blue area presents the localization errors of the predicted boxes (Loc) and indicates that the PR curve is a 0.1 IOU value. The white area shows the area under the precision–recall curve, which is comprised of the prediction with IOU above 0.75 (C75). Lastly, the grey area represents the detections with IOU above 0.5 (C50). The brown area (Sim) is the PR curve after the super-category false positives are removed. Green area (Oth) is the PR after all class confusions are removed. As this research does not include a super-category or any other category, these curves do not exist in the provided plots.

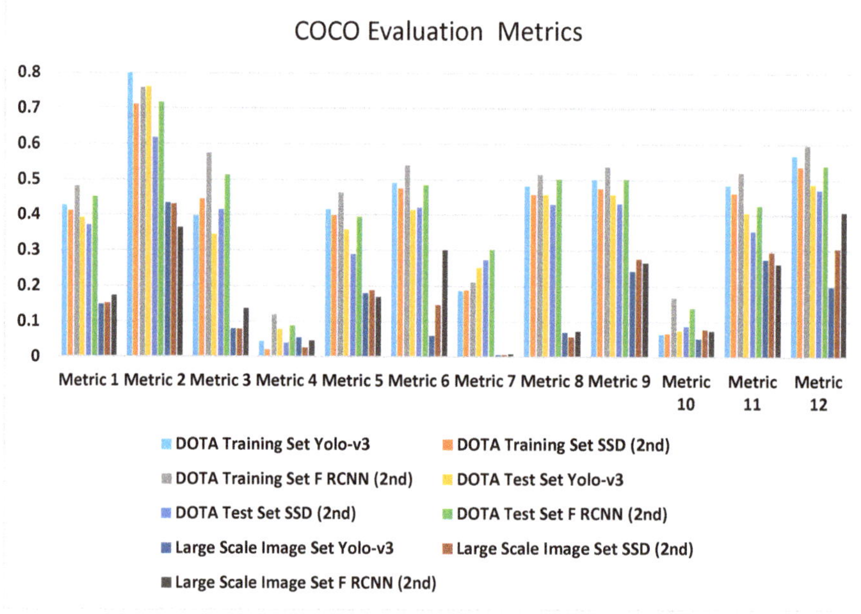

Figure 9. Graphic representation of COCO evaluation metrics.

When the PR plots were investigated together with the AP metrics, which is presented in Table 5, it was observed that the large-sized aircrafts were detected better for the DOTA test and large-scale image set. Additionally, it can be asserted that all of the networks detect better with the IOU above 0.5 when the margin area was compared with IOU above 0.75. The localization error for the DOTA test set is smaller compared with large-scale images. For the Yolo-v3 network, nine optimum anchor sizes were selected by clustering the whole DOTA training samples according to the object sizes; however, the pixel sizes of objects are much smaller in the large-scale image set. Besides, the number of objects in the DOTA training set is much more than the large-scale image set, which possibly resulted in unbalanced object sizes between the two datasets. Lastly, the sizes of the anchor boxes that were selected with the k-means algorithm in the training phase did not match with the optimum size for the large-scale image dataset. This condition could be an explanation for the higher localization errors observed for the large-scale image set.

Table 5. Average precision (AP) metrics of all test sets for all networks according to COCO metric API.

Dataset	DOTA Test Set				Large Scale Image Set			
AP	C75	C50	Loc	Bg	C75	C50	Loc	Bg
Yolo-v3 All	0.34	0.76	0.78	0.79	0.08	0.43	0.79	0.79
Yolo-v3 Large	0.36	0.80	0.82	0.83	0.02	0.22	0.69	0.69
Yolo-v3 Medium	0.33	0.67	0.70	0.71	0.10	0.51	0.85	0.85
Yolo-v3 Small	0	0.25	0.25	0.25	0.04	0.11	0.48	0.48
SSD All (2nd)	0.41	0.61	0.64	0.67	0.07	0.43	0.74	0.78
SSD Large (2nd)	0.48	0.68	0.70	0.70	0.06	0.41	0.76	0.77
SSD Medium (2nd)	0.29	0.52	0.59	0.63	0.11	0.51	0.77	0.80
SSD Small (2nd)	0	0.13	0.13	0.25	0	0.15	0.41	0.62
Faster R-CNN All (2nd)	0.51	0.71	0.72	0.73	0.13	0.36	0.81	0.81
Faster R-CNN Large (2nd)	0.56	0.77	0.78	0.79	0.22	0.52	0.82	0.82
Faster R-CNN Medium (2nd)	0.42	0.60	0.61	0.61	0.13	0.37	0.83	0.83
Faster R-CNN Small (2nd)	0.06	0.17	0.17	0.25	0.04	0.14	0.55	0.59

Although the SSD network provided the worst performance for the test sets, it is much effective in the localization of the objects when compared with the other networks. Moreover, the Yolo-v3 network provided better detection of the small objects with 0.5 IOU. Additionally, Faster R-CNN can detect the small objects of the DOTA test set with 6% AP, while the other networks cannot, and for the small objects of the large-scale image set, it has a similar performance with the Yolo-v3 (Figure 10).

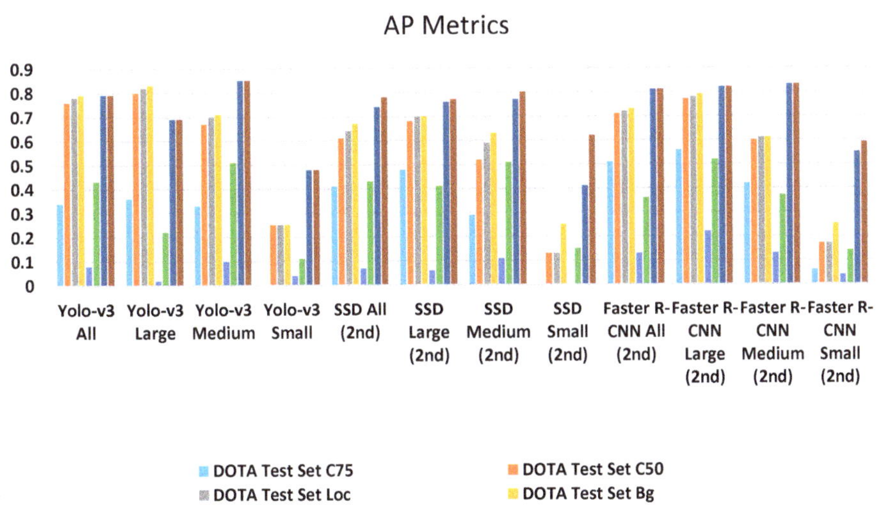

Figure 10. Graphic representation of COCO AP metric results.

3.3. Evaluation with Accuracy Metrics

As the last evaluation step, the precision, recall, and F1 scores were calculated for all networks and all datasets with a 0.5 IOU threshold. To observe how the models can generalize the training data, these metrics were calculated for the training data as well. Moreover, the first attempt of the training for SSD and Faster R-CNN were added to the evaluation in this step to observe the improvements gained by second training with modified parameters for these models. According to the results presented in Table 6, the Faster R-CNN with the second training parameter set provided the highest precision, recall, and F1 scores for both the DOTA and large-scale test sets. Moreover, it took second place after YOLO-v3 with slight differences for the DOTA training set, which indicates good generalization and learning through the training phase. The YOLO-v3 performance is ranked as second for both test sets, with comparatively low recall values, which is a sign of an increment in non-detected objects. SSD with the second training parameter set provided the lowest scores for test sets as well as the training set, which indicates a low level of generalization and learning process (Figure 11). When the results of SSD and Faster R-CNN with the first training parameter set were compared with the second parameter set, an obvious improvement was observed with the modified parameters, indicating the importance of parameter selection in the training phase.

Table 6. Precision, recall, and F1 score of all datasets.

Dataset	DOTA Training Set			DOTA Test Set			Large Scale Image Set		
Method/Metric	Prec.	Rec.	F1	Prec.	Rec.	F1	Prec.	Rec.	F1
Yolo-v3	0.99	0.95	0.97	0.96	0.89	0.92	0.97	0.87	0.91
SSD (1st)	0.99	0.44	0.61	0.96	0.43	0.59	0.65	0.36	0.46
SSD (2nd)	0.89	0.73	0.80	0.86	0.68	0.76	0.87	0.65	0.74
Faster R-CNN (1st)	0.99	0.51	0.67	0.99	0.47	0.63	0.74	0.41	0.52
Faster R-CNN (2nd)	0.97	0.92	0.95	0.98	0.89	0.93	0.98	0.92	0.94

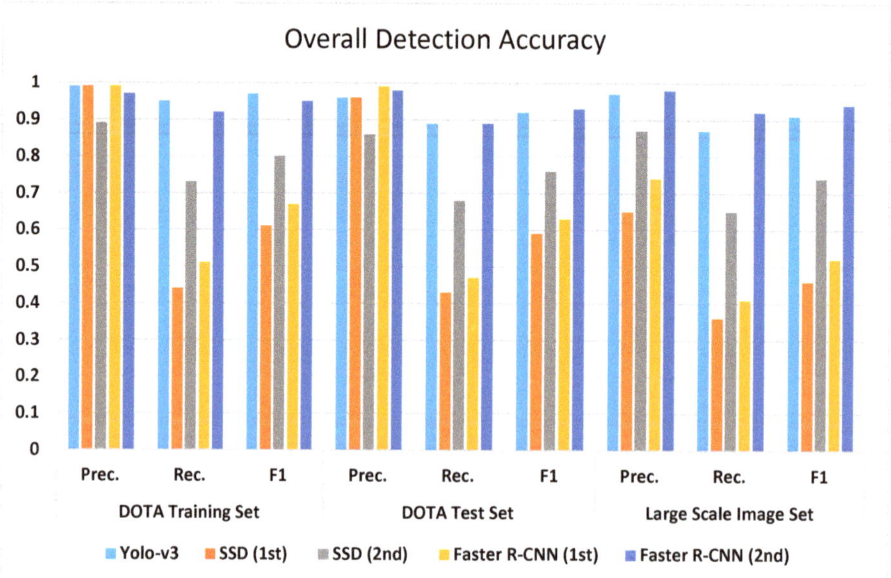

Figure 11. Graphic representation of the precision, recall, and F1 scores.

3.4. Visual Evaluation

The detection results from the DOTA test set and Pleiades large-scale image set were interpreted visually to assess the performance of algorithms. According to the detection results of the DOTA test set, Yolo-v3 is more successful than the other networks. Although the selected samples provided in Figure 12 include different sized aircrafts, and the image patches have illuminance differences, background complexities, and different band information, the Yolo-v3 provided a lesser amount of missing objects, while SSD provided the worst results.

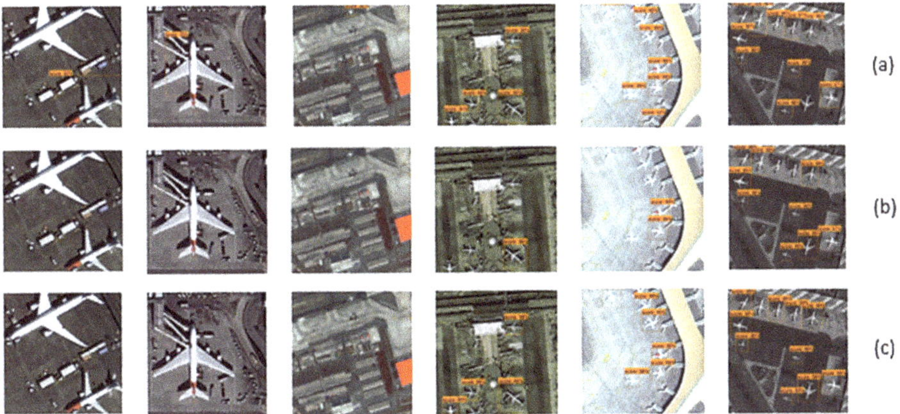

Figure 12. Detection results of some of the DOTA test set patches (**a**) Yolo-v3, (**b**) SSD, (**c**) Faster R-CNN.

The aircraft detection from the large-scale Pleiades image set, which covers a 53-km^2 area in total, lasted around 37 s for SSD, 97 s for Yolo v3, and 102 s for the Faster R-CNN with the proposed detection flow approach. The results from Sabiha Gokcen Airport and Antalya Airport are provided in Figures A4 and A5, respectively. In Figure A4, non-detected objects are observable in the center and northern part of the image for SSD. YOLO-v3 missed only two airplanes for that image scene, however, it faced multiple detections at the bottom left part of the image scene where several airplanes are grouped. Faster R-CNN provided a balanced performance with a high detection rate and good localization of the objects. For Figure A5, similar results were achieved, and some false detections were also observed in the SSD case.

4. Conclusions

This article presented a comparative evaluation of state-of-the-art CNN-based object detection models for determining airplanes from satellite images. The networks were trained with the DOTA dataset and the performance of them was evaluated with both the DOTA dataset and independent Pleiades satellite images. The best results were obtained with the Faster R-CNN network according to the COCO metrics and F1 scores. The Yolo-v3 architecture also provided promising results with a lower processing time, but SSD could not converge the training data well with low iterations. All of the networks tended to learn more with different parameters and more iterations. It can be asserted that Yolo-v3 has a faster convergence capability when compared with the other networks; however, the optimization methods also play an important role in the process. Although SSD provided the worst detection performance, it was better in object localization. The imbalance between the object sizes and the diversities also affected the results. In the training of deep learning architectures, imbalances should be avoided, or the categories should be divided into finer grains, such as airplanes, gliders, small planes, jet planes, and warplanes. In summary, transfer learning and parameter tuning approaches on pre-trained object detection networks provided promising results for airplane detection from satellite images. Besides, the proposed slide and detect and non-maximum suppression-based detection flow enabled algorithms to be run on full-sized (large-scale) satellite images.

For future work, the anchor box sizes can be defined by weighted clustering according to the sample size of the datasets. Moreover, all of the networks can be used together to define the offsets of the bounding boxes by averaging the predicted bounding boxes, to prevent false positives and increase the recall ratio. In this way, the localization errors could be decreased as well. Finding a way to use the ensemble learning methods for object detection architectures could be another improvement. In addition, the object detection networks often use R, G, and B bands, as they are mostly developed for natural images. However, satellite imageries can contain more spectral bands. Therefore, further studies are planned to integrate the additional spectral bands of the satellite images, to increase the number of labels and train the model more accurately.

Author Contributions: Formal analysis, M.S.; Methodology, U.A., M.S. and E.S.; Project administration, E.S.; Software, M.S.; Validation, U.A. and E.S.; Visualization, U.A. and M.S.; Writing—original draft, U.A., M.S. and E.S. All authors have read and agree to the published version of the manuscript.

Funding: This research received no external funding.

Acknowledgments: Authors acknowledge the support of Istanbul Technical University—Center for Satellite Communications and Remote Sensing (ITU-CSCRS) by providing the Pleiades satellite images.

Conflicts of Interest: The authors declare no conflict of interest.

Appendix A

   Algorithm A1 Non-Maximum Suppression

```
1:   procedure NMS
2:       input:
3:           d = {b_d = [l_x, l_y, l_w, l_h], c_d}: bounding box offsets
4:           and confidence scores of detection list,
5:           t: iou threshold,
6:           t_s: score threshold.
7:       output:
8:           f_d = {b_f = [l_x, l_y, l_w, l_h], c_f}: final detection list.
9:           if (size of d < 2)
10:              return
11:          %sort detections in descending order according to c_d
12:          d ← sort(c_d)
13:      f_d = d[0]
14:          for all b_d, c_d in d do
15:          %calculate iou between b_d, b_f
16:              iou = iou (b_d, b_f)
17:          idxs ← where (iou < t) in d
18:              if (size of idxs == 0)
19:                  if (c_d > t_s)
20:                      f_d = stack (f_d, {b_d, c_d})
21:          end for
22:      end procedure
```

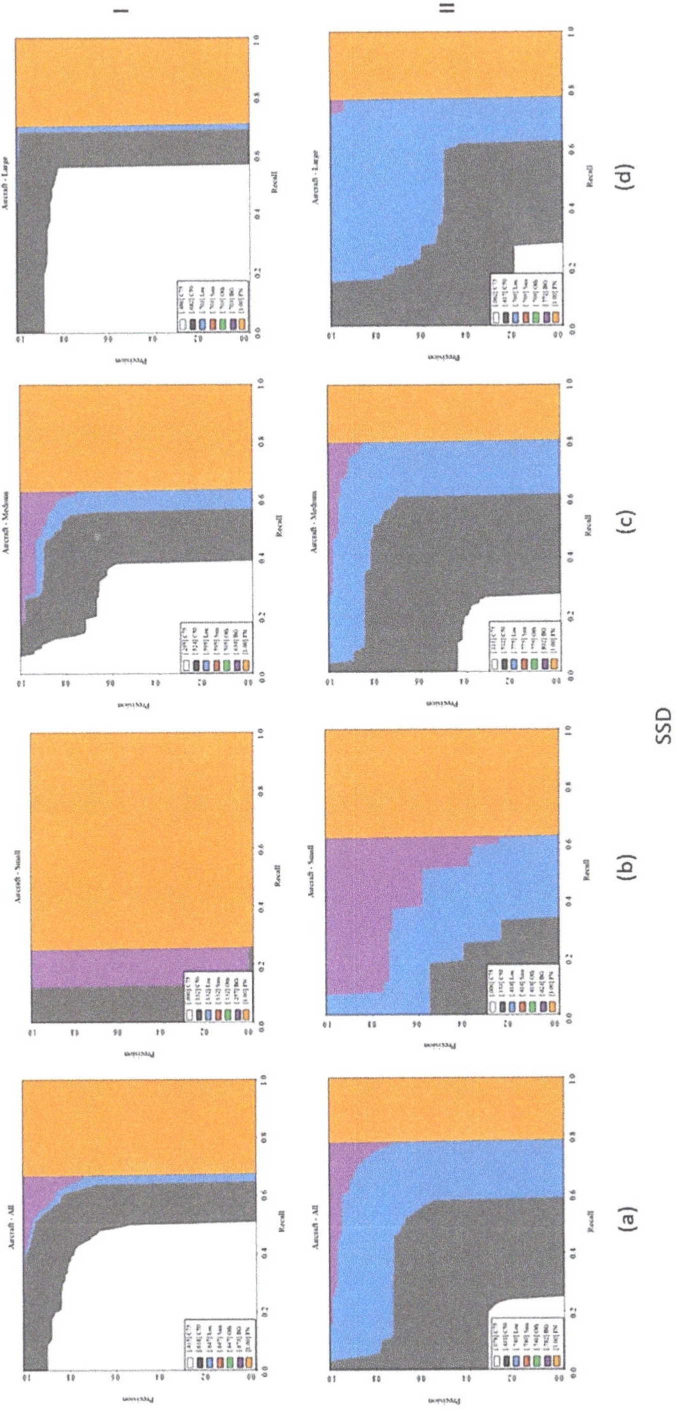

Figure A1. SSD Precision–recall curves for, (**a**) all object sizes, (**b**) small, (**c**) medium, (**d**) large object size obtained from, (I) DOTA test data and (II) Pleiades large-scale test image data.

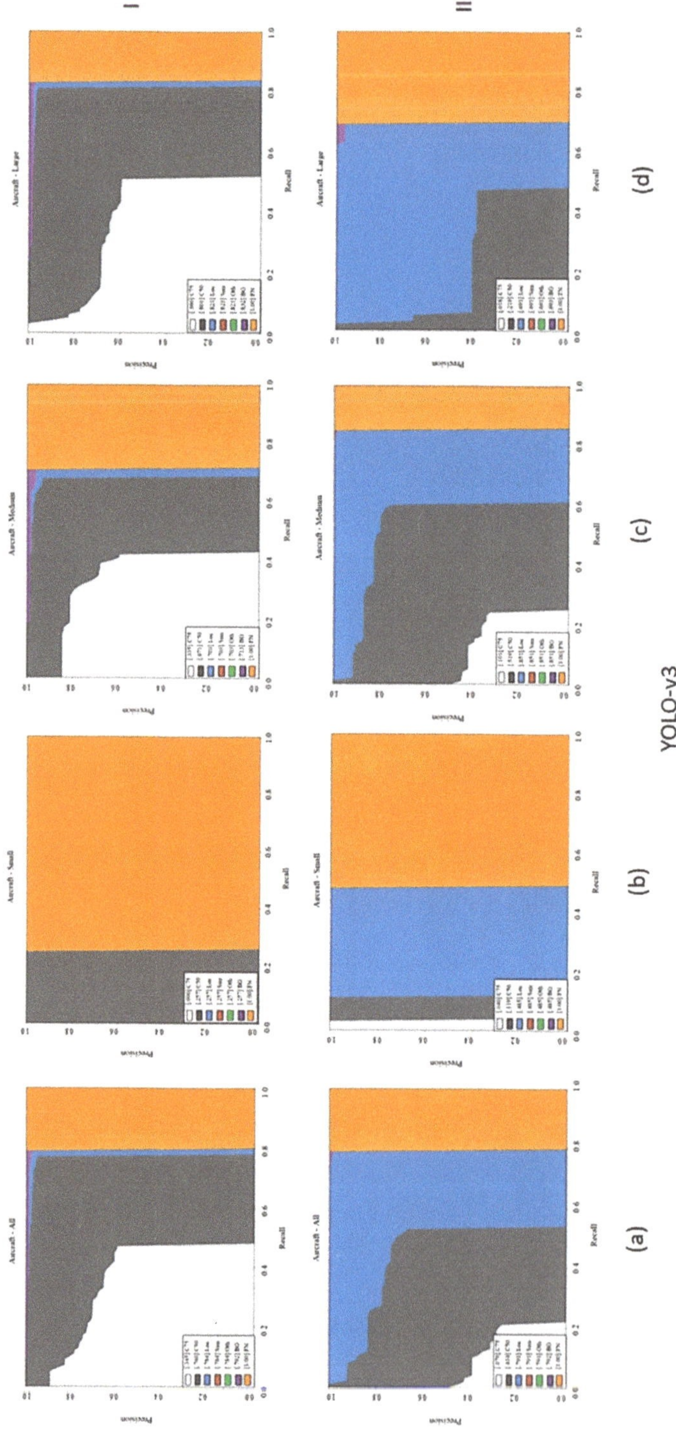

Figure A2. YOLO—V3 Precision–recall curves for, (**a**) all object sizes, (**b**) small, (**c**) medium, (**d**) large object size obtained from, (I) DOTA test data and (II) Pleiades large-scale test image data.

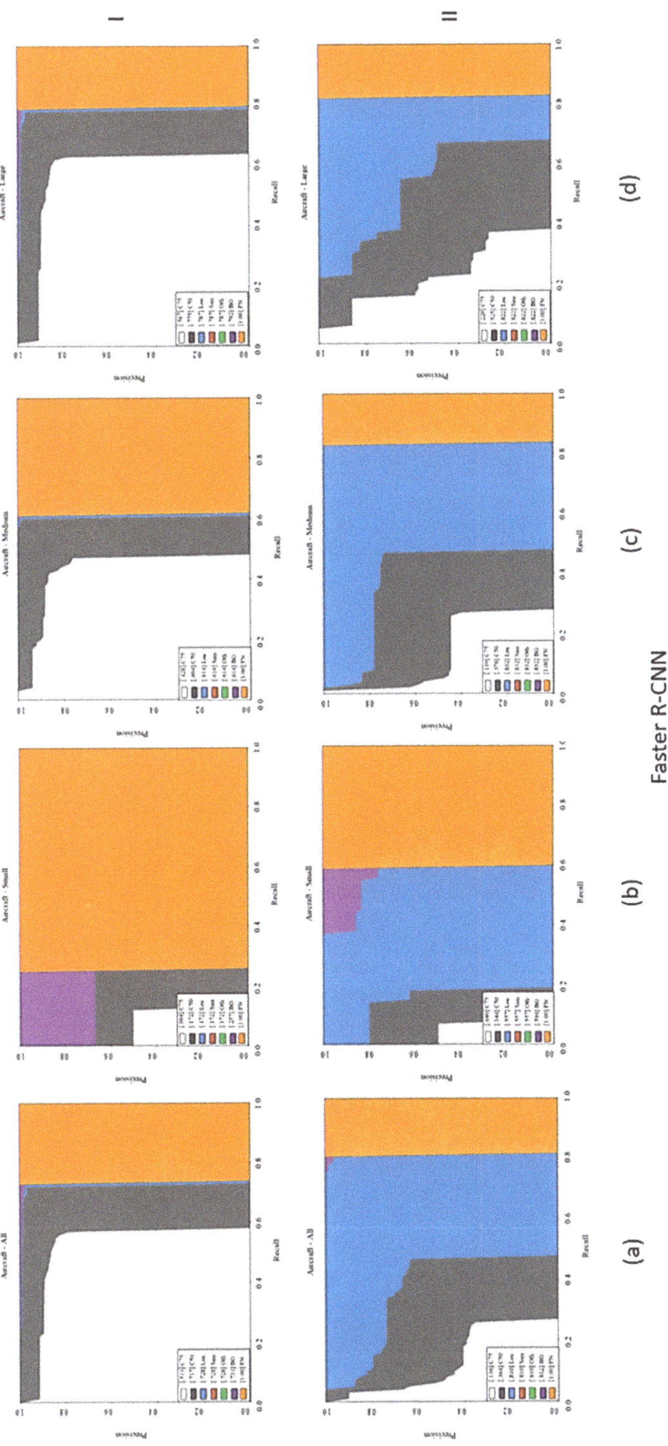

Figure A3. Faster R-CNN Precision–recall curves for, (**a**) all object sizes, (**b**) small, (**c**) medium, (**d**) large object size obtained from, (I) DOTA test data and (II) Pleiades large-scale test image data.

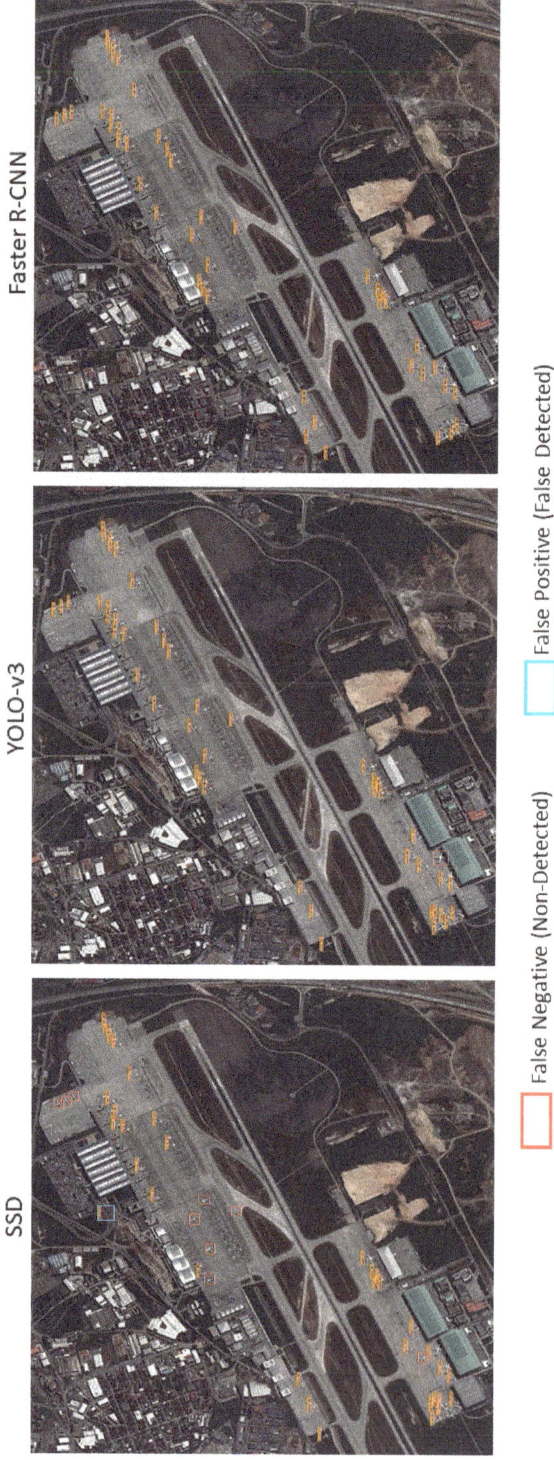

Figure A4. The aircraft detection results of Sabiha Gokcen Airport from Pleiades image data.

Figure A5. The aircraft detection results of Antalya Airport from Pleiades image data.

References

1. Cheng, G.; Han, J. A survey on object detection in optical remote sensing images. *ISPRS J. Photogramm. Remote Sens.* **2016**, *117*, 11–28. [CrossRef]
2. Svatonova, H. Analysis of visual interpretation of satellite data. *Int. Arch. Photogramm. Remote Sens Spat Inf. Sci.* **2016**, *41*, 675–681. [CrossRef]
3. Wu, H.; Zhang, H.; Geofan, J.; Xu, F. Typical target detection in satellite images based on convolutional neural networks. In Proceedings of the 2015 IEEE International Conference on Systems, Man, and Cybernetics, Hong Kong, China, 9–12 October 2015; pp. 2956–2961.
4. Zhou, P.; Cheng, G.; Liu, Z.; Bu, S.; Hu, X. Weakly supervised target detection in remote sensing images based on transferred deep features and negative bootstrapping. *Multidim. Syst. Sign. Process.* **2015**, *27*, 925–944. [CrossRef]
5. Gidaris, S.; Komodakis, N. Locnet: Improving localization accuracy for object detection. In Proceedings of the IEEE Conference on Computer Vision and Pattern Recognition (CVPR), Las Vegas, NV, USA, 26 June–1 July 2016; pp. 789–798. Available online: http://openaccess.thecvf.com/content_cvpr_2016/html/Gidaris_LocNet_Improving_Localization_CVPR_2016_paper.html (accessed on 10 January 2020).
6. Tello, M.; Martinez, C.L. A novel algorithm for ship detection in sar imagery based on the wavelet transform. *IEEE Trans. Geosci. Remote Sens.* **2005**, *2*, 201–205. [CrossRef]
7. Sirmacek, B.; Unsalan, C. Urban-area and building detection using sift keypoints and graph theory. *IEEE Trans. Geosci. Remote Sens.* **2009**, *47*, pp–1156. [CrossRef]
8. Han, J.; Zhang, D.; Cheng, G.; Guo, L.; Ren, J. Object detection in optical remote sensing images based on weakly supervised learning and high-level feature learning. *IEEE Trans. Geosci. Remote Sens.* **2014**, *53*, 3325–3337. [CrossRef]
9. Ma, L.; Li, M.; Ma, X.; Cheng, L.; Du, P.; Liu, Y. A review of supervised object-based land-cover image classification. *ISPRS J. Photogramm. Remote Sens.* **2017**, *130*, 277–293. [CrossRef]
10. Sun, H.; Sun, X.; Wang, H.; Li, Y.; Li, X. Automatic target detection in high-resolution remote sensing images using spatial sparse coding bag-of-words model. *IEEE Trans. Geosci. Remote Sens.* **2012**, *9*, 109–113. [CrossRef]
11. Polat, E.; Yildiz, C. Stationary aircraft detection from satellite images. *Istanb. Univ. J. Electr. Electron. Eng.* **2012**, *12*, 1523–1528.
12. Cheng, G.; Han, J.; Guo, L.; Qian, X.; Zhou, P.; You, X.; Hu, X. Object detection in remote sensing imagery using a discriminatively trained mixture model. *ISPRS J. Photogramm. Remote Sens.* **2013**, *85*, 32–43. [CrossRef]
13. Cheng, G.; Guo, L.; Zhao, T.; Han, J.; Li, H.; Fang, J. Automatic landslide detection from remote-sensing imagery using a scene classification method based on BoVW and pLSA. *Int. J. Remote Sens.* **2012**, *34*, 45–59. [CrossRef]
14. Han, J.; Zhou, P.; Zhang, D.; Cheng, G.; Guo, L.; Liu, Z.; Bu, S.; Wu, J. Efficient, simultaneous detection of multi-class geospatial targets based on visual saliency modeling and discriminative learning of sparse coding. *ISPRS J. Photogramm. Remote Sens.* **2014**, *89*, 37–48. [CrossRef]
15. Krizhevsky, A.; Sutskever, I.; Hinton, G.E. Imagenet classification with deep convolutional neural networks. In Proceedings of the Advances in Neural Information Processing Systems 25 (NIPS 2012), Lake Tahoe, NV, USA, 3–6 December 2012; pp. 1097–1105. Available online: http://papers.nips.cc/paper/4824-imagenet-classification-with-deep-convolutional-neural-networ (accessed on 10 January 2020).
16. Deng, J.; Dong, W.; Socher, R.; Li, L.J.; Li, K.; Li, F.F. Imagenet: A large-scale hierarchical image database. In Proceedings of the 2009 IEEE Conference on Computer Vision and Pattern Recognition, Miami, FL, USA, 20–25 June 2009; pp. 248–255. Available online: https://ieeexplore.ieee.org/abstract/document/5206848 (accessed on 10 January 2020).
17. Simonyan, K.; Zisserman, A. Very deep convolutional networks for large-scale image recognition. *arXiv* **2014**, arXiv:1409.1556.
18. Szegedy, C.; Liu, W.; Jia, Y.Q.; Sermanet, P.; Reed, S.; Anguelov, D.; Erhan, D.; Vanhoucke, V.; Rabinovich, A. Going deeper with convolutions. In Proceedings of the IEEE Conference on Computer Vision and Pattern Recognition (CVPR), Boston, MA, USA, 7–12 June 2015; pp. 1–9. Available online: https://www.cv-foundation.org/openaccess/content_cvpr_2015/html/Szegedy_Going_Deeper_With_2015_CVPR_paper.html (accessed on 10 January 2020).

19. He, K.; Zhang, X.; Ren, S.; Sun, J. Deep residual learning for image recognition. In Proceedings of the IEEE Conference on Computer Vision and Pattern Recognition (CVPR), Boston, MA, USA, 7–12 June 2015; pp. 1–9.

20. Hu, F.; Xia, G.S.; Hu, J.; Zhang, L. Transferring deep convolutional neural networks for the scene classification of high-resolution remote sensing imagery. *Remote Sens.* **2015**, *7*, 14680–14707. [CrossRef]

21. Scott, G.J.; England, M.R.; Starms, W.A.; Marcum, R.A.; Davis, C.H. Training deep convolutional neural networks for land-cover classification of high-resolution imagery. *IEEE Geosci. Remote Sens. Lett.* **2017**, *14*, pp–549. [CrossRef]

22. Cheng, G.; Han, J.; Lu, X. Remote sensing image scene classification: Benchmark and state of the art. *Proc. IEEE* **2017**, *105*, 1865–1883. [CrossRef]

23. Zhong, Y.; Fei, F.; Liu, Y.; Zhao, B.; Jiao, H.; Zhang, L. SatCNN: Satellite image dataset classification using agile convolutional neural networks. *Remote Sens. Lett.* **2016**, *2*, 136–145. [CrossRef]

24. Papadomanolaki, M.; Vakalopoulou, M.; Zagoruyko, S.; Karantzalos, K. Benchmarking deep learning frameworks for the classification of very high resolution satellite multispectral data. *ISPRS Ann. Photogramm. Remote Sens. Spat. Inf. Sci.* **2016**, *3*, 83–88. [CrossRef]

25. Girshick, R.; Donahue, J.; Darrell, T.; Malik, J. Rich feature hierarchies for accurate object detection and semantic segmentation. In Proceedings of the IEEE Conference on Computer Vision and Pattern Recognition (CVPR), Colombus, OH, USA, 24–27 June 2014; pp. 580–587.

26. He, K.; Zhang, X.; Ren, S.; Sun, J. Spatial pyramid pooling in deep convolutional networks for visual recognition. *IEEE Trans. Pattern Anal. Mach. Intell.* **2015**, *37*, 1904–1916. [CrossRef]

27. Girshick, R. Fast R-CNN. In Proceedings of the IEEE International Conference on Computer Vision, Santiago, Chile, 11–18 December 2015; pp. 1440–1448.

28. Ren, S.; He, K.M.; Girshick, R.; Sun, J. Faster R-CNN: Towards real-time object detection with region proposal networks. In Proceedings of the Advances in Neural Information Processing Systems 28 (NIPS 2015), Montreal, QC, Canada, 7–12 December 2015; pp. 91–99.

29. Hu, G.; Yang, Z.; Han, J.; Huang, L.; Gong, J.; Xiong, N. Aircraft detection in remote sensing images based on saliency and convolution neural network. *EURASIP J. Wirel. Commun. Netw.* **2018**, *26*, 1–16. [CrossRef]

30. Tang, T.; Zhou, S.; Deng, Z.; Zou, H.; Lei, L. Vehicle detection in aerial images based on region convolutional neural networks and hard negative example mining. *Sensors* **2017**, *17*, 336. [CrossRef]

31. Zhang, R.; Yao, J.; Zhang, K.; Feng, C.; Zhang, J. S-cnn-based ship detection from high-resolution remote sensing images. *Int. Arch. Photogramm. Remote Sens. Spat. Inf. Sci.* **2016**, *41*, 423–430. Available online: https://www.int-arch-photogramm-remote-sens-spatial-inf-sci.net/XLI-B7/423/2016/ (accessed on 10 January 2020). [CrossRef]

32. Liu, Z.; Yuan, L.; Weng, L.; Yang, Y. A high resolution optical satellite image dataset for ship recognition and some new baselines. In Proceedings of the 6th International Conference on Pattern Recognition Applications and Methods, Porto, Portugal, 24–26 February 2017; pp. 324–331. [CrossRef]

33. Chen, C.; Gong, W.; Hu, Y.; Chen, Y.; Ding, Y. Learning oriented region-based convolutional neural networks for building detection in satellite remote sensing images. *Int. Arch. Photogramm. Remote Sens. Spat. Inf. Sci.* **2017**, *42*, 461–464. [CrossRef]

34. Cheng, G.; Zhou, P.; Han, J. Learning rotation-invariant convolutional neural networks for object detection in VHR optical remote sensing images. *IEEE Trans. Geosci. Remote Sens.* **2016**, *54*, 7405–7415. [CrossRef]

35. Redmon, J.; Divvala, S.; Girshick, R.; Farhadi, A. You only look once: Unified, real-time object detection. In Proceedings of the IEEE Conference on Computer Vision and Pattern Recognition (CVPR), Las Vegas, NV, USA, 26 June–1 July 2016; pp. 779–788. Available online: https://www.cv-foundation.org/openaccess/content_cvpr_2016/html/Redmon_You_Only_Look_CVPR_2016_paper.html (accessed on 10 January 2020).

36. Liu, W.; Anguelov, D.; Erhan, D.; Szegedy, C.; Reed, S.; Fu, C.Y.; Berg, A.C. SSD: Single Shot MultiBox Detector. In Proceedings of the European Conference on Computer Vision, Amsterdam, The Netherlands, 11–14 October 2016; pp. 21–37. [CrossRef]

37. Everingham, M.; Gool, L.V.; Williams, C.K.I.; Winn, J.; Zisserman, A. The PASCAL visual object classes (VOC) challenge. *Int. J. Comput. Vis.* **2010**, *88*, 303–338. [CrossRef]

38. Lin, T.Y.; Marie, M.; Belongie, S.; Hays, J.; Perona, P.; Ramanan, D.; Dollár, P.; Zitnick, C.L. Microsoft COCO: Common Objects in Context. In Proceedings of the European Conference on Computer Vision, Zurich, Switzerland, 6–12 September 2014; pp. 740–755.

39. Radovic, M.; Adarkwa, O.; Wang, Q. Object recognition in aerial images using convolutional neural networks. *J. Imaging* **2017**, *3*, 21. [CrossRef]

40. Nie, G.H.; Zhang, P.; Niu, X.; Dou, Y.; Xia, F. Ship detection using transfer learned single shot multi box detector. In Proceedings of the 4th Annual International Conference on Information Technology and Applications (ITA 2017), Guangzhou, China, 26–28 May 2017; pp. 1006–1012. [CrossRef]

41. Wang, Y.; Wang, C.; Zhang, H. Combining single shot multibox detector with transfer learning for ship detection using Sentinel-1 images. In Proceedings of the BIGSARDATA, Beijing, China, 13–14 November 2017; pp. 1–4. [CrossRef]

42. Szegedy, C.; Vanhoucke, C.; Ioffe, S.; Shlens, J.; Wojna, Z. Rethinking the inception architecture for computer vision. In Proceedings of the IEEE Conference on Computer Vision and Pattern Recognition (CVPR), Las Vegas, NV, USA, 26 June–1 July 2016; pp. 2818–2826. Available online: https://www.cv-foundation.org/openaccess/content_cvpr_2016/papers/Szegedy_Rethinking_the_Inception_CVPR_2016_paper.pdf (accessed on 10 January 2020).

43. Wojek, C.; Dorko', G.; Schulz, A.; Schiele, B. Sliding-Windows for Rapid Object Class Localization: A Parallel Technique. In *Joint Pattern Recognition Symposium*; Springer: Berlin/Heidelberg, Germany, 2008. [CrossRef]

44. Neubeck, A.; Van Gool, L. Efficient Non-Maximum Suppression. In Proceedings of the 18th International Conference on Pattern Recognition (ICPR'06), Hong Kong, China, 20–24 August 2006. [CrossRef]

45. Redmon, J.; Farhadi, A. YOLOv3: An Incremental Improvement. *arXiv* **2018**, arXiv:1804.02767.

46. Abadi, M.; Barham, P.; Chen, J.; Chen, Z.; Davis, A.; Dean, J.; Devin, M.; Ghemawat, S.; Irving, G.; Isard, M.; et al. TensorFlow: A System for Large-Scale Machine Learning. In Proceedings of the 12th USENIX Symposium on Operating Systems Design and Implementation, Savannah, GA, USA, 2–4 November 2016; pp. 265–283.

47. Common Objects in Context Challenge. Available online: http://cocodataset.org/#detection-eval (accessed on 12 January 2020).

 *remote sensing* 

Article

A Coarse-to-Fine Network for Ship Detection in Optical Remote Sensing Images

Yue Wu [1], Wenping Ma [2], Maoguo Gong [3,*], Zhuangfei Bai [1], Wei Zhao [2], Qiongqiong Guo [2], Xiaobo Chen [2] and Qiguang Miao [1]

[1] Key Laboratory of Big Data and Intelligent Vision, School of Computer Science and Technology, Xidian University, Xi'an 710071, China; ywu@xidian.edu.cn (Y.W.); zfbai@stu.xidian.edu.cn (Z.B.); qgmiao@xidian.edu.cn (Q.M.)
[2] Key Laboratory of Intelligent Perception and Image Understanding of Ministry of Education, School of Articial Intelligence, Xidian University, Xi'an 710071, China; wpma@mail.xidian.edu.cn (W.M.); weizhao_90@stu.xidian.edu.cn (W.Z.); qqiongguo@stu.xidian.edu.cn (Q.G.); xiaobochen_1@stu.xidian.edu.cn (X.C.)
[3] School of Electronic Engineering, Key Laboratory of Intelligent Perception and Image Understanding of Ministry of Education, Xidian University, Xi'an 710071, China
* Correspondence: gong@ieee.org

Received: 20 November 2019; Accepted: 27 December 2019; Published: 10 January 2020

Abstract: With the increasing resolution of optical remote sensing images, ship detection in optical remote sensing images has attracted a lot of research interests. The current ship detection methods usually adopt the coarse-to-fine detection strategy, which firstly extracts low-level and manual features, and then performs multi-step training. Inadequacies of this strategy are that it would produce complex calculation, false detection on land and difficulty in detecting the small size ship. Aiming at these problems, a sea-land separation algorithm that combines gradient information and gray information is applied to avoid false alarms on land, the feature pyramid network (FPN) is used to achieve small ship detection, and a multi-scale detection strategy is proposed to achieve ship detection with different degrees of refinement. Then the feature extraction structure is adopted to fuse different hierarchical features to improve the representation ability of features. Finally, we propose a new coarse-to-fine ship detection network (CF-SDN) that directly achieves an end-to-end mapping from image pixels to bounding boxes with confidences. A coarse-to-fine detection strategy is applied to improve the classification ability of the network. Experimental results on optical remote sensing image set indicate that the proposed method outperforms the other excellent detection algorithms and achieves good detection performance on images including some small-sized ships and dense ships near the port.

Keywords: convolutional neural networks (CNNs); feature fusion; ship detection; optical remote sensing images

---

1. Introduction

Ship detection in optical remote sensing image is a challenging task and has a wide range of applications such as ship positioning, maritime traffic control and vessel salvage [1]. Differing from natural image that taken in close-range shooting with horizontal view, remote sensing image acquired by satellite sensor with a top-down perspective is vulnerable to the factor such as weather. Offshore and inland river ship detection has been studied on both synthetic aperture radar (SAR) and optical remote sensing imagery. Some alternative methods of machine learning approaches have also been proposed [2–5]. However, the classic ship detection methods based on SAR images will cause a high false alarm ratio and be influenced by the sea surface model, especially on inland rivers and in offshore areas. Schwegmann et al. [6] used deep highway networks to avoid the vanishing gradient problem. They developed their own three-class SAR dataset that allows for more meaningful analysis of ship discrimination performances. They used data from Sentinel-1 (Extra Wide Swath), Sentinel-3 and RADARSAT-2 (Scan-SAR Narrow). They used Deep Highway Networks 2, 20, 50, 100 with 5-fold cross-validation and obtained an accuracy of 96% outperforming classical techniques such as SVM, Decision Trees, and Adaboost. Carlos Bentes et al. [7] used a custom CNN with TerraSAR-X Multi Look Ground Range Detected (MGD) images to detect ships and iceberg. They compared their results with SVM and PCA+SVM, and showed that the proposed model outperforms these classical techniques. The classic detection methods based on SAR images do not perform well on small and gathering ships. And with the increasing resolution and quantity of optical remote sensing images, ship detection in optical remote sensing images has attracted a lot of research interests.This paper mainly discusses ship detection in optical remote sensing images. In the object detection task, natural image is mainly used to front-back object detection. By contrast, remote sensing image is mainly used to left-right object detection [8]. Ship detection in remote sensing image is immensely affected by the viewpoint changes, cloud occlusion, wave interference, background clutter. Of these, the characteristics of optical remote sensing image such as the diversity of target size, high complexity of background and small targets makes ship detection particularly difficult.

In recent years, ship detection methods in optical remote sensing image mainly adopt a coarse-to-fine detection strategy which is based on two-stage [9,10]. The first step is the ship candidate extraction stage. All candidate regions that possibly contain ship targets are searched out in the entire image by some region proposal algorithms, which is a coarse extraction process. In this case, the information of image such as color, texture and shape is usually taken into account [1,11]. Region proposal algorithms include the sliding windows-based method [12], the image segmentation-based method [10], and the saliency analysis-based method [13,14]. These methods can preliminarily extract the candidate region of ships, then the ship candidate region is filtered and merged according to shape, scale information and neighborhood analysis methods [15]. Selective search algorithm (SS) [16] is a representative algorithm for candidate region extraction and is widely used in object detection task.

The second step is the ship recognition stage. The ship candidate regions are classified by a binary classifier which distinguishes whether the ship target is located in the candidate region [17]. It is a fine recognition process. The features of ships are extracted and then candidate regions are classified. Many traditional methods extract low-level features, such as scale-invariant feature transform (SIFT) [18], histogram of oriented gradients (HOG) [19], deformed part mode (DPM) feature [20] and structure-local binary patterns (LBP) feature [21,22] to classify candidate regions. With the popularization of deep learning, some methods use convolution neural network (CNN) to extract the features of ships, which are the high-level feature with more semantic information. These extracted features combine with a classifier to classify all candidate regions to distinguish the ship from the background. Many excellent classifiers such as the support vector machine (SVM) [1,23], AdaBoost [24], and unsupervised discrimination strategy [13] are adopted to recognize the candidate regions.

Although the traditional method has achieved considerable detection results in clear and calm ocean environments, there still have many deficiencies. Yao et al. [25] found that the traditional methods have some shortcomings. First, the extracted candidate regions have a large amount of redundancy, which leads to expensive calculation. Second, the manual feature focuses on the shape or texture of ships, which requires manual observation of all ships. The complex background and variability in ship size will lead to poor detection robustness and low detection speed. Most important of all, when the size of the ships is very small or the ships are concentrated at the port, the extraction of the ship candidate region is particularly difficult. Therefore, the accuracy and efficiency of ship detection are greatly reduced.

Recently, convolutional neural networks (CNN) with good feature expression capability have widely used in image classification [26], object detection [27,28], semantic segmentation [29], image segmentation [30], image registration [31,32]. Object detection based on deep convolution neural network has achieved good performance on large scale natural image data set. These methods are mainly divided into two main categories: two-stage method and one-stage method. Two-stage method originated from R-CNN [33], then successively arise Fast R-CNN [34] and Faster R-CNN [28]. R-CNN is the first object detection framework based on deep convolutional neural networks [35], which uses the selective search algorithm (SS) to extract the candidate regions and computes features by CNN. A set of class-specific linear SVMs [36] and regressors are used to classify and fine-tune the bounding boxes, respectively. Fast R-CNN is improved on the basis of R-CNN to avoid repeated calculations of candidate region features. Faster R-CNN proposes a region proposed network (RPN) instead of the selective search method (SS) to extract candidate regions, which improves the computational efficiency by sharing the features between the RPN and the object detection network. One-stage methods, such as YOLO [27] and SSD [37], solve the detection problem as a regression problem and achieve an end-to-end mapping directly from image pixels to bounding box coordinates by a full convolutional network. SSD detects objects on multiple feature maps with different resolutions from a deep convolutional network and achieves better detection results than YOLO.

In recent years, many ship detection algorithms [25] based on deep convolutional neural networks have been proposed. These methods intuitively extract features of images through CNN, avoiding complex shape and texture analysis, which significantly improve the detection accuracy and efficiency of ships in optical remote sensing images. Zhang et al. [38] proposed S-CNN, which combines CNN with the designed proposals extracted from two ship models. Zou et al. [23] proposed the SVD Networks, which use CNN to adaptively learn the features of the image and adopt feature pooling operation and the linear SVM classifier to determine the position of the ship. Hou et al. [39] proposed the size-adapted CNN to enhance the performance of ship detection for different ship sizes, which contains multiple fully convolutional networks of different scales to adapt to different ships sizes. Yao et al. [25] applied a region proposal network (RPN) [28] to discriminate ship targets and regress the detection bounding boxes, in which the anchors are designed by intrinsic shape of ship targets. Wu et al. [40] trained a classification network to detect the locations of ship heads, and adopted an iterative multitask network to perform bounding-box regression and classification [41]. But these methods must first perform feature region extraction operations, so the efficiency of the algorithm is reduced. The most important is that these methods can produce more false detection on land and small ship cannot be detected.

This paper includes three main contributions:

(1) Aiming at the false detection on land, we use a sea-land separation algorithm [42] which combines gradient information and gray information. This method uses gradient and gray information to achieve preliminary separation of land and sea, and then eliminates non-connected regions through a series of morphological operations and ignoring small area operations.

(2) About small ship cannot be detected, we used The Feature Pyramid Network (FPN) [43] and a multi-scale detection strategy to solve this problem. The Feature Pyramid Network (FPN) proposes a top-down path that combines a horizontally connected structure that combines low resolution, strong semantic features with high resolution, weak semantic features to effectively solve small target detection problem. The multi-scale detection strategy is proposed to achieve ship detection with different degrees of refinement.

(3) We designed a two-stage inspection network for ship detection in optical remote sensing images. It can obtain the position of the predicted ship directly from the image without additional candidate region extraction operations, which greatly improves the efficiency of ship detection. Finally, we propose a coarse-to-fine ship detection network (CF-SDN) which has the feature extraction structure with the form of feature pyramid network, achieving end-to-end mapping directly from image pixels to bounding boxes with confidence scores. The CF-SDN contains multiple detection layers with a coarse-to-fine detection strategy employed at each detection layer.

The remainder of this paper is organized as follows. In Section II, we introduce our method including procedure of optical remote sensing image preprocessing including the sea-land separation algorithm, the multi-scale detection strategy, two strategies to eliminate the influence of cutting image, and the structure of the coarse-to-fine ship detection network (CF-SDN), including the feature extraction structure, the distribution of anchor, the coarse-to-fine detection strategy, the details of training and testing.. Section III describes the experiments performed on optical remote sensing image data set and Section IV presents conclusions.

2. Methodology

In this section, we will introduce the procedure of optical remote sensing image preprocessing including the sea-land separation algorithm, the multi-scale detection strategy, two strategies to eliminate the influence of cutting image, and the structure of the coarse-to-fine ship detection network (CF-SDN), including the feature extraction structure, the distribution of anchor, the coarse-to-fine detection strategy, the details of training and testing. The procedure of optical remote sensing image preprocessing is shown in Figure 1.

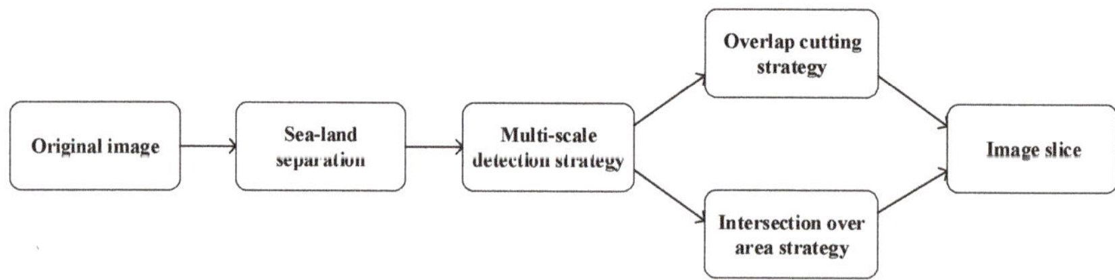

Figure 1. Flow diagram of the overall detection process.

2.1. Sea-land Separation Algorithm

Optical remote sensing images are obtained by satellites and aerial sensors. So the area that the image covered is wide and the geographical background is complex. In ship detection task, ships are usually scattered in water area (sea area) or in inshore area. In generally, the land and ship area present a relatively high gray level and have much complex texture, which are contrary to the situation in the sea area. Due to the complexity of the background in optical remote sensing images, the characteristics of some land areas are very similar to those of ships. This can easily lead to the detection of ship on land, which is called false alarm. Therefore, it is necessary to use sea-land separation algorithms to distinguish the sea area (or water area) from the land area before formal detection.

The sea-land separation algorithm [42] used in this paper considers the gradient information and the gray information of the optical remote sensing image comprehensively, combines some typical image morphology algorithms, and finally generates a binary image. In the process of sea-land separation, the algorithm that only considers the gradient information of the image performs well when the sea surface is relatively calm and the land texture is complex. However, the algorithm is difficult to achieve sea-land separation when the sea surface texture is complicated. The algorithm considering gray-scale information of the image is suitable for processing uniform texture images, but is difficult to process a complex image region. Therefore, the advantages of these two algorithms can be complemented with each other. The combination of gradient information and gray scale information can adapt to the complex situation of the optical remote sensing images, and can overcome the problem of poor sea-land separation performance caused by considering single information. The sea-land separation process is shown in Figure 2. The specific implementation details of the algorithm are as follows:

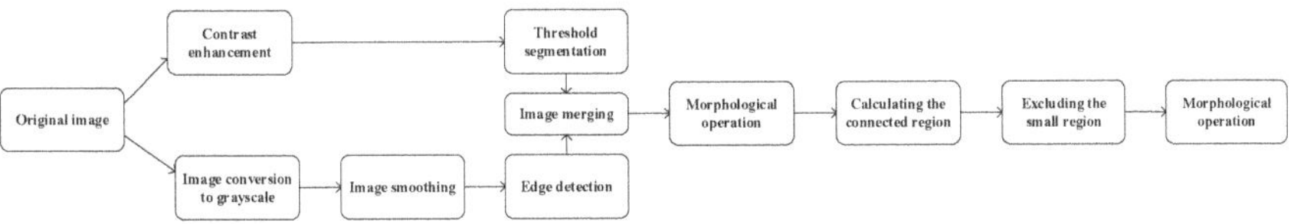

Figure 2. Flow diagram of the proposed sea-land separation algorithm.

(1) Threshold segmentation and edge detection are performed on the original image respectively. Before the threshold segmentation, the contrast of the image is enhanced to highlight the regions where the pixel values have large difference. Similarly, the image should be smoothed before performing edge detection. The traditional edge detection methods produce a lot of subtle wavy textures on the sea surface, which can be eliminated by filters. Here, we enhance the contrast of the image by histogram equalization, and perform threshold segmentation by the Otsu algorithm. At the same time, the median filter is used to smooth the image and the median filter size that selected in our experiment is 5×5, because the median filter is a nonlinear filtering that can not only remove noise but also preserve the edge information of the image when the image background is complex. Then the canny operator is used to detect the image edges, and we set the low and high thresholds to 10% of the maximum and 20% of the maximum, respectively.

(2) The threshold segmentation results and the edge detection results are combined by logical OR operation, then a binary image is generated to highlight non-water areas, which is regarded as the preliminary sea-land separation result. In the binary image, The pixel value of the land area is set to 1, and the pixel value of the water area is set to 0. The final result (such as IMAGES3) is shown in Figure 3.

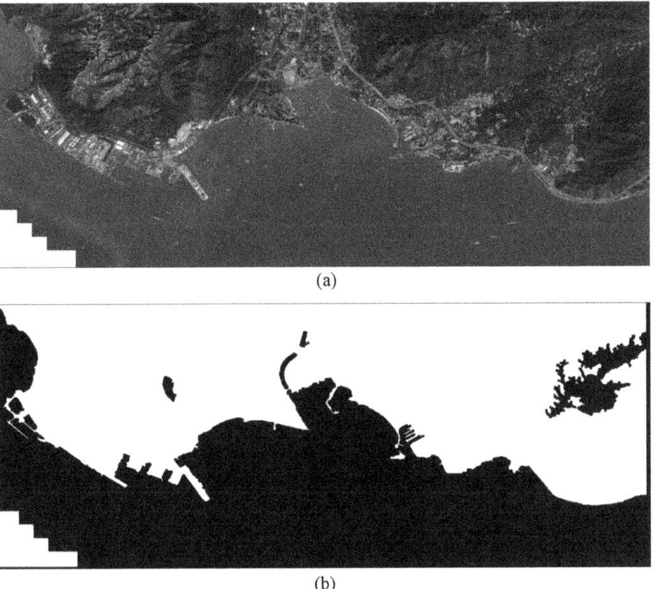

(a)

(b)

Figure 3. (**a**) The testing optical remote-sensing image IMAGE3. (**b**) The sea–land separation result corresponding to IMAGE3. It is a binary image, where the value of the position corresponding to the sea (or water) area is 0 (it is shown in black in the figure), and the position corresponding to the land region is 1(it is shown in white in the figure).

(3) Finally, a series of specific morphological operations are performed on this binary image. The basic specific morphological operation algorithms include dilation operation, erosion operation, open operation and close operation. Among them, the dilation operation and the close operation can fill gaps in the land contours of the binary map and remove small holes, while the erosion operations and the open operations can eliminate some small protrusions and narrow sections in the land area. Here, we first perform dilation operation and close operation on the binary image to eliminate the small holes in the land area. Then we calculate the connected regions for the processed binary image and exclude the small regions (corresponding to the ship or the small island at sea). The bumps on the land edges are eliminated by the erosion operation and the opening operation. The above specific morphological operation can be repeated to ensure the sea and land areas are completely separated. The size and shape of structuring elements is determined by analyzing the characteristic of non-connected areas on land from every experiment. The shape of structuring elements that selected in our experiment is disk, and the size of disk is 5 and 10. Figure 4 gives the intermediate results of a typical image slice in the sea-land separation process.

During test, only the area that contains the water area is sent into CF-SDN to detect ships and the pure land area is ignored.

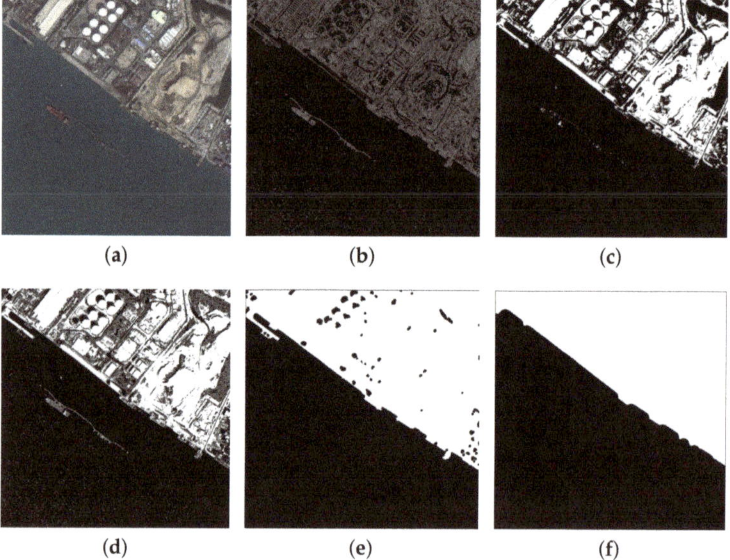

(a)   (b)   (c)

(d)   (e)   (f)

Figure 4. The intermediate results of one typical image slice in the sea-land separation process. The intermediate results of one typical image slice in the sea–land separation process. (**a**) The original image. (**b**) The edge detection result of the original image. (**c**) The threshold segmentation result of the original image. (**d**) The result after logical OR operation. (**e**) The result after preliminary morphological filtering. (**f**) The final sea–land separation result of the testing image.

Figure 4 gives the intermediate results of a typical image slice in the sea-land separation process. It can be found that the results of edge detection and threshold segmentation can complement each other to highlight non-water areas more completely. When only use threshold segmentation method, the area with low gray values on land may be classified as sea areas. Edge detection highlights the areas with complex textures and complements the results of threshold segmentation. We perform the expansion filtering and closing operations on the combined results in sequence. Then the connected regions are calculated and the small regions are removed. The final sea-land separation results highlight the land area and ships on the surface are classified as sea area.

2.2. Multi-Scale Detection Strategy

Generally the optical remote sensing images size is very large. The length and width of the image is usually several thousand pixels, the ship targets seem to be very small on the entire image. Therefore, it is necessary to cut the entire remote sensing image into some image slices and detect it separately. These image slices are normalized to the fixed size (320 × 320) in a certain proportion. Then the coarse-to-fine ship detection network outputs the detection results of these image slices. Here, the outputs of network are scaled according to the corresponding proportion. Finally, these detection results are mapped back to the original image according to the cutting position.

The sea-land separation results obtained in the previous subsection will also be applied in this subsection. Most ship detection methods set the pixel value of the land area in the remote sensing image to zero or the image's mean value to achieve the purpose of shielding land during the detection process. However, roughly removing original pixel values of the land area can easily lead to miss detection of ships at boundary between sea and land. If separation results are not accurate enough, detection performance will be greatly reduced. In this paper, we use a threshold to quickly exclude the areas that only contain land, and detect ships in areas that contain water (include the boundary between the sea and land). The specific method is as follows:

First, when the testing optical remote sensing image is cut, the corresponding sea-land separation result (a binary image) will be cut into some binary image slices with the same cutting method. And In the cut image, the ratio of ship area to slice area will become larger. Through a lot of experimental and statistical analysis, we found that when the average value of each binary image slice is less than a certain threshold, the water area in the image slice does not appear the ships. So each remote sensing image slice corresponds to a binary image slice. Figure 5 lists 3 examples. We calculate the average value of each binary image slice, and determine whether the image contains water. If the value is greater than the set threshold (0.8), we can think the corresponding remote sensing image slice almost does not contain water area, so we skip it and do not detect it.

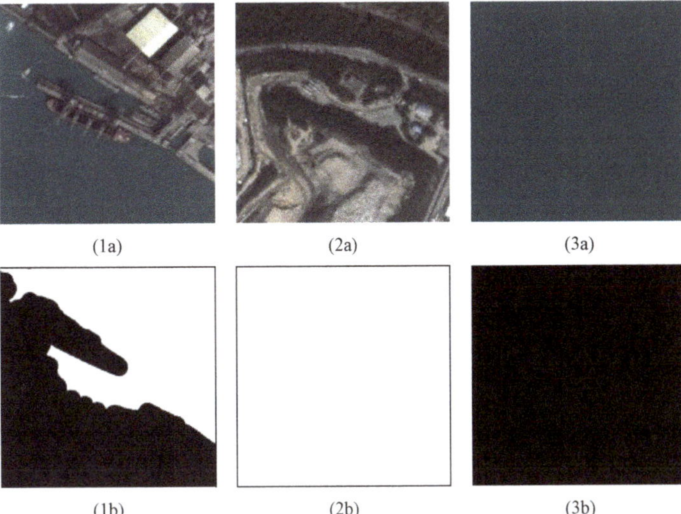

(1a)   (2a)   (3a)

(1b)   (2b)   (3b)

Figure 5. The top part are 3 remote sensing images slices, and the bottom part are the corresponding binary image slices. (**1b**) The mean value of the binary image slice is 0.52, which is smaller than the threshold, so the image slice in (**1a**) should be sent to the ship detection network. (**2b**) The mean value of the binary image slice is 1.0, which means that the image slice in (**2a**) only contains land, and can be skipped directly. (**3b**) The mean value of the binary image slice is 0, which means that the image slice in (**3a**) only contains water.

All mentioned above is the method using a single cutting size to cut and detect the testing optical remote sensing image. However, the scale distribution range of ships is wide. The size of small ship is only dozens of pixels, while the size of large ships is tens of thousands of pixels. It is difficult to determine the cutting size to ensure that ships at all scales can be accurately predicted. If the cutting size is small, many large ships will be cut off, which leads to miss detection. If the cutting size is large, many small ships will look smaller, which are difficult to detect. We propose a multi-scale detection strategy shown in Figure 6 to solve this dilemma.

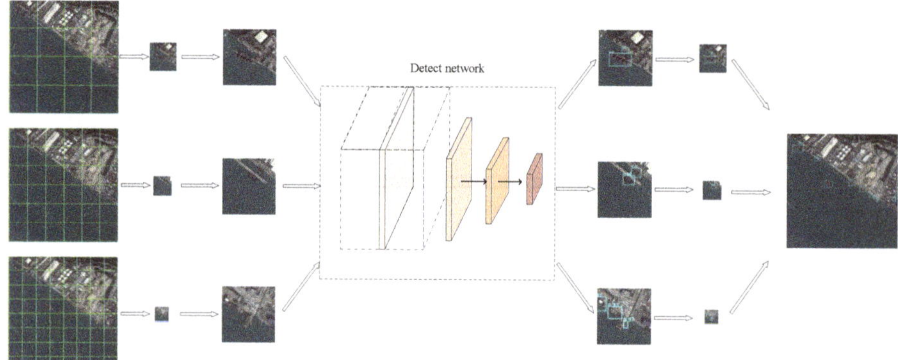

Figure 6. Flow diagram of the proposed multi-scale Detection.

The multi-scale detection strategy is that multiple cutting sizes are used to cut the testing optical remote sensing image into multiple different scales image slices in the test process. The testing optical remote sensing image is detected with multiple cutting sizes to achieve different degrees of refinement detection. And the detection results at each cutting size are combined to make the ship detection in optical remote sensing image more detailed and accurate.

In the experiment, we do a lot of tests and statistical analysis on the data set used in the experiments, and we find that the maximum length of the ship in the data does not exceed 200 pixels, the maximum width does not exceed 60 pixels, the minimum length is greater than 30 pixels, and the minimum width is greater than 10 pixels. Finally, the image slices can achieve satisfactory results when we choose the three cutting scales , 300×300, 400×400 and 500×500 respectively. And then image slices of each scale are detected separately. The detection results at multiple cutting sizes are combined and most of the redundant bounding boxes are deleted by non-maximal suppression (NMS), then we obtain the final detection results.

2.3. Elimination of Cutting Effect

Because the optical remote sensing images need to be cut during the detection process, many ships are easy to be cut off. This results in some bounding boxes which are output by the network only containing a portion of the ship. We adopt two strategies to eliminate the effect of cutting.

(1) We slice the image by overlap cutting. The overlap cutting is a strategy to ensure each ship appears completely at least once in all cutting image slices. This strategy produces overlapping slices by moving stride smaller than the slice size. For example, when the slice size is 300*300, the stride must be less than 300, and the produced slices certainly have overlapping parts. Moreover, different cutting scales are used in the test process. The ship which is cut off at one scale may completely appear at another scale. These bounding boxes detected from each image slice are mapped back to the original image according to the cutting position, which ensure that at least one of the bounding boxes of the same ship can completely contain the ship. The overlap cutting size used in experiment is 100 and the stride is 100.

(2) Suppose there are two bounding boxes *A* and *B*, shown in Figure 7a. The original NMS method calculates the Intersection over Union (IoU) of the two bounding boxes and compares it with the threshold to decide whether to delete the bounding box with lower confidence. However, optical remote sensing image ship detection is special. As shown in Figure 7b, it is assumed that the bounding box A only contains a part of a ship, and the bounding box B completely contains the same ship, so most of the area A is contained in B. But according to the above calculation method, the IOU between A and B may not exceed the threshold, so the bounding box A is retained and becomes a redundant bounding box.

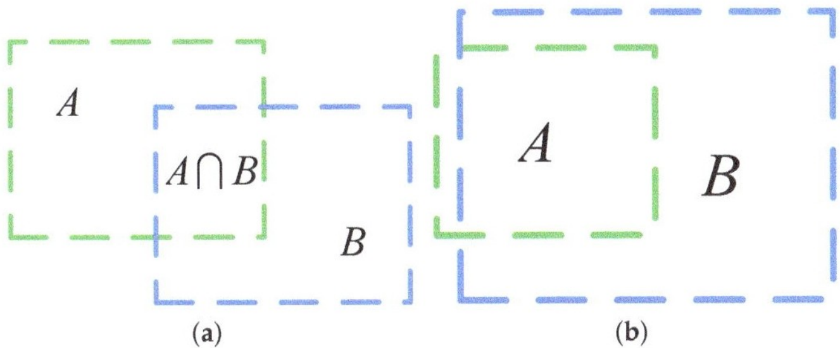

Figure 7. Two bounding boxes with overlapping areas.

In order to solve this situation, a new metric named IOA (intersection over area) is used in the NMS to determine whether to delete the bounding box. We define IOA between box *A* and box *B* as:

$$IOA = \frac{area(A \cap B)}{area(B)} \tag{1}$$

Here, assuming that the confidence of B is lower than A (if the confidence of the two boxes is equal, then box B is the smaller one.) and $area(A \cap B)$ refers to the area of the overlap between box *A* and box *B*.

During the test, we first perform non-maximum suppression on all detection results, which calculates the value of IOU between overlapping bounding boxes (the threshold is 0.5) to remove some redundant bounding boxes. For the remaining bounding boxes, the IOA between the overlapping bounding boxes are calculated. If the IOA between the two bounding boxes exceeds the threshold which is set to 0.8 in the experiments, the bounding box with lower confidence is removed. The remaining bounding boxes are the final detection results.

2.4. The Feature Extraction Structure

Using deep convolutional neural networks for target detection have an important problem. It is that the feature map output by the convolutional layer becomes smaller as the network deepens, and the information of the small target is also lost. This causes low detection accuracy for small target. Considering that shallow feature maps have higher resolution and deep feature maps contain more semantic information, we used FPN [43] to solve this problem. This structure can fuse features of different layers and independently predict object position of each feature layer. Therefore, the CF-SDN not only can preserve the information of small ship, but also have more semantic information. The input of the network is an image slice which is cut from optical remote sensing images, and the output is the predicted bounding boxes and the corresponding confidences. The feature extraction structure of the CF-SDN is shown in the Figure 8.

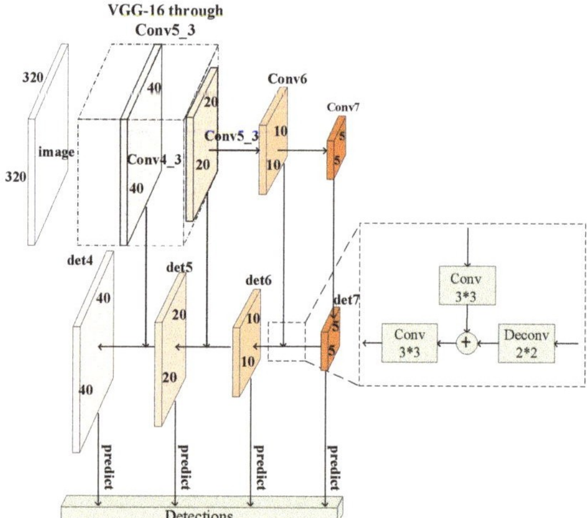

Figure 8. The feature extraction structure of the coarse-to-fine ship detection network.

We select the first 13 convolutional layers and the first 4 max pooling layers of VGG-16 which is pre-trained with ImageNet dataset [44] as the basic network, and add 2 convolutional layers (*conv6* and *conv7*) at the end of the network. The two convolutional layers (*conv6* and *conv7*) reduce the resolution of the feature map to half in sequence. With the deepening of the network, the features are continuously sampled by the max pooling layer, and the resolution of the output feature map get smaller, but the semantic information is more abundant. This is similar to the bottom-up process in FPN networks(A deep convnet computes an inherent multi-scale and pyramidal shape feature hierarchy). We select four different resolution feature maps that output from *conv4_3*, *conv5_3*, *conv6* and *conv7* , as shown in Figure 8. The strides of the selected feature maps are 8, 16, 32 and 64. The input size of this network is 320×320 pixels and the resolutions of the selected feature map are 40×40 (*conv4_3*), 20×20 (*conv5_3*), 10×10 (*conv6*) and 5×5 (*conv7*).

We set four detection layers in the network, and generate four feature maps of corresponding size through the selected feature maps. Then these feature maps are used as the input of four detection layers respectively. The deepest feature map (5×5) output by *conv7* is directly considered as the feature map of the last detection layer input, which is named det7. The feature maps used as inputs of the remaining detection layers are generated sequentially from the back to front in a lateral connections manner. The dotted line in the Figure 8 demonstrates the lateral connections manner. The deconvolution layer doubles the resolution of the deep feature map, while the convolution layer only changes the channel number of the feature map without changing the resolution. Feature maps are fused by element addition, and a 3×3 convolutional layer is added to decrease the aliasing effect caused by up-sampling. The fusion feature map serves as the input of the detection layer.

2.5. The Distribution of Anchors

In this subsection, we design the distribution of anchors at each detection layer. Anchors [28] are a set of reference boxes at each feature map cell, which tile the feature map in a convolutional manner. At each feature map cell, we predict the offsets relative to the anchor shapes in the cell and the confidence that indicate the presence of ship in each of those boxes. In optical remote sensing images, the scale distribution of ships is discrete, and ships usually have diverse aspect ratio depending on different orientations. So anchors with multiple sizes and aspect ratios are set at each detection layer to increase the number of matched anchors.

Feature maps from different detection layer have different resolutions and receptive field sizes introduce two types of receptive fields in CNN [45,46] , one is the theoretical receptive field which indicates the input region that theoretically affects the value of this unit, the other is the effective receptive field which indicates the input region has effective influence on the output value. Zhang et al. [47] points out that the effective receptive field is smaller than the theoretical receptive field, and anchors should be significantly smaller than theoretical receptive field in order to match the effective receptive field. At the same time, the article states that the stride size of a detection layer determines the interval of its anchor on the input image.

As listed in the second and third column of Table 1, the stride size and the size of theoretical receptive field at each detection layer are fixed. Considering that the anchor size set for each layer should be smaller than the calculated theoretical receptive field, we design the anchor size of each detection layer as shown in the fourth column of Table 1. The anchors of each detect layer have two scales and five aspect ratios. The aspect ratios are set to $\{\frac{1}{3}, \frac{1}{2}, 1, 2, 3\}$, so there are $2 \times 5 = 10$ anchors at each feature map cell on each detection layer.

Table 1. The distribution of the anchors.

Detect Layer	Stride	Theoretical Receptive Field Size	Anchor Size
conv4_3	8	92^2	$\{32^2, 64^2\}$
conv5_3	16	196^2	$\{64^2, 96^2\}$
conv6	32	404^2	$\{128^2, 160^2\}$
conv7	64	404^2	$\{192^2, 224^2\}$

2.6. The Coarse-to-Fine Detection Strategy

The structure of the detection layer is shown in Figure 9. We set up three parallel branches at each detection layer, two for classification and the other for bounding box regression. In Figure 9, the branches from top to bottom are coarse classification network, fine classification network and bounding box regression network, respectively. At each feature map cell, the bounding box regression network predicts the offsets relative to the anchor shapes in the cell, and the coarse classification network predicts the confidence which indicates the presence of ship in each of those boxes. This is a coarse detection process which obtains some bounding boxes with confidences. Then, the image block contained in the bounding box which has a confidence higher than the threshold (set to 0.1) is further classified (ship or background) by the fine classification network to obtain the final detection result. This is a fine detection process.

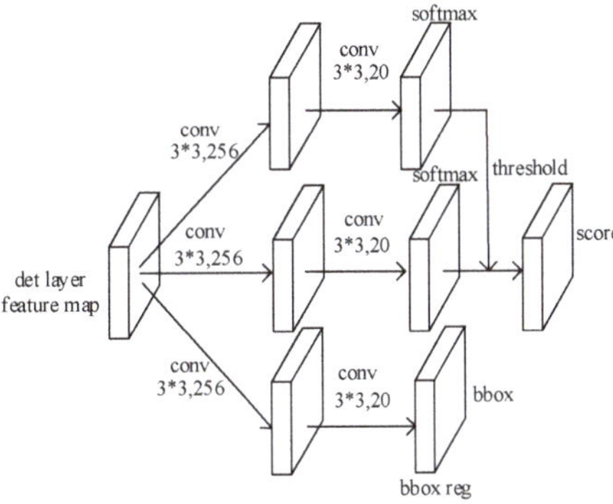

Figure 9. The structure of the detection layer.

2.6.1. Loss Function

Aiming at the structure of the detection layer, the multi-task loss L are used to jointly optimize model parameters:

$$L = \alpha \frac{1}{N_{cls_1}} \sum_i L_{cls}(p_i, p_i^*) + \beta \frac{1}{N_{reg}} \sum_i p_i^* L_{reg}(t_i, t_i^*) + \gamma \frac{1}{N_{cls_2}} \sum_j L_{cls}(p_j, p_j^*) \tag{2}$$

In Equation (2) i is the index of an anchor from the coarse classification network and the bounding box regression network in a batch, and p_i is the predicted probability that the anchor i is a ship. If the anchor is positive, the ground truth label p_i^* is 1, and p_i^* is 0 conversely. t_i is a vector representing the 4 parameterized coordinates of the predicted bounding box, and t_i^* is that of the ground-truth box associated with a positive anchor. The term $p_i^* L_{reg}$ means the regression loss is activated only for positive anchors and disabled otherwise. j is the index of an anchor from the fine classification network in a mini-batch, and the meaning of p_j and p_j^* is similar to p_i and p_i^*. The three terms are normalized by N_{cls_1}, N_{reg} and N_{cls_2} and weighted by the balancing parameter α, β and γ. N_{cls_1} represents the number of positive and negative anchors from the coarse classification network in the batch. N_{reg} represents the number of positive anchors from the bounding box regression network in the batch, and N_{cls_2} represents the number of positive and negative anchors from the fine classification network in the batch. In our experiment, we set $\alpha = \beta = \gamma = \frac{1}{3}$.

In Equation (2) the classification loss L_{cls} is the log loss from the coarse classification network:

$$L_{cls}(p_i, p_i^*) = -log[p_i^* p_i + (1 - p_i)(1 - p_i^*)] \tag{3}$$

the regression loss L_{reg} is the smooth L1 loss from the bounding box regression network:

$$L_{reg}(t_i, t_i^*) = R(t_i - t_i^*) \tag{4}$$

R is smooth L1 function:

$$smooth_{L1}(x) = \begin{cases} 0.5x^2 & if\ |x| < 1, \\ |x| - 0.5 & otherwise \end{cases} \tag{5}$$

2.6.2. Training Phase

In the training phase, these three branches are trained at the same time. A binary class label is set for each anchor in each branch.

(1) For coarse classification network and bounding box regression network, the anchors assigned positive label must satisfy one of the following two conditions: (i) match a ground truth box with the highest Intersection-over-Union (IoU) overlap. (ii) match a ground-truth box with an IoU overlap higher than 0.5. The anchors which have IoU overlap lower than 0.3 for all ground-truth boxes are assigned as negative label. The SoftMax layer outputs the confidences of each anchor at each cell on the feature map. Anchors whose confidence higher than 0.1 are selected as the train samples of the fine classification network.

(2) For fine classification network, the anchors selected from the previous step are further given positive and negative label. Here, the IoU overlap threshold for selecting the positive anchor is raised from 0.5 to 0.6. The larger threshold means that the positive anchor selected is closer to the ground truth box, which makes the classification more precise. Since the number of negative samples in remote sensing images is much larger than the number of positive samples, we randomly select negative samples to ensure that the ratio between positive and negative samples in each mini-batch is 1:3. If the number of positive samples is 0, the number of negative samples is set to 256.

2.6.3. Testing Phase

In the testing phase, firstly the bounding box regression network outputs the coordinate offsets to each anchor at each feature map cell. Then we adjust the position of each anchor by the box regression strategy and to get the bounding boxes. The outputs of the two classification networks are the confidence scores s1 and s2 corresponding to each bounding box. The confidence scores encode the probability of the ship appearing in the bounding box. First, if s1 output from the coarse classification network is lower than 0.1, the corresponding bounding box is removed. Then the confidence corresponding to the remaining bounding box is determined as the product of s1 and s2. The bounding box with the confidence larger than 0.2 is selected. Finally, non-maximum suppression (NMS) is applied to get final detection results.

3. Experiments and Results

In this section, the details of the experiments are described and the performances of the proposed method are studied. First, we introduce the data set used in the experiment. Then we introduce evaluation metrics used in the experiments. Finally, we conduct multiple sets of experiments to evaluate the performance of our methods and compare it with three excellent detection methods.

3.1. Data Set

Due to the lack of public data sets intended for ship detection in optical remote sensing image, we collected seven typical and representative images from different geographic conditions in Google Earth. The resolution of these images is 0.5 meter per pixel. The number of ships contained in each image range from dozens to hundreds and the ship size varies from 10×10 pixels to 400×400 pixels. Among these images, we selected 4 images for training and 3 images were remained for testing. The position of each ship in training images were labeled, including the coordinates of the center point, length and width of the ship. The data set we used is shown in Figure 10. Table 2 introduces the three images IMAGE1, IMAGE2, and IMAGE3 of the testing set.

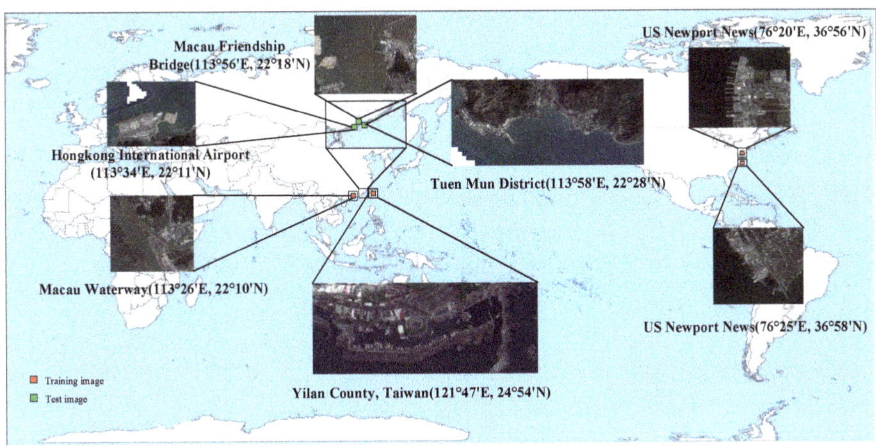

Figure 10. The data set we used

For training set images, the center of each ship was regarded as the center of image slice and some image slices were cut out as the train samples with the size of 300×300, 400×400, and 500×500. Data augmentation was achieved through translation, rotation, image brightness, contrast changes and so on. After data augmentation, 30000 image slices with different sizes composed the training data set for CF-SDN. Each ship is completely contained in at least an image slice and the corresponding position information constituted the training label set.

Table 2. The information of the testing images.

	Image Size	Numbers of Ships	Main Feature of Ships
IMAGE1	7948 × 11,289	190	small
IMAGE2	5726 × 4267	67	dense
IMAGE3	10,064 × 23,168	560	small and dense

3.2. Evaluations Metrics

The precision-recall curve (PRC) and average precision (AP) are used to quantitatively evaluate the performance of an object detection system.

3.2.1. Precision-Recall Curve

The precision-recall curve reflects the trend in precision and recall. The precision rate represents the proportion of the real target in the predicted target, and the recall rate represents the proportion of the correctly detected targets in the actual real targets. The precision and recall metrics are computed as follows:

$$precision = \frac{N_{tp}}{N_{tp} + N_{fp}} \tag{6}$$

$$recall = \frac{N_{tp}}{N_{tp} + N_{fn}} \tag{7}$$

Here, N_{tp} represents the number of true positives, which indicates the number of the correctly detected targets. N_{fp} represents the number of false positives, which indicates the number of the error detected targets(misjudge the background as a target). N_{fn} represents the number of false negatives, which indicates the number of miss detected targets. If the IoU between the predicted bounding box and the ground truth bounding box exceeds 0.5, the detection is regarded as true positive, otherwise, as a false positive. If there are multiple predicted bounding boxes overlap the same ground truth bounding box, then only one is considered as true positive, while others are considered as false positive.

The higher precision rate and recall rate, the better detection performance. But the precision rate is usually balanced against the recall rate. When the recall rate increases, the precision rate will decrease accordingly. Therefore, we calculate the average precision of the P-R curve to reflect the detection performance.

3.2.2. Average Precision

The average precision is the area under the precision-recall curve. Here, the average precision is obtained by calculating the average value of the corresponding precision when the recall rate changes from 0 to 1. In this paper, the average precision is calculated by the method used in the PASCAL VOC Challenge, which calculates the average precision by taking the mean of the precision rate of the points at all different recall rates on the P-R curve.

3.3. Implementation Details

Our experiments are implemented in Caffe, in a hardware environment consisting of HP-Z840 Workstation with an TITAN X12-GB GPU.

In the training of CF-SDN, the layers from VGG-16 are initialized by pre-training a model for ImageNet classification [48], which is a common technique used in deep neural networks. All other new layers are initialized by drawing weights from a zero-mean Gaussian distribution with standard deviation 0.01. The whole network is trained end-to-end by back propagation algorithm and SGD. The initial learning rate is set to 0.001 and we use it for 30k iterations; then we continue training for 30k iterations with 0.0005. The batch size is set to 20, and the total number of positive anchors and negative anchors in a batch is 256. The momentum is set to 0.9 and the weight decay is set to 0.0005.

3.4. Experimental Results and Analysis

3.4.1. Performance on the Testing Data Set

Using the trained CF-SDN, we perform ship detection on the testing data set which contains three optical remote sensing images with different scenes. The sea-land separation algorithm is used to obtain a binary image of the testing image, which is used to remove the image slices that only contain land. The multi-scale detection strategy is used to achieve different degrees of refinement detection. Figures 11–13 shows the detection results of CF-SDN on IMAGE1, IMAGE2 and IMAGE3 respectively, in which the true positives, false positives and false negatives are indicated by red, green and blue rectangles. The top left corner of the rectangle shows the confidence. Due to the large size of the testing image, we only take some representative areas to show the details.

As shown in Figure 11, the proposed method exhibits good detection performance for small size ships. Despite some ships on the sea are fuzzy which caused by cloud occlusion and wave interference, the proposed method has successfully detected most of these ships. As shown in Figure 12 and Figure 13, the proposed method has accurately located the scattered ships on the sea. Many ships on the land boundary also can be well detected, though they are easily be confused with the land features. For the dense ships in the port, as shown in Figure 13, our method can also detect most of the ships.

Figure 11. The detection results of IMAGE1. The true positives, false positives and false negatives are indicated by red, green and blue rectangles. The top left corner of the rectangle shows the confidence.

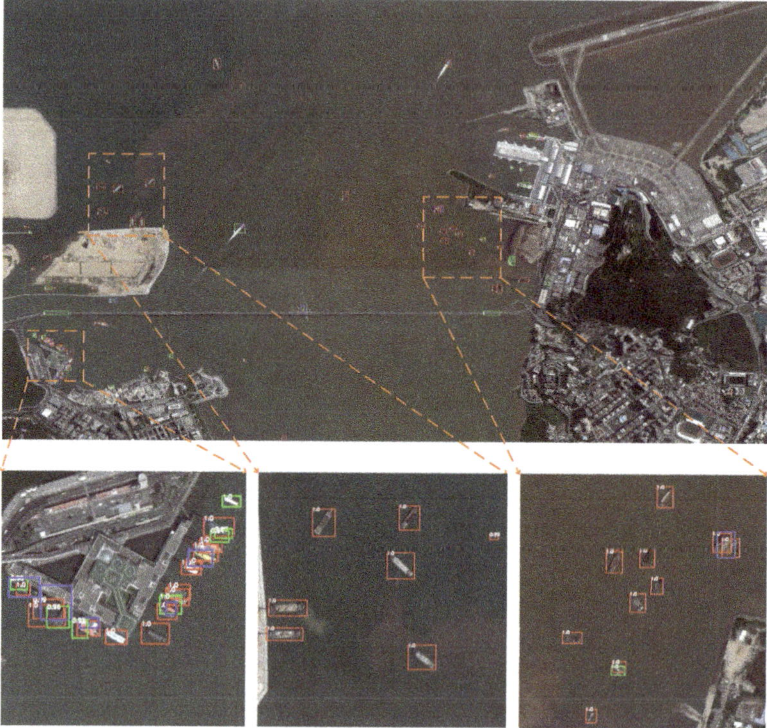

Figure 12. The detection results of IMAGE2. The true positives, false positives and false negatives are indicated by red, green and blue rectangles. The top left corner of the rectangle shows the confidence.

Figure 13. The detection results of IMAGE3. The true positives, false positives and false negatives are indicated by red, green and blue rectangles. The top left corner of the rectangle shows the confidence.

3.4.2. Comparison with other detection algorithms

In order to quantitatively demonstrate the superiority of our approach, we compared it with the other object detection algorithms. We choose R-CNN [33], Faster R-CNN [28], SSD [37] and the latest ship detection algorithms [49] as the comparison algorithm. R-CNN is an object detection model based on deep convolutional neural network and has been widely used in object detection of remote sensing images. Faster R-CNN is the representative two-stage object detection model and is improved from R-CNN. SSD is the representative one-stage object detection model, which is the same as CF-SDN and achieves an end-to-end mapping directly from image pixels to bounding box coordinates. The latest ship detection algorithm is a R-CNN based ship detection algorithm. Figure 14 is shown the specific example from six different methods.

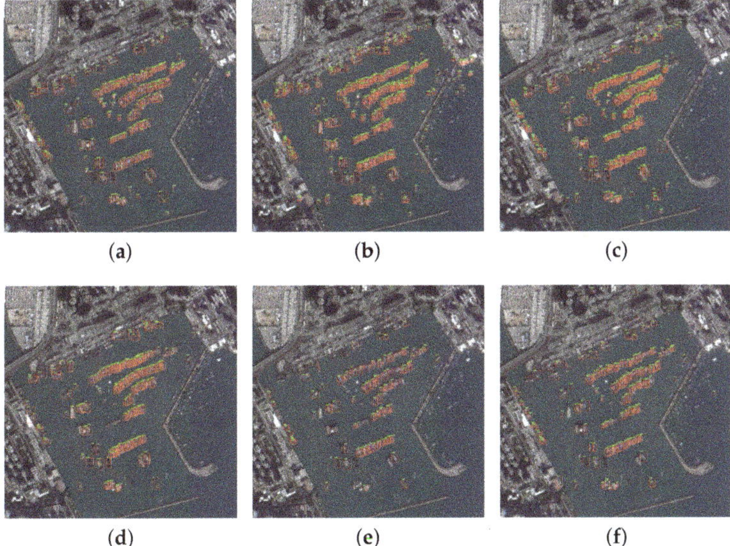

Figure 14. The specific example from six different methods. (**a**) The CF-SDN detection result. (**b**) The C-SDN detection result. (**c**) The Faster-RCNN detection result. (**d**) The SSD detection result. (**e**) The RCNN detection result. (**f**) The RCNN-based ship detection result.

In addition, to further validate the effectiveness of the proposed feature extraction structure and the coarse-to-fine detection strategy, we compare the proposed CF-SDN with CF-SDN without fine classification. In this experiment, C-SDN represents the CF-SDN without fine classification network, which has the same feature extraction structure as CF-SDN, but only contains a coarse classification network and a bounding box regression network at the detection layer. CF-SDN represents the complete CF-SDN, which adopts the coarse-to-fine detection strategy in detection, and predicts the boundary box that may contain ships through a coarse classification network and a bounding box regression network, and further finely classifies the detection results through a fine classification network.

For all test methods, the sea-land separation algorithm was implemented to remove the image slice that only contains land. We used the overlap cutting to slice the images the cutting size used in test is 400 (the overlap cutting size is 100 and the stride is 100). In addition, the detection results of the whole testing images are processed by NMS, and the IOA threshold is set to 0.5.

Tables 3 and 4 and Figure 15 show the quantitative comparison results of these methods on testing data set. As can be seen, the proposed CF-SDN exceed all other methods for all images in terms of AP. Compared with R-CNN, SSD, Faster R-CNN, C-SDN and R-CNN Based Ship Detection, the proposed CF-SDN acquires 27.3%, 9.2%, 4.8%, 2.7%, 22.4% performance gains in terms of AP on entire data set, respectively. Among them, the performance of C-SDN is second only to CF-SDN. Compared with R-CNN, SSD, Faster R-CNN and R-CNN Based Ship Detection, the CF-SDN without fine classification (C-SDN) acquires 24.6%, 6.5%, 2.1%, 21.7% performance gains in terms of AP on entire data set, respectively. This benefits from the proposed feature extraction structure which fuses different hierarchical features to improve the representation of features. Through the comparion between the C-SDN and CF-SDN, we can find the superiority of the coarse-to-fine detection strategy. Many false alarms are removed and the average precision is improved by the further fine classification.

Table 3. Performance comparison of the six methods on the testing set in terms of AP.

	IMAGE1	IMAGE2	IMAGE3	Comprehensive
R-CNN	0.389	0.442	0.475	0.415
SSD	0.504	0.691	0.625	0.596
Faster R-CNN	0.590	0.695	0.645	0.640
R-CNN Based Ship Detection	0.549	0.581	0.411	0.464
C-SDN	0.607	0.706	0.668	0.661
CF-SDN	0.610	0.742	0.706	0.688

Table 4. Performance comparison of the five methods on the testing set in terms of time(unit: second).

	R-CNN	SSD	Faster R-CNN	R-CNN Based Ship Detection	CF-SDN
IMAGE1	42.432	19.584	29.376	31.469	13.661
IMAGE2	11.232	5.184	7.776	9.159	4.218
IMAGE3	65.208	30.096	47.652	56.326	26.752
Total	119.872	54.864	84.804	99.69	44.631

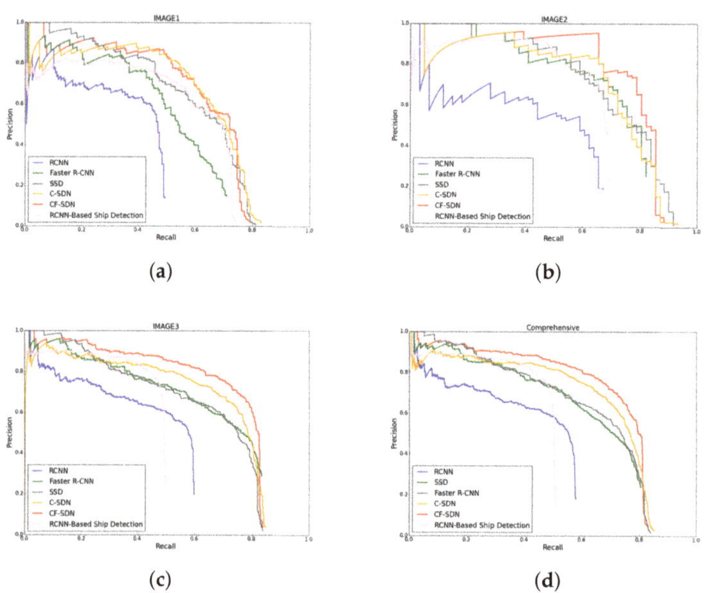

Figure 15. Performance comparison of the six methods on the testing set in terms of the P-R Curves. (a) Comparison of detection performance on IMAGE1. (b) Comparison of detection performance on IMAGE2. (c) Comparison of detection performance on IMAGE3. (d) Comparison of detection performance on the whole testing data set.

3.4.3. Sea-Land Separation to Improve the Detection Accuracy

In order to validate the effectiveness of the sea-land separation algorithm, we compared the detection results with and without sea-land separation during the test. We choose SSD and CF-SDN as the detection model. SSD-I and C-SDN indicate that the sea-land separation method was not used during the test. SSD-II and CF-SDN indicate that the proposed sea-land separation method is used to remove the areas which only contain land during the test. The cutting size is 400 (the overlap is 100). The detection result of the whole testing image is processed by NMS, and the IoU threshold is set to 0.5.

Table 5 shows the quantitative comparison results of the experiments. Table 6 shows the time spent in two different phases during the test. It can be observed that SSD-II acquires 19.6% performance gains in terms of AP in entire data set compared with SSD-I, while CF-SDN acquires 2.1% performance gains compared with C-SDN. As shown in Figure 16, the method that use sea-land separation has achieved higher accuracy when the recall rate is almost equal. This demonstrates that the sea-land separation can avoid some false alarms and improve the detection accuracy. The detection performance of SSD is more affected by sea-land separation than that of CF-SDN, which confirms that CF-SDN can extract features better and generate fewer false alarms.

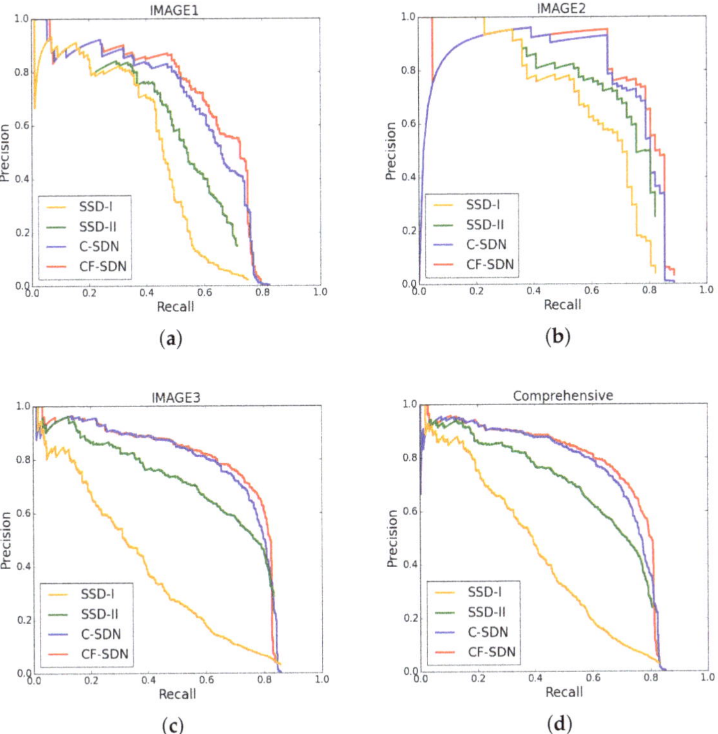

Figure 16. Performance comparison of with and without sea-land separation on the testing set in terms of the P-R Curves. (**a**) Comparison of detection performance on IMAGE1. (**b**) Comparison of detection performance on IMAGE2. (**c**) Comparison of detection performance on IMAGE3. (**d**) Comparison of detection performance on the whole testing data set.

Table 5. Performance comparison of with sea-land separation on the testing set and without sea-land separation on the testing set in terms of AP.

	SSD-I	SSD-II	C-SDN	CF-SDN
IMAGE1	0.434	0.504	0.586	0.610
IMAGE2	0.621	0.691	0.720	0.742
IMAGE3	0.370	0.625	0.690	0.706
Comprehensive	0.400	0.596	0.667	0.688

Table 6. Time spent at different phases during the test(unit: second).

	IMAGE1	IMAGE2	IMAGE3	Total
CF-SDN	13.661	4.218	26.752	44.631
Threshold segmentation	20.264	5.969	53.797	80.03
Edge detection	3.997	0.940	10.249	15.366
Morphological operation	2.858	0.804	7.764	11.426
Excluding the small region	7.843	3.416	19.496	30.755

3.4.4. Multi-Scale Detection Strategy Improves Performance

In order to validate the effectiveness of the multi-scale detection strategy, we compare the detection performance of using single cutting size and using multi-scale detection strategy during the test. For the experiment that using single cutting size, we adopt three different cutting sizes of 300×300, 400×400 and 500×500 respectively. For the experiment that using multi-scale detection strategy, we combine the detection results of three single cutting size and use NMS to remove some redundant bounding boxes. The detection model used in these experiments is the CF-SDN, and both of them use the sea-land separation algorithm to remove the area that only contains land.

Table 7 and Figure 17 show the quantitative comparison results of using each single cutting sizes and using the multi-scale detection strategy. As can be seen from them, the highest detection accuracy is obtained by using the multi-scale detection strategy. When we only adopt a single cutting size, the cutting scale of 400×400 demonstrates the best detection performance on the testing data set. Compared with the single detection scale of 300, 400, 500, the combined result acquired 4.4%, 3.9%, 13.8% performance gains in terms of AP in entire data set. Combined with the detection results at different cutting sizes, the multi-scale detection strategy shows the outstanding advantages. The combination of multiple detection with different refinement degree effectively improves the accuracy and the recall of ship detection.

Table 7. Performance comparison of using each single cutting size and using the multi-scale detection strategy (combined) on the testing set in terms of AP.

	300×300	400×400	500×500	Combined
IMAGE1	0.668	0.610	0.579	0.705
IMAGE2	0.757	0.742	0.710	0.745
IMAGE3	0.683	0.706	0.590	0.735
Comprehensive	0.683	0.688	0.589	0.727

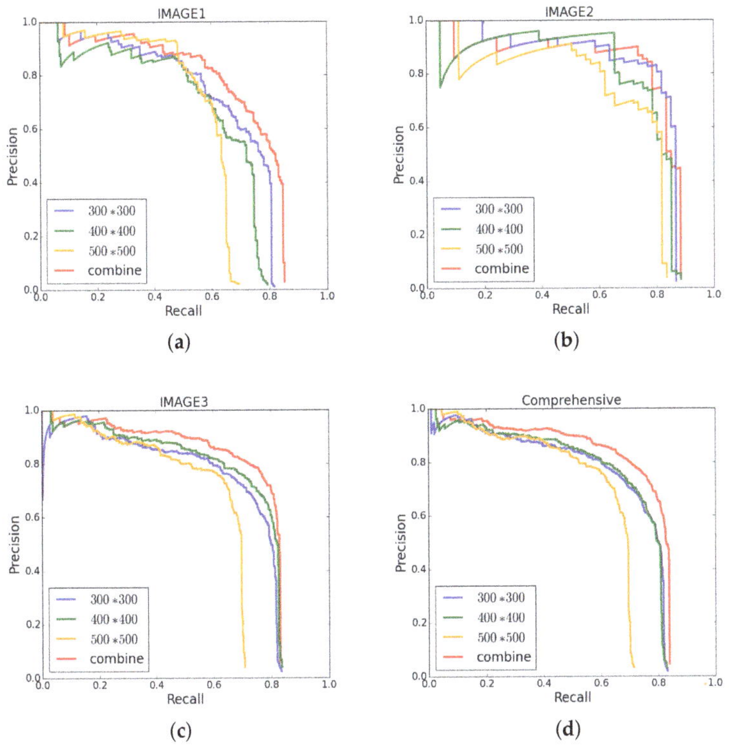

Figure 17. Performance comparison of using each single cutting size (300 × 300, 400 × 400 and 500 × 500) and using the multi-scale detection strategy (combined) on the testing set in terms of the P-R Curves. (**a**) Comparison of detection performance on IMAGE1. (**b**) Comparison of detection performance on IMAGE2. (**c**) Comparison of detection performance on IMAGE3. (**d**) Comparison of detection performance on the whole testing data set.

4. Conclusions

This paper presents a coarse-to-fine ship detection network (CF-SDN) which includes a sea-land separation algorithm, a coarse-to-fine ship detection network and a multi-scale detection strategy. The sea-land separation algorithm can avoid false alarms on land. The coarse-to-fine ship detection network do not need to use the region proposal algorithm and directly achieves an end-to-end mapping directly from image pixels to bounding boxes with confidences. The multi-scale detection strategy can achieve ship detection with different degrees of refinement. It effectively improves the accuracy and speed of ship detection.

Experimental results on optical remote sensing data set show that the proposed method outperforms other excellent detection algorithms and achieves good detection performance on the data set including some small-sized ships. For the dense ships near the port, our method can locate most of the ships well, although produce a little false alarms and miss detections at the same time. The main reason for the missing detection is that many bounding boxes with high overlap are removed by NMS. In fact, the overlaps between the ground truth of dense ships is very high. Therefore, our future work will focus on the two aspects: (1) The orientation angle information is taken into account when determining the position of the ship, which can effectively reduce the overlap between the bounding boxes of the dense ships. (2) Combined with the characteristics of remote sensing images, the select strategy of positive and negative samples are considered in the network to improve the classification and location ability of the detection network.

Author Contributions: Y.W. and W.Z. conceived and designed the experiments; X.C. and Z.B. performed the experiments; Q.G. and X.C. analyzed the data;W.M., M.G. and Q.M. contributed materials; Z.B. wrote the paper. Y.W. and W.M. supervised the study and reviewed this paper. All authors have read and agreed to the published version of the manuscript.

Funding: This work was supported in part by National Natural Science Foundation of China (No. 61702392).

Acknowledgments: The authors would like to thank the anonymous reviewers for their very competent comments and helpful suggestions.

Conflicts of Interest: The authors declare no conflict of interest.

References

1. Zhu, C.; Zhou, H.; Wang, R.; Guo, J. A Novel Hierarchical Method of Ship Detection from Spaceborne Optical Image Based on Shape and Texture Features. *IEEE Trans. Geosci. Remote Sens.* **2010**, *48*, 3446–3456. [CrossRef]

2. Lang, F.; Yang, J.; Yan, S.; Qin, F. Superpixel Segmentation of Polarimetric Synthetic Aperture Radar (SAR) Images Based on Generalized Mean Shift. *Remote Sens.* **2018**, *10*, 1592. [CrossRef]

3. Ciecholewski, M. River Channel Segmentation in Polarimetric SAR Images: Watershed Transform Combined with Average Contrast Maximisation. *Expert Syst. Appl.* **2017**, *82*, 196–215. [CrossRef]

4. Braga, A.M.; Marques, R.C.; Rodrigues, F.A.; Medeiros, F.N. A Median Regularized Level Set for Hierarchical Segmentation of SAR Images. *IEEE Geosci. Remote Sens. Lett.* **2017**, *14*, 1171–1175. [CrossRef]

5. Jin, R.; Yin, J.; Zhou, W.; Yang, J. Level Set Segmentation Algorithm for High-resolution Polarimetric SAR Images Based on a Heterogeneous Clutter Model. *IEEE J. Sel. Top. Appl. Earth Obs. Remote Sens.* **2017**, *10*, 4565–4579. [CrossRef]

6. Schwegmann, C.P.; Kleynhans, W.; Salmon, B.P.; Mdakane, L.W.; Meyer, R.G. Very deep learning for ship discrimination in synthetic aperture radar imagery. In Proceedings of the 2016 IEEE International Geoscience and remote-sensing Symposium (IGARSS), Beijing, China, 10–15 July 2016; Volume 10, pp. 104–107.

7. Bentes, C.; Frost, A.; Velotto, D.; Tings, B. Ship-iceberg Discrimination with Convolutional Neural Networks in High Resolution SAR Images. In Proceedings of the 11th European Conference on Synthetic Aperture Radar, Hamburg, Germany, 6–9 June 2016; Volume 6, pp. 1–4.

8. Zhao, B.; Zhong, Y.; Zhang, L. A Spectral–structural Bag-of-features Scene Classifier for Very High Spatial Resolution remote-sensing Imagery. *ISPRS J. Photogramm. Remote Sens.* **2016**, *116*, 73–85. [CrossRef]

9. Yang, G.; Li, B.; Ji, S.; Gao, F.; Xu, Q. Ship Detection from Optical Satellite Images Based on Sea Surface Analysis. *IEEE Geosci. Remote Sens. Lett.* **2013**, *11*, 641–645. [CrossRef]

10. Bi, F.; Zhu, B.; Gao, L.; Bian, M. A Visual Search Inspired Computational Model for Ship Detection in Optical Satellite Images. *IEEE Geosci. Remote Sens. Lett.* **2012**, *9*, 749–753.

11. Corbane, C.; Najman, L.; Pecoul, E.; Demagistri, L.; Petit, M. A Complete Processing Chain for Ship Detection Using Optical Satellite Imagery. *Int. J. Remote Sens.* **2010**, *31*, 5837–5854. [CrossRef]

12. Soofbaf, S.; Sahebi, M.; Mojaradi, B. A Sliding Window-based Joint Sparse Representation (SWJSR) Method for Hyperspectral Anomaly Detection. *Remote Sens.* **2018**, *10*, 434. [CrossRef]

13. Qi, S.; Ma, J.; Lin, J.; Li, Y.; Tian, J. Unsupervised Ship Detection Based on Saliency and S-HOG Descriptor from Optical Satellite Images. *IEEE Geosci. Remote Sens. Lett.* **2015**, *12*, 1451–1455.

14. Ding, Z.; Yu, Y.; Wang, B.; Zhang, L. An Approach for Visual Attention Based on Biquaternion and Its Application for Ship Detection in Multispectral Imagery. *Neurocomputing* **2012**, *76*, 9–17. [CrossRef]

15. Yang, F.; Xu, Q.; Li, B. Ship Detection from Optical Satellite Images Based on Saliency Segmentation and Structure-LBP Feature. *IEEE Geosci. Remote Sens. Lett.* **2017**, *14*, 602–606. [CrossRef]

16. Uijlings, J.R.; Van De Sande, K.E.; Gevers, T.; Smeulders, A.W. Selective Search for Object Recognition. *Int. J. Comput. Vis.* **2013**, *104*, 154–171. [CrossRef]

17. Miyamoto, H.; Uehara, K.; Murakawa, M.; Sakanashi, H.; Nasato, H.; Kouyama, T.; Nakamura, R. Object Detection in Satellite Imagery Using 2-Step Convolutional Neural Networks. In Proceedings of the IEEE International Geoscience and remote-sensing Symposium, Valencia, Spain, 22–27 July 2018; pp. 1268–1271.

18. Chang, H.H.; Wu, G.L.; Chiang, M.H. Remote-sensing Image Registration Based on Modified SIFT and Feature Slope Grouping. *IEEE Geosci. Remote Sens. Lett.* **2019**, *16*, 1363–1367. [CrossRef]
19. Dong, C.; Liu, J.; Xu, F.; Liu, C. Ship Detection from Optical remote-sensing Images Using Multi-Scale Analysis and Fourier HOG Descriptor. *Remote Sens.* **2019**, *11*, 1529. [CrossRef]
20. Li, Z.; Yang, D.; Chen, Z. Multi-layer Sparse Coding Based Ship Detection for remote-sensing Images. In Proceedings of the IEEE International Conference on Information Reuse and Integration, San Francisco, CA, USA, 13–15 August 2015; pp. 122–125.
21. Haigang, S.; Zhina, S. A Novel Ship Detection Method for Large-scale Optical Satellite Images Based on Visual LBP Feature and Visual Attention Model. In Proceedings of the International Archives of Photogrammetry, remote-sensing and Spatial Information Sciences, Prague, Czech Republic, 12–19 July 2016; Volume 41, pp. 917–921.
22. Yang, F.; Xu, Q.; Gao, F.; Hu, L. Ship Detection from Optical Satellite Images Based on Visual Search Mechanism. In Proceedings of the IEEE International Geoscience and remote-sensing Symposium (IGARSS), Milan, Italy, 26–31 July 2015; pp. 3679–3682.
23. Zou, Z.; Shi, Z. Ship Detection in Spaceborne Optical Image with SVD Networks. *IEEE Trans. Geosci. Remote Sens.* **2016**, *54*, 5832–5845. [CrossRef]
24. Shi, Z.; Yu, X.; Jiang, Z.; Li, B. Ship Detection in High-resolution Optical Imagery Based on Anomaly Detector and Local Shape Feature. *IEEE Trans. Geosci. Remote Sens.* **2013**, *52*, 4511–4523.
25. Yao, Y.; Jiang, Z.; Zhang, H.; Zhao, D.; Cai, B. Ship Detection in Optical remote-sensing Images Based on Deep Convolutional Neural Networks. *J. Appl. Remote Sens.* **2017**, *11*, 042611. [CrossRef]
26. Krizhevsky, A.; Sutskever, I.; Hinton, G.E. Imagenet Classification with Deep Convolutional Neural Networks. In *Advances in Neural Information Processing Systems*; NIPS: Denver, CO, USA, 2012; pp. 1097–1105.
27. Redmon, J.; Divvala, S.; Girshick, R.; Farhadi, A. You Only Look Once: Unified, Real-time Object Detection. In Proceedings of the IEEE Conference on Computer Vision and Pattern Recognition, Las Vegas, NV, USA, 27–30 June 2016; pp. 779–788.
28. Ren, S.; He, K.; Girshick, R.; Sun, J. Faster R-CNN: Towards Real-Time Object Detection with Region Proposal Networks. *IEEE Trans. Pattern Anal. Mach. Intell.* **2017**, *39*, 1137–1149. [CrossRef]
29. Long, J.; Shelhamer, E.; Darrell, T. Fully Convolutional Networks for Semantic Segmentation. In Proceedings of the IEEE Conference on Computer Vision and Pattern Recognition, Boston, MA, USA, 7–12 June 2015; pp. 3431–3440.
30. Wu, Y.; Ma, W.; Gong, M.; Li, H.; Jiao, L. Novel Fuzzy Active Contour Model with Kernel Metric for Image Segmentation. *Appl. Soft Comput.* **2015**, *34*, 301–311. [CrossRef]
31. Wu, Y.; Ma, W.; Gong, M.; Su, L.; Jiao, L. A Novel Point-matching Algorithm Based on Fast Sample Consensus for Image Registration. *IEEE Geosci. Remote Sens. Lett.* **2014**, *12*, 43–47. [CrossRef]
32. Wu, Y.; Ma, W.; Miao, Q.; Wang, S. Multimodal Continuous Ant Colony Optimization for Multisensor Remote Sensing Image Registration with Local Search. *Swarm Evol. Comput.* **2017**, *47*, 89–95. [CrossRef]
33. Girshick, R.; Donahue, J.; Darrell, T.; Malik, J. Rich Feature Hierarchies for Accurate Object Detection and Semantic Segmentation. In Proceedings of the IEEE Conference on Computer Vision and Pattern Recognition, Columbus, OH, USA, 23–28 June 2014; pp. 580–587.
34. Girshick, R. Fast R-CNN. In Proceedings of the IEEE International Conference on Computer Vision, Santiago, Chile, 7–13 December 2015; pp. 1440–1448.
35. Ren, Y.; Zhu, C.; Xiao, S. Small Object Detection in Optical remote-sensing Images Via Modified Faster R-CNN. *Appl. Sci.* **2018**, *8*, 813. [CrossRef]
36. Gallego, A.J.; Pertusa, A.; Gil, P. Automatic Ship Classification from Optical Aerial Images with Convolutional Neural Networks. *Remote Sens.* **2018**, *10*, 511. [CrossRef]
37. Liu, W.; Anguelov, D.; Erhan, D.; Szegedy, C.; Reed, S.; Fu, C.Y.; Berg, A.C. SSD: Single Shot Multibox Detector. In Proceedings of the European Conference on Computer Vision, Amsterdam, The Netherlands, 11–14 October 2016; pp. 21–37.
38. Zhang, R.; Yao, J.; Zhang, K.; Feng, C.; Zhang, J. S-CNN-Based Ship Detection from High-resolution remote-sensing Image. *Int. Arch. Photogramm. Remote Sens. Spat. Inf. Sci.* **2016**, *41*, 917–921.
39. Hou, X.; Xu, Q.; Ji, Y. Ship Detection from Optical remote-sensing Image based on Size-Adapted CNN. In Proceedings of the Fifth International Workshop on Earth Observation and Remote Sensing Applications (EORSA), Xi'an, China, 18–20 June 2018; pp. 1–5.

40. Wu, F.; Zhou, Z.; Wang, B.; Ma, J. Inshore Ship Detection Based on Convolutional Neural Network in Optical Satellite Images. *IEEE J. Sel. Top. Appl. Earth Obs. Remote Sens.* **2018**, *11*, 4005–4015. [CrossRef]

41. Cheng, G.; Han, J.; Zhou, P.; Xu, D. Learning Rotation-invariant and Fisher Discriminative Convolutional Neural Networks for Object Detection. *IEEE Trans. Image Process.* **2018**, *28*, 265–278. [CrossRef]

42. Ma, L.; Soomro, N.Q.; Shen, J.; Chen, L.; Mai, Z.; Wang, G. Hierarchical sea–land Segmentation for Panchromatic remote-sensing Imagery. *Math. Probl. Eng.* **2017**, *2017*, 1–8. [CrossRef]

43. Lin, T.Y.; Dollár, P.; Girshick, R.; He, K.; Hariharan, B.; Belongie, S. Feature Pyramid Networks for Object Detection. In Proceedings of the IEEE Conference on Computer Vision and Pattern Recognition, Honolulu, HI, USA, 21–26 July 2017; pp. 2117–2125.

44. Simonyan, K.; Zisserman, A. Very Deep Convolutional Networks for Large-scale Image Recognition. *arXiv* **2014**, arXiv:1409.1556.

45. He, K.; Zhang, X.; Ren, S.; Sun, J. Spatial Pyramid Pooling in Deep Convolutional Networks for Visual Recognition. *IEEE Trans. Pattern Anal. Mach. Intell.* **2015**, *37*, 1904–1916. [CrossRef] [PubMed]

46. Luo, W.; Li, Y.; Urtasun, R.; Zemel, R. Understanding the Effective Receptive Field in Deep Convolutional Neural Networks. In *Advances in Neural Information Processing Systems*; NIPS: Denver, CO, USA, 2016; pp. 4898–4906.

47. Zhang, S.; Zhu, X.; Lei, Z.; Shi, H.; Wang, X.; Li, S.Z. S3FD: Single Shot Scale-invariant Face Detector. In Proceedings of the IEEE International Conference on Computer Vision, Venice, Italy, 22–29 October 2017; pp. 192–201.

48. Russakovsky, O.; Deng, J.; Su, H.; Krause, J.; Satheesh, S.; Ma, S.; Huang, Z.; Karpathy, A.; Khosla, A.; Bernstein, M.; et al. Imagenet large-scale Visual Recognition Challenge. *Int. J. Comput. Vis.* **2015**, *115*, 211–252. [CrossRef]

49. Zhang, S.; Wu, R.; Xu, K.; Wang, J.; Sun, W. R-CNN-Based Ship Detection from High Resolution Remote-sensing Imagery. *Remote Sens.* **2019**, *11*, 631. [CrossRef]

 *remote sensing*

Article

Improved Remote Sensing Image Classification Based on Multi-Scale Feature Fusion

Chengming Zhang [1,2,3,*,†], Yan Chen [1,3], Xiaoxia Yang [1,†], Shuai Gao [4], Feng Li [5], Ailing Kong [1], Dawei Zu [1] and Li Sun [1]

1 College of Information Science and Engineering, Shandong Agricultural University, 61 Daizong Road, Taian 271000, China
2 Key Open Laboratory of Arid Climate Change and Disaster Reduction of CMA, 2070 Donggangdong Road, Lanzhou 730020, China
3 Shandong Technology and Engineering Center for Digital Agriculture, 61 Daizong Road, Taian 271000, China
4 Chinese Academy of Sciences, Institute of Remote Sensing and Digital Earth, 9 Dengzhuangnan Road, Beijing 100094, China
5 Shandong Provincal Climate Center, No. 12 Wuying Mountain Road, Jinan 250001, China
* Correspondence: chming@sdau.edu.cn; Tel.: +86-139-5382-3659
† These authors are co-first authors as they contributed equally to this work.

Received: 25 November 2019; Accepted: 6 January 2020; Published: 8 January 2020

Abstract: When extracting land-use information from remote sensing imagery using image segmentation, obtaining fine edges for extracted objects is a key problem that is yet to be solved. In this study, we developed a new weight feature value convolutional neural network (WFCNN) to perform fine remote sensing image segmentation and extract improved land-use information from remote sensing imagery. The WFCNN includes one encoder and one classifier. The encoder obtains a set of spectral features and five levels of semantic features. It uses the linear fusion method to hierarchically fuse the semantic features, employs an adjustment layer to optimize every level of fused features to ensure the stability of the pixel features, and combines the fused semantic and spectral features to form a feature graph. The classifier then uses a Softmax model to perform pixel-by-pixel classification. The WFCNN was trained using a stochastic gradient descent algorithm; the former and two variants were subject to experimental testing based on Gaofen 6 images and aerial images that compared them with the commonly used SegNet, U-NET, and RefineNet models. The accuracy, precision, recall, and F1-Score of the WFCNN were higher than those of the other models, indicating certain advantages in pixel-by-pixel segmentation. The results clearly show that the WFCNN can improve the accuracy and automation level of large-scale land-use mapping and the extraction of other information using remote sensing imagery.

Keywords: convolutional neural network; image segmentation; multi-scale feature fusion; semantic features; Gaofen 6; aerial images; land-use; Tai'an

---

1. Introduction

Remote sensing images have become the main data source for obtaining land-use information at broad spatial scales. The most common method first assigns each pixel to a category using image segmentation and subsequently generates land-use information according to the pixel-by-pixel classification result [1,2]. Since the accuracy of the final extraction is determined by the accuracy of the pixel-by-pixel classification, improving the segmentation accuracy is a common research focus [3]. The pixel feature extraction method and the classifier performance both have a decisive influence on segmentation results [4,5].

Researchers have proposed a variety of methods to extract ideal features. For example, spectral indexes have been widely applied in the classification of low- and medium-spatial-resolution remote sensing images, as they can accurately reflect statistical information regarding pixel spectral values. Commonly used indexes include the vegetation index [6–13], water index [14], normalized difference building index [15], ecological index [16], normalized difference vegetation index (NDVI) [10,13], and derivative indexes such as the re-normalized difference vegetation index [12] or the growing season normalized difference vegetation index [13]. Taking account of the advantage of the short time period of low- and medium-spatial-resolution remote sensing images, some researchers have applied spectral indexes to time-series images [17,18]. However, such indexes are mainly used to express common information within the same land-use type through simple band calculations. When these are applied to remote sensing images with rich detail and high spatial resolution, it becomes difficult to extract features with good discrimination, limiting the application of spectral indexes to the classification of such images.

For high-spatial-resolution remote sensing images with a high level of detail, researchers initially proposed the use of texture features with the gray matrix method [19]. Subsequently, other researchers have proposed a series of methods for extracting more abundant texture features, including the Gabor filter [20], the Markov random field model [21], the Gibbs random field model [22], and the wavelet transform [23]. Compared with spectral index features, texture features can better express the spatial correlation between pixels, improve the ability to distinguish between features, and effectively improve the accuracy of pixel classification.

Although the combination of spectral and texture features has greatly promoted the development of remote sensing image segmentation technology and significantly improved the accuracy of pixel classification results, the ongoing improvement in remote sensing technology has resulted in increasing image resolutions and higher levels of detail. Traditional texture feature extraction techniques now struggle with high-resolution images, such that new methods are needed to obtain effective feature information from such images [24,25].

The developing field of machine learning is currently being applied to pixel feature extraction, with early applications to image processing, including neural networks [26,27], support vector machines [28,29], decision trees [30,31], and random forests [32,33]. These methods use pixel spectral information as inputs and achieve the desired feature results through complex calculations. Although these methods can fully explore the relationship between channels and obtain some effective feature information, these features only express the information of a single pixel rather than the spatial relationship between pixels, limiting these methods' application in the processing of remote sensing images.

Convolutional neural networks (CNNs) use pixel blocks as inputs and compute them with a set of convolution kernels to obtain more features with a stronger distinguishing ability [34–37]. The biggest advantage of CNNs lies in their ability to simultaneously extract specific pixel features as well as spatial features of pixel blocks. This approach can reasonably set the structure of the convolutional layer according to the characteristics of the image and the extraction target, so as to extract features that meet the requirements, achieving good results in image processing [34,36]. The most widely used CNNs in the field of camera image processing include fully convolutional networks (FCNs) [38], SegNet [39], DeepLab [40], RefineNet [41], and U-NET [42]. DeepLab expands the convolution for rich features in camera images, enabling the model to effectively expand the receptive field without increasing the calculations required. SegNet and UNET can establish a symmetrical network structure and segment camera images, resulting in an efficient utilization of high-level semantic features. RefineNet uses a multipath structure to combine coarse high-level semantic features and relatively fine low-level semantic features by equal-weight fusion.

As CNNs have outstanding advantages in feature extraction, they have been widely used in other fields, such as real-time traffic sign recognition [43], pedestrian recognition [44], apple target recognition [45], plant disease detection [46], and pest monitoring [47]. Researchers have established a method to extract coordinate information of an object from street view imagery [48] using CNNs.

Compared to camera images, remote sensing images have fewer details and more mixed pixels. When using a CNN to extract information from remote sensing images, the influence of the convolution structure on feature extraction must be considered [49]. Existing CNN structures are mainly designed for camera images with a high level of detail; therefore, a more ideal result can be obtained through adjustments that consider the specific characteristics of remote sensing images [50,51]. Researchers have proposed a series of such adjustments for applying CNNs to remote sensing image processing [3] and some classic CNNs have been widely applied in this field [52,53]. Based on the characteristic analysis of target objects and specific remote sensing imagery, researchers have established a series of CNNs such as two-branch CNN [49], WFS-NET [54], patch-based CNN [55], and hybrid MLP-CNN [56]. Researchers have also used remote sensing images to create many benchmark datasets, such as EuroSAT [57] or the Inria Aerial Image dataset [58], to test the performance of CNNs.

In this study, we established a CNN structure based on variable weight fusion, the weight feature value convolutional neural network (WFCNN), and experimentally assessed its performance. The main contributions of this work are as follows.

- Based on the analysis of the data characteristics of remote sensing images, fully considering the impact of image spatial resolution on feature extraction, we establish a suitable network convolution structure;
- The proposed approach can effectively fuse low-level semantic features with high-level semantic features and fully considers the data characteristics of adjacent areas around objects on remote sensing images. Compared with the strategies adopted by other models, our approach is more conducive to effective features.

2. Datasets

We employed two sets of datasets to test the performance of the model. We created the GF-6 images dataset that contains 587 image-label pairs. The aerial image labeling dataset was the benchmark dataset [58].

2.1. The GF-6 Images Dataset

2.1.1. Study Area

Tai'an is a prefecture-level city covering ~7761 km^2 in Shandong Province, eastern China (116°20–117°59′E, 35°38′–36°28′N; Figure 1); it has jurisdiction over the districts of Taishan and Daiyue, the cities of Feicheng and Xintai, and the counties of Ningyang and Dongping. Its terrain is highly variable, including mountains, hills, plains, basins, and lakes (Figure 1c). The mountains are concentrated in the east and north, including Mount Tai (one of the Five Great Mountains of China). The hills are mainly distributed in southwestern Xintai, eastern Ningyang, the northwestern city suburbs, southern Feicheng, and northern Dongping. The major basin lies within Dongping and contains Dongping Lake, Daohu Lake, and Zhoucheng Lake.

In the study area, crops are divided into summer and autumn crops, according to the growing season. Summer crops mainly refers to winter wheat, and its growth period begins in the autumn and reaches the early summer of the second year. Autumn crops mainly refer to corn, millet, and potato. The growth period generally ranges from early summer to early winter. Therefore, crop planting areas are usually divided into winter wheat and farmland.

There are eight main land-use types in the study area: developed land, water bodies, agricultural buildings, roads, farmland, winter wheat, woodland, and others. Developed land includes residential and factory areas, agricultural buildings refer to buildings in crop planting areas, and others refer to areas not used. These diverse landforms and land-uses render the study area representative of many different regions and suitable as an experimental area for this study.

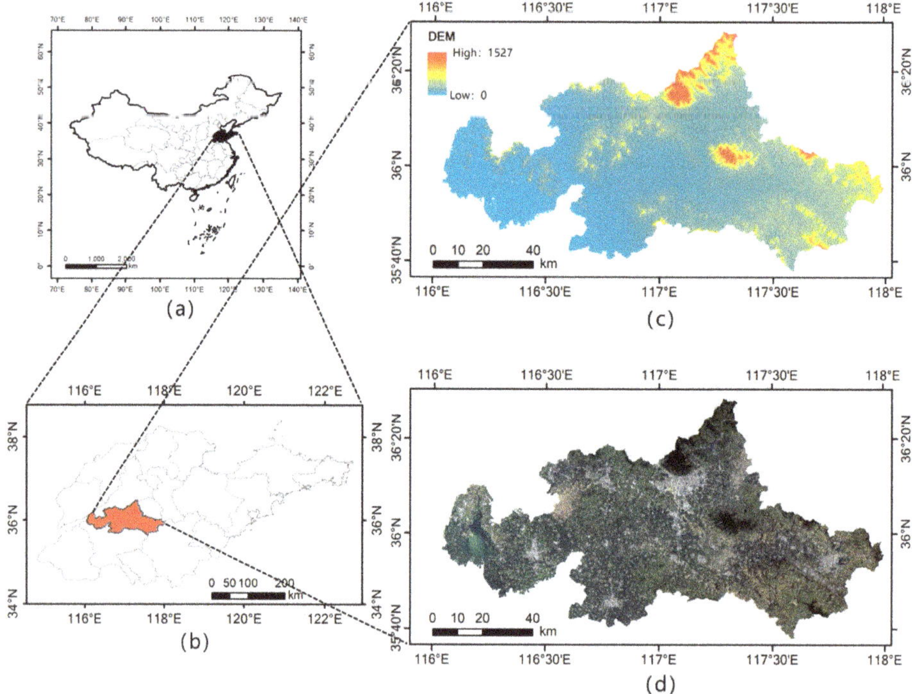

Figure 1. Geographic location of the city of Tai'an (red boundary) within Shandong Province, China:
(**a**) Geographic location of China; (**b**) Geographic location of Shandong Province; (**c**) Terrain of Tai'an
and (**d**) remote sensing images used in this study.

2.1.2. Remote Sensing Data

Gaofen 6 (GF-6) is a low-orbit optical remote sensing satellite launched by China in 2018, which
provides images with high resolution, wide coverage, high quality, and high efficiency. GF-6 has a
design life of eight years and two cameras: a full-color 8 m (high-resolution) multi-spectral camera
with an image swath of 90 km and a 16 m (medium-resolution) multi-spectral wide-format camera
with an image swath of 800 km. The GF-6 satellite will operate with the Gaofen 1 satellite network,
reducing the time resolution of remote sensing data acquisition from 4 d to 2 d. In-depth analysis of
these remote sensing images and the establishment of appropriate image segmentation methods will
allow the acquisition of high-precision global land-use information, which is of great significance for
improving the application level of GF-6.

We selected images from different seasons to increase the anti-interference abilities of the WFCNN
and mitigate potential complications, such as the change in seasons, and thus enhance applicability. We
collected a total of fifty-two GF-6 remote sensing images (Figure 1d). These images were divided into
four groups according to the image acquisition time. The first image group was captured in autumn
2018, the second in winter 2018, the third in spring 2019, and the fourth in summer 2019. Together, the
images of each group covered the study area. Specifications are given in Table 1.

Table 1. Specifications for Gaofen 6 satellite imagery used in this study.

Band	Range (μm)	Spatial Resolution (m)	Width (km)
Panchromatic	0.45–0.90	2	>90
B1	0.45–0.52	8	>90
B2	0.52–0.60	8	>90
B3	0.43–0.69	8	>90
B4	0.76–0.90	8	>90

Image preprocessing included geometric correction, radiation correction, and image fusion. We used Python to develop a program for geometry correction that captured control points from geometrically corrected Gaofen 2 (GF-2) remote sensing images. These geographically corrected GF-2 images have a spatial resolution of 1 meter, which is suitable for selecting control points from them.

Atmospheric correction was performed using the fast line-of-sight atmospheric analysis of spectral hypercubes module in the environment for visualizing images (ENVI) software. The multi-spectral and panchromatic images were fused using the Pan-sharping module in ENVI. The resulting image included four bands (blue, green, red, and near-infrared) with a 2-m spatial resolution.

2.1.3. Dataset Creation

In the images captured in winter or spring, developed land, water bodies, agricultural buildings, roads, bare fields, and farmland can be directly distinguished by visual interpretation within ENVI. To accurately distinguish winter wheat and woodland, 353 sample points were obtained through ground surveys (118 woodland and 235 winter wheat, Figure 2). Woodland areas had a rough texture, a large color change, and an irregular shape, while winter wheat areas were finer, smoother, and generally regular in shape. Defining these features helped to improve the accuracy of visual interpretation.

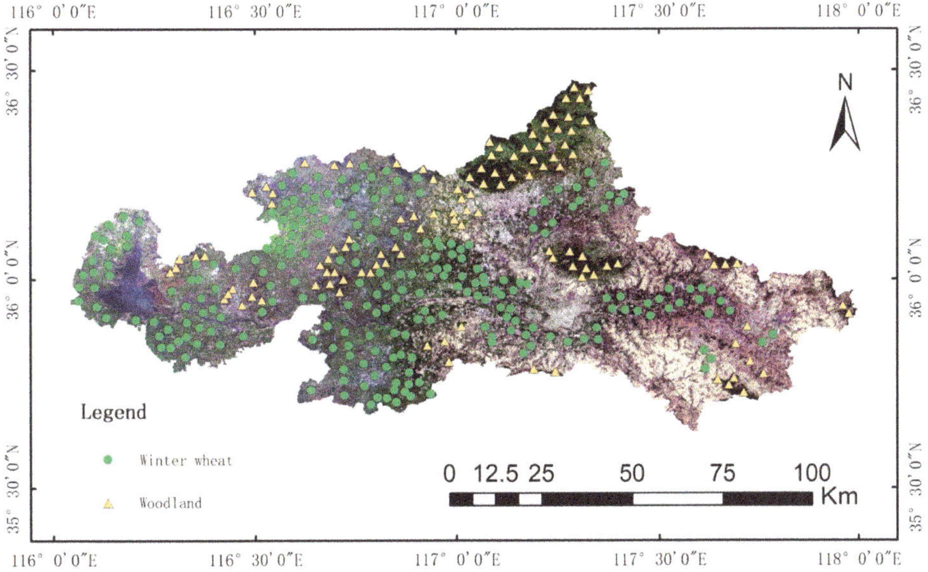

Figure 2. Geographic location of ground survey sample points used to distinguish winter wheat (green) and woodland (yellow).

The ENVI software was used to fuse the images into a complete mosaic, covering the whole study area. This image was then segmented into patches of 480 × 360 pixels, from which 587 images were selected for manual classification based on the eight land-use types defined above. Numerical codes were then assigned to these land-use types; an example is provided in Figure 3. All labeled image patches and their labeled files formed data sets for training and testing the WFCNN model.

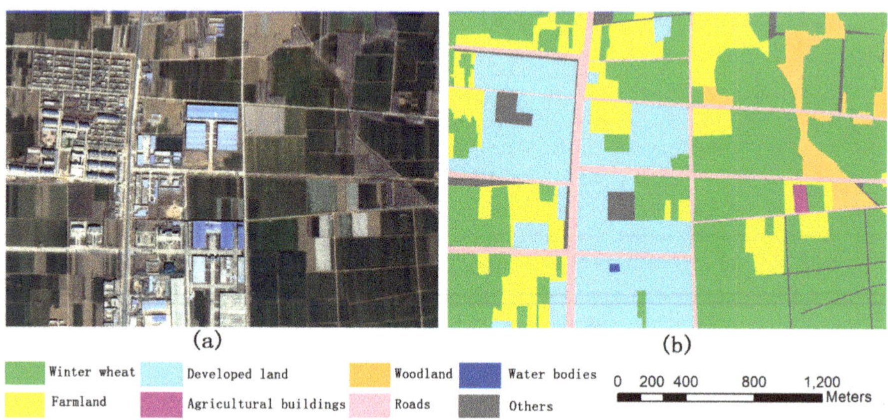

Figure 3. Example of land-use classification: (**a**) original image and (**b**) classified image.

2.2. The Aerial Image Labeling Dataset

The aerial image labeling dataset was downloaded from https://project.inria.fr/aerialimagelabeling/ [58]. The images cover dissimilar urban settlements, ranging from densely populated areas (e.g., the financial district of San Francisco) to alpine towns (e.g., Lienz in the Austrian Tyrol). The dataset features were as follows:

Coverage of 810 km^2

Aerial orthorectified color imagery with a spatial resolution of 0.3 m

Ground truth data for two semantic classes: building and no building

The original aerial image labeling dataset contained 180 color image tiles, 5000 × 5000 pixels in size, which we cropped into small image patches, each with a size of 500 × 500 pixels. An image-label pair example is provided in Figure 4.

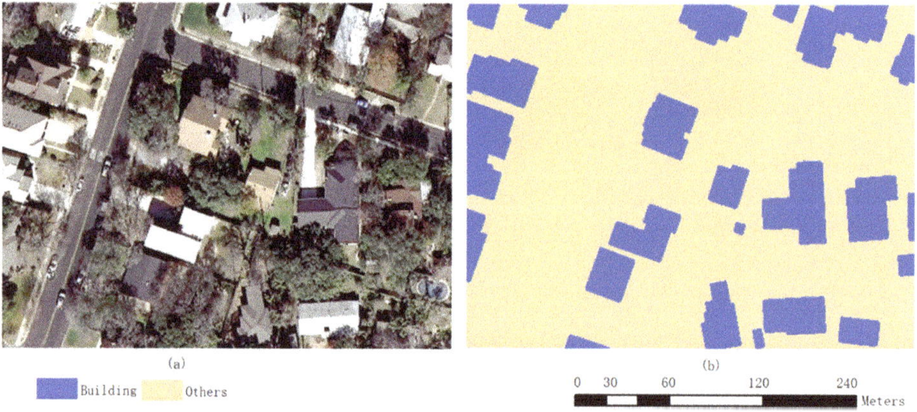

Figure 4. Example of image-label pair: (**a**) original image and (**b**) classified image.

3. Methods

3.1. Structure of the WFCNN Model

The WFCNN model includes an encoder, a decoder, and a classifier (Figure 5). The encoder is used to extract pixel-by-pixel features, while the decoder is used to fuse the coarse high-level semantic features and fine low-level semantic features. The classifier is used to complete the pixel-by-pixel classification. When training the model, the image patch and the corresponding label file are used as inputs. Once the model is successfully trained, the image to be segmented is used as the input, with the output being pixel-by-pixel label files.

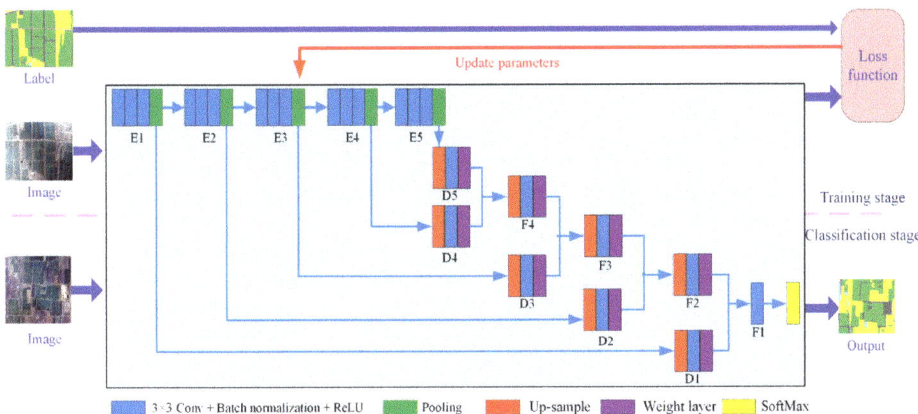

Figure 5. Structure of the weight feature value convolutional neural network model.

3.1.1. Encoder

The encoder consists of five serial connection feature extraction units that can extract five levels of semantic features for each pixel. Each unit consists of three convolutional layers, a batch normalization layer, an activation layer, and a pooling layer. The encoder contains a total of 15 convolutional layers with differing amounts of convolution kernels (Table 2). The advantage of this structural design is that the influence of the image's spatial resolution and geographical coverage is fully considered. Ensuring that sufficient semantic features can be extracted avoids the risk that the excessively deep convolution structure may cause noise in extracted feature values. This is beneficial to the classifier when performing pixel-by-pixel classification.

Table 2. Number of convolution kernels for each convolutional layer.

Layer	Number of Convolution Kernels
1, 2, 3	64
4, 5, 6,	128
7, 8, 9	256
10, 11, 12, 13, 14, 15	512

The activation layer uses the widely used rectified linear unit function as the activation function. Commonly, the pooling operation can accelerate feature aggregation and eliminate feature values with poor discrimination, which is beneficial to the operation of the classifier. However, the pooling operation generally leads to a reduced feature map resolution, affecting segmentation accuracy. When pooling the edge area of two object types, the fact that two adjacent pixels often belong to different categories makes it is easy to mistakenly apply the feature values of one category to another category.

When the pooling operation is carried out on images with a high level of detail or in the shallow convolutional structure, the influence of this problem is not obvious, but if this operation is carried out in the deep convolutional structure, the problem has a serious influence on feature extraction.

WFCNN uses a new pooling strategy to solve this problem. In the E1, E2, and E3 feature extraction units, a 2×2 pool core is used and the pooling step length is 2, encouraging the advantages of the pooling operation in accelerating feature aggregation. In the later stages, due to the large difference between the current resolution and that of original images, the step length of the pooling layer in E4 and E5 cells is adjusted to 1. Here, the pool kernel size is still 2×2, but when the size of the function block is smaller than that of the kernel participating in the pool operation, the size of the kernel pool is adjusted to match the function block. This ensures that a valid pool value is obtained, so the size of the feature graph output through E4 and E5 remains the same.

3.1.2. Decoder

The decoder is composed of five decoding units (D5, D4, D3, D2, and D1) that correspond to and decode the encoding units with the same number (Figure 4). The fusion units F4, F3, F2, and F1 are used to fuse the feature values obtained by the decoding unit. Other than F1, which only contains one pooling layer, the structure of the other units is the same. Each unit includes one up-sampling layer, several convolution layers, and one weight layer.

The up-sampling layer is a deconvolution layer, which is used to restore the size of the feature graph. WFCNN adopts a gradual recovery strategy, in which the number of rows and columns are respectively doubled during each adjustment; the number of rows and columns of the feature graph are eventually restored to be consistent with the original image.

The convolutional layer adjusts the feature values after up-sampling to ensure the consistency of the structure of the feature graphs involved in fusion. The adjustment strategy adopted by WFCNN reduces the depth of the feature graph (Table 3). The decoder finally generates a feature vector consisting of 64 elements for each pixel.

Table 3. Number of feature layers per decoding unit.

Unit	Layers before Adjustment	Layers after Adjustment
D5	512	512
D4	512	512
F4	512	256
D3	256	256
F3	256	128
D2	128	128
F2	128	64
D1	64	64
F1	64	64

Each weight layer contains only one convolution kernel of type $1 \times 1 \times h$, which is used to unify the feature values for stretching or narrowing transformation before fusion. The essence of the weight layer is to uniformly multiply a certain coefficient for a certain feature layer and convert it into a convolutional operation, to ensure that the model can be trained end-to-end.

3.1.3. Classifier

The Softmax model is a widely used classifier in FCN, SegNet, DeepLab, RefineNet, UNET, and other models. In WFCNN, Softmax uses the 64-layer feature graph generated by the decoder as input, calculates the probability of the pixel belonging to each category pixel-by-pixel, and organizes it into a category probability vector as the output. The WFCNN uses the category corresponding to the maximum probability value as the pixel category.

3.2. Training WFCNN

3.2.1. Loss Function

WFCNN defines the loss function based on the cross entropy of the sample:

$$H(p,q) = -\sum_{i=1}^{m} q_i \log(p_i) \tag{1}$$

where m is the number of categories, p is the category probability vector with a length of m output by WFCNN, and q is the real probability distribution, generated according to the manual label. In q, the components are 0 except for that corresponding to the pixel's category, where it is 1. In the WFCNN, each pixel is regarded as an independent sample, allowing the loss function to be defined as:

$$loss = -\frac{1}{total} \sum \sum_{i=1}^{m} q_i \log(p_i) \tag{2}$$

where *total* represents the sample count.

3.2.2. Training Algorithm

WFCNN is used end-to-end, with a stochastic gradient descent algorithm [50] used as the training algorithm with the following steps:

1. The hyperparameters in the training process are determined and the parameters of the model are initialized.
2. The selected image-label pairs are input into the model as training data.
3. The model carries out a forward calculation on the current training data.
4. Equation (2) is used to calculate the loss function of the real probability distribution and the predicted probability distribution.
5. The random gradient descent algorithm is used to update the parameters of the model and complete the training process.
6. Steps (3), (4), and (5) are repeated until the loss function is less than the specified expected value.

3.3. Experimental Setup

We used the SegNet, U-Net, and RefineNet models for comparison with WFCNN, as their structures and working principles are similar, providing the best test for WFCNN. The models were set up based on previous research, with SegNet containing 13 convolutional layers [39], U-Net containing 10 convolutional layers [42], and RefineNet containing 101 convolutional layers [41]. We implemented WFCNN based on the TensorFlow Framework, using the Python language. To better assess its performance, we also tested two variants, termed WFCNN-1 and WFCNN-2 (Table 4).

Table 4. Models used in the comparative experiment.

Name	Description
WFCNN	
SegNet	Similar to WFCNN, the classifier used only high-level semantic features.
U-Net	Similar to WFCNN, the classifier used fused features.
RefineNet	Similar to WFCNN, the linear model was also adopted for feature fusion, but the parameters were all fixed as 1.
WFCNN-1	The decoding unit was modified to use an adjustment strategy for the feature map depth ascending scale. The length of the feature vector generated by the decoder was 512 for each pixel.
WFCNN-2	The decoding unit was modified to remove the weight layer.

All experiments were conducted on a graphics workstation with a 12 GB NVIDIA graphics card and the Linux Ubuntu 16.04 operating system, using the data set defined in Section 2.

To increase the number and diversity of samples, each image in the training data set was processed using color adjustment, horizontal flip, and vertical flip steps. The color adjustment factors included brightness, saturation, hue, and contrast, and each image was processed 10 times. These enhanced images were only used as training data.

We used cross-validation for comparative experiments. When using the GF-6 images dataset, 157 images were randomly selected as test data every training round, and the other images were used as training data until all images were tested once. When using the aerial image labeling dataset, 3000 images were used as test data every training session.

4. Results

Overall, 10 results tested on the GF-6 images dataset were randomly selected from all models (Figure 6), and 10 results were tested on the aerial image labeling dataset (Figure 7). As can be seen from Figures 6 and 7, WFCNN performed best in all cases.

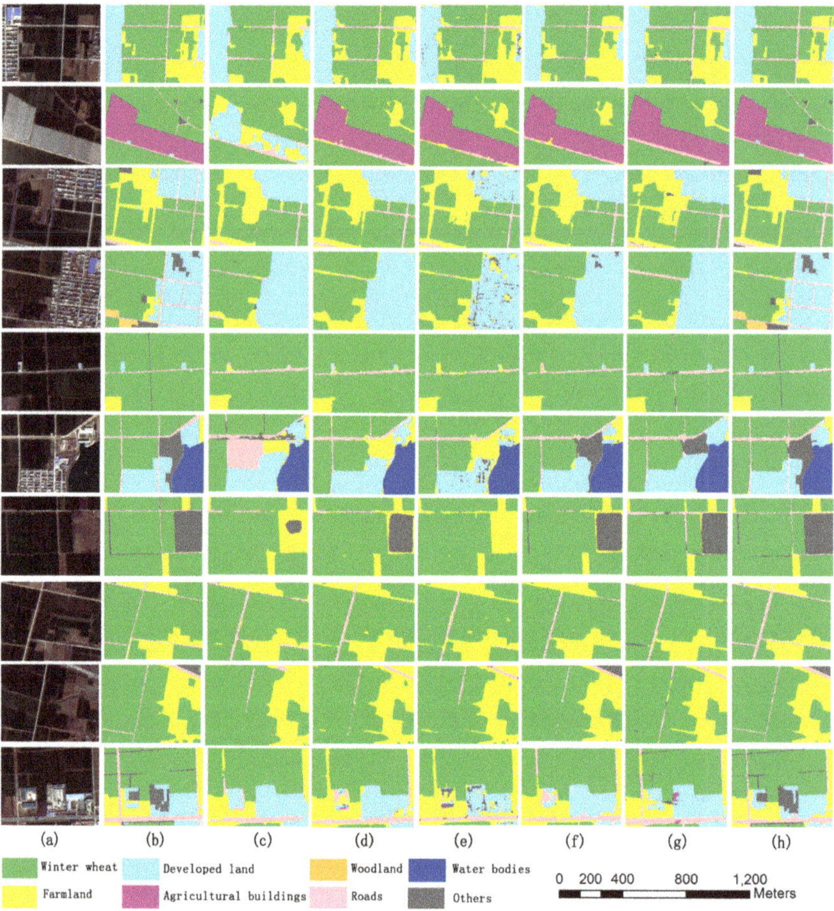

Figure 6. Comparative experimental results for 10 selected images from the GF-6 images dataset: (**a**) original GF-6 image, (**b**) manual classification, (**c**) SegNet, (**d**) U-Net, (**e**) RefineNet, (**f**) WFCNN-1, (**g**) WFCNN-2, and (**h**) WFCNN.

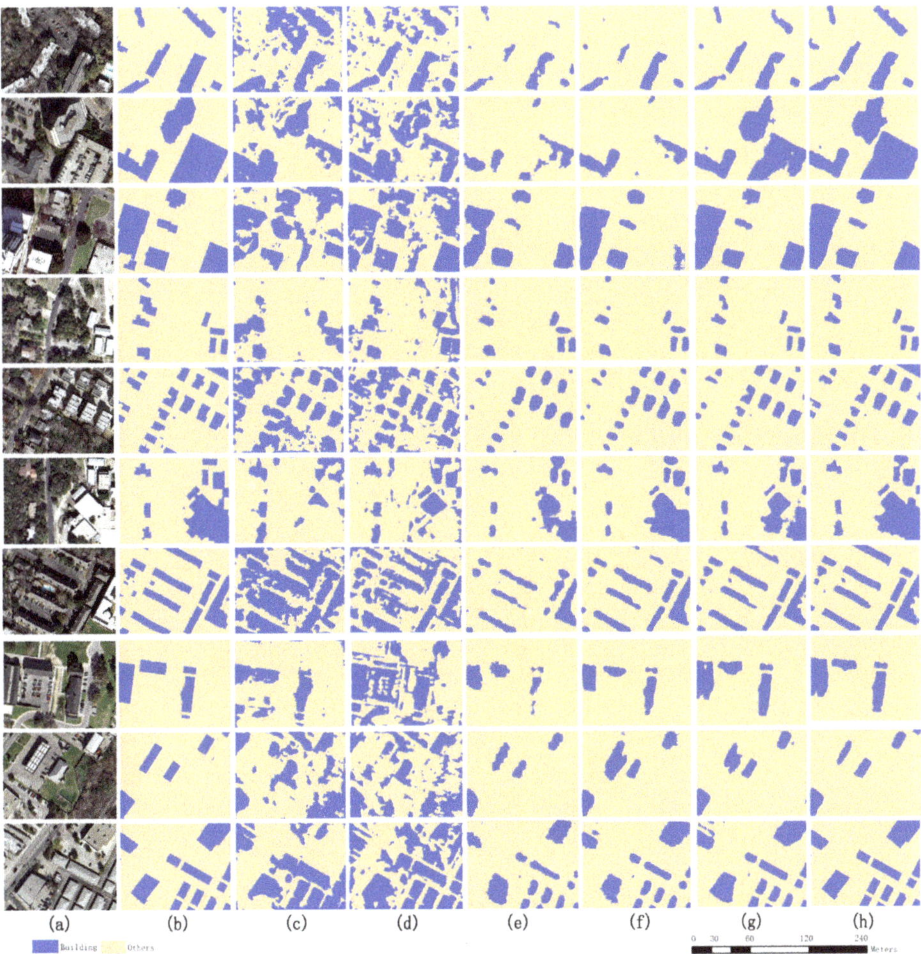

Figure 7. Comparative experimental results for 10 selected images from the aerial image labeling dataset: (**a**) original image, (**b**) manual classification, (**c**) SegNet, (**d**) U-Net, (**e**) RefineNet, (**f**) WFCNN-1, (**g**) WFCNN-2, and (**h**) WFCNN.

SegNet exhibited most errors and the distribution was relatively scattered, with more misclassified pixels in both the edge and inner areas. This shows that it is more reasonable to combine high semantic features with low semantic features than to use only high semantic features.

The number of misclassified pixels produced by all three variants of WFCNN was lower for RefineNet and U-net. The excellent performance of RefineNet and U-Net in the camera image indicates that the network structure should be determined in accordance with the spatial resolution of the image.

WFCNN performed better than WFCNN-1. This excellent performance indicates that the feature vector dimension was too high and is not conducive to improving the accuracy of the classifier. The result that WFCNN performed better than WFCNN-2 shows that the weight layer played a role.

By comparing the performance of U-Net, RefineNet, WFCNN-1, WFCNN-2, and WFCNN, it indicates that different feature fusion methods differ in their contributions to improving accuracy, so it is necessary to choose an appropriate feature fusion method for a given situation.

Figures 8 and 9 present confusion matrices for the different models, which again demonstrate that WFCNN had the best segmentation results. Comparing Figures 8 and 9, it can be found that the performance of each model on the aerial image labeling dataset is better than that on the GF-6 images dataset, indicating that the spatial resolution has a certain impact on the performance of the model.

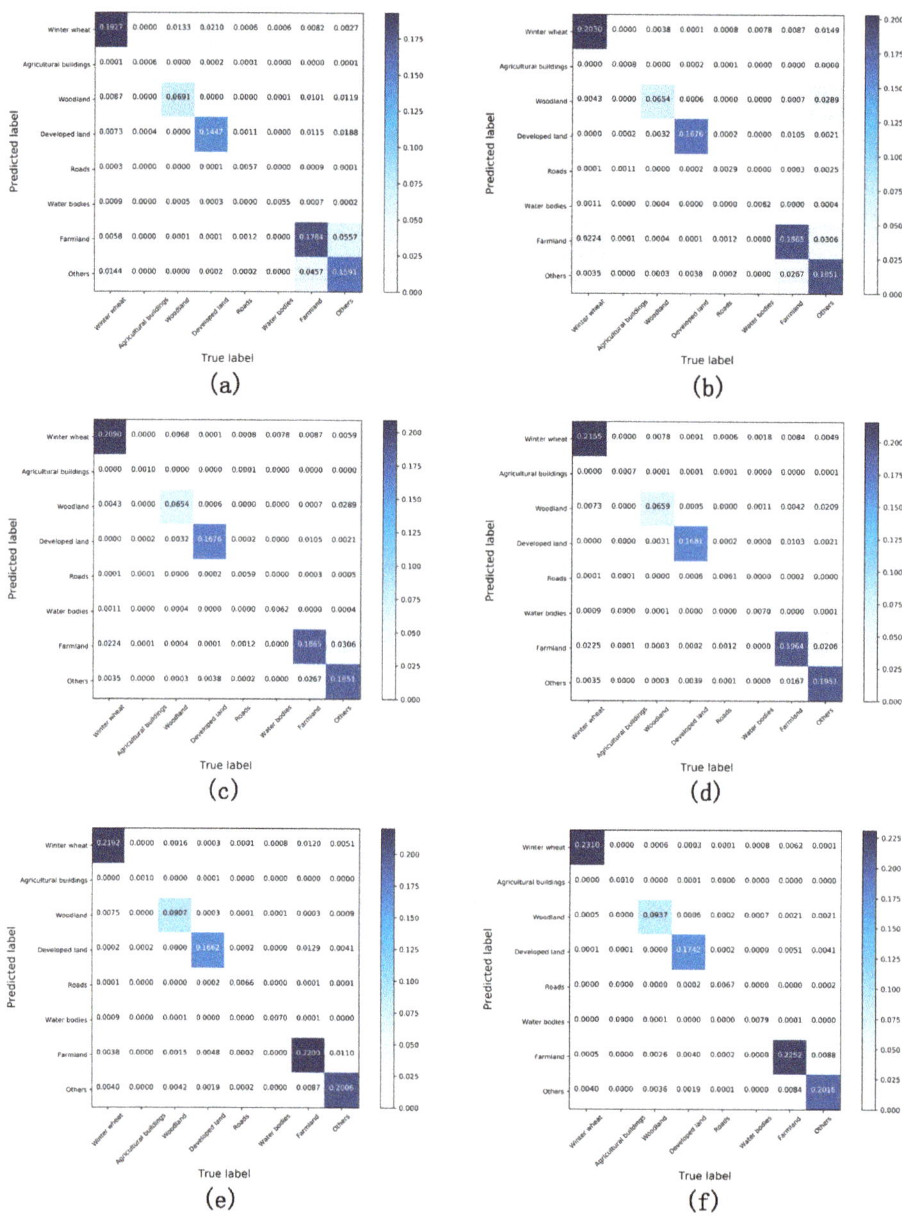

Figure 8. Confusion matrix for different model on the GF-6 images dataset: (**a**) SegNet, (**b**) U-Net, (**c**) RefineNet, (**d**) WFCNN-1, (**e**) WFCNN-2, and (**f**) WFCNN.

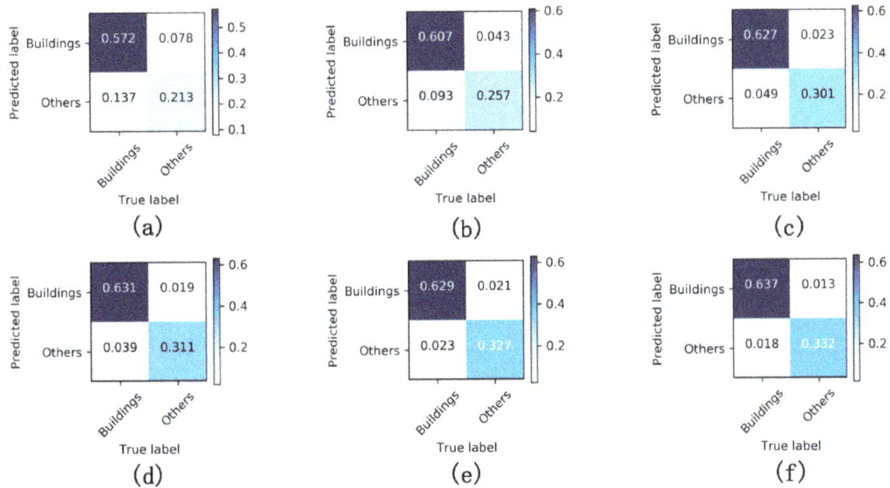

Figure 9. Confusion matrix for different model on the aerial image labeling dataset: (**a**) SegNet, (**b**) U-Net, (**c**) RefineNet, (**d**) WFCNN-1, (**e**) WFCNN-2, and (**f**) WFCNN.

For comparison, we use Table 5 to summarize the data given in Figure 8, and use Table 6 to summarize the data given in Figure 9.

Table 5. Comparison of model performance statistics on the GF-6 images dataset.

Indicator	SegNet	U-Net	RefineNet	WFCNN-1	WFCNN-2	WFCNN
A	75.58%	81.75%	82.67%	85.48%	91.13%	94.13%
B	24.42%	18.25%	17.33%	14.52%	8.87%	5.87%

A denotes the proportion of correctly classified pixels; B denotes the proportion of misclassified pixels.

Table 6. Comparison of model performance statistics on the aerial image labeling dataset.

Indicator	SegNet	U-Net	RefineNet	WFCNN-1	WFCNN-2	WFCNN
A	78.50%	86.40%	92.80%	94.20%	95.60%	96.90%
B	21.50%	13.60%	7.20%	5.80%	4.40%	3.10%

A denotes the proportion of correctly classified pixels; B denotes the proportion of misclassified pixels.

We used accuracy, precision, recall, F1-Score, intersection over union (IoU), and Kappa coefficient as indicators to further evaluate the segmentation results for each model (Tables 7 and 8). The F1-Score is defined as the harmonic mean of precision and recall. IoU is defined as the number of pixels labeled as the same class in both the prediction and the reference, divided by the number of pixels labeled as the class in the prediction or the reference.

The average accuracy of WFCNN was 18.48 % higher than SegNet, 11.44% higher than U-Net, 7.78% higher than RefineNet, 6.68% higher than WFCNN-1, and 2.15% higher than WFCNN-2.

Table 7. Comparison of model indicators on the GF-6 images dataset comparison.

Indicator	SegNet	U-Net	RefineNet	WFCNN-1	WFCNN-2	WFCNN
Accuracy	75.58%	81.75%	82.67%	85.48%	91.13%	94.13%
Precision	75.05%	69.62%	76.07%	81.65%	89.90%	91.93%
Recall	72.20%	74.16%	82.02%	81.72%	90.71%	94.14%
F1-Score	0.7101	0.7797	0.7904	0.8231	0.8906	0.9271
IoU	0.7360	0.7181	0.7893	0.8169	0.9031	0.9302
Kappa coefficient	0.5826	0.5652	0.6514	0.6905	0.8232	0.8693

Table 8. Comparison of model indicators on the aerial image labeling dataset.

Indicator	SegNet	U-Net	RefineNet	WFCNN-1	WFCNN-2	WFCNN
Accuracy	78.50%	86.40%	92.80%	94.20%	95.60%	96.90%
Precision	76.94%	86.19%	92.83%	94.21%	95.22%	96.74%
Recall	74.43%	83.41%	91.23%	92.97%	95.10%	96.43%
F1-Score	0.6117	0.7362	0.8522	0.8794	0.9077	0.9340
IoU	0.7566	0.8478	0.9202	0.9358	0.9516	0.9658
Kappa coefficient	0.6122	0.7355	0.8520	0.8793	0.9080	0.9341

5. Discussion

5.1. Effect of Features on Accuracy

At present, remote sensing image classification mainly relies on three feature types: spectral features mainly express information for individual pixels, textural features mainly express the spatial correlation between adjacent pixels, and semantic features mainly express the relationship between pixels in a specific region in a more abstract way.

CNNs can simultaneously extract these three feature types by reasonably organizing the convolution kernel. For example, the features extracted by using a 1×1 convolution kernel are equivalent to spectral features, those obtained by shallow convolutional layers can be regarded as textural features, and those obtained by convolutional layers of different depths can be regarded as semantic. Therefore, advanced semantic features can be considered as including all three feature types. However, in CNNs, advanced semantic features are obtained by deepening the network structure and expanding the receptive field. The spatial resolution of the image, the number of pixels covered by the object area, and other factors will affect the feature extraction results.

In a high-resolution and detailed camera image, an object tends to occupy a larger area, and it is advantageous to use advanced features. When SegNet only uses high-level semantic features to process an image with a high level of detail, a certain number of misclassified pixels occur at the edge, but few occur inside the object.

In the experiments of this study, the results of SegNet contained more misclassified pixels both at the edge and the inner area. Comparing objects of different sizes showed that smaller objects had more other types of objects adjacent to them and more pixels filled in. This is because the receptive field of high-level semantic features is generally large, and when the type distribution of objects in the receptive field is messy, the extracted feature values will deviate greatly from those of other pixels of the same kind, thus, affecting the accuracy of the segmentation results.

Unlike SegNet, the other models tested here combined low-level semantic features with high-level semantic features, with more accurate results. When segmenting remote sensing images with high spatial resolution, such as GF-6, the different levels of semantic features should be fused and used for classification, which is more reasonable than using advanced semantic features alone.

5.2. Effect of Up-Sampling on Accuracy

The purpose of up-sampling is to encrypt the rough high-level semantic feature graph to generate a feature vector for each pixel. In this study, WFCNN, U-Net, and RefineNet generated feature maps with the same size as compared to the input image through multi-step sampling. However, WFCNN used deconvolution to complete up-sampling, while RefineNet and U-Net used deconvolution to perform up-sampling first and chose linear interpolation for the final up-sampling. The segmentation results for these three models had few mis-segmented pixels within the object, but at the edges, WFCNN's results were significantly better. When RefineNet and U-Net processed camera images, the edges of the object were very fine. We believe that the reason for this phenomenon is that the structure of a remote sensing image is considerably different from that of a camera image, such that bilinear interpolation does not achieve good results when processing the former. Since the resolution of a camera image is generally high, when two objects are adjacent to each other, the pixel changes across the boundary are usually gentle and the difference between adjacent pixels is small, such that up-sampling using bilinear interpolation is more effective. In contrast, the pixel changes in a remote sensing image are usually sharp across the boundary between two objects and the difference between adjacent pixels is large, resulting in a poor bilinear interpolation effect. Unlike RefineNet and U-Net, WFCNN uses deconvolution for up-picking and correction and all parameters are obtained through learning samples, allowing the model to adapt to the unique characteristics of remote sensing images and achieve good results.

5.3. Effect of Feature Fusion on Accuracy

WFCNN, WFCNN-1, and WFCNN-2 all used feature fusion methods to generate feature vectors for each pixel. Since the convolution kernels used in different convolutional layer levels may be different, potentially causing differences in the feature map depths, it is necessary to adjust the latter before fusion. Choosing an appropriate fusion strategy can help improve the accuracy of the results.

WFCNN used a dimensional reduction strategy while WFCNN-1 used a dimensional increase. Although both can achieve consistent depth adjustments of the feature graph, our results showed that WFCNN performed better, indicating that dimensional reduction is more effective. Our analysis suggests that the effect of the classifier may be influenced by a high dimensionality. In a future study, we intend to test additional fusion strategies to further improve the accuracy of the segmentation results.

Like WFCNN, WFCNN-2 used dimensional reduction, but the former adjusted the characteristics of the participation fusion by using the weight layer beforehand, while the latter directly used dimensional reduction. Once the resulting feature map was fused, it was clear that the weight layer used by WFCNN improved the accuracy of the segmentation results.

6. Conclusions

This study tested a new approach to obtain high-precision land-use information from remote sensing images, using a new CNN-based model (WFCNN) to obtain multi-scale image features. Experimental comparisons with the currently used SegNet and RefineNet models, along with two variants on the original WFCNN model, demonstrated the advantages of this approach. This paper discusses the influence of classification feature selection, up-sampling method, and fusion strategy on segmentation accuracy, indicating that WFCNN can also be applied to other high spatial resolution remote sensing images.

Analyzing the data characteristics of high-resolution remote sensing images allowed the effects of spatial resolution on feature extraction to be fully considered, leading to the adoption of a hierarchical pooling strategy. In the initial stage of feature extraction, a large cell length step was used in WFCNN because the feature values were relatively scattered, allowing the pooling operation to accentuate the advantages of feature aggregation. In the subsequent stage, a smaller pooling step size was adopted due to the large difference between the current resolution and the original image. The experimental results showed that this strategy was more conducive to generating features with good discrimination.

The data characteristics of adjacent regions within remote sensing images were fully considered to effectively combine low-level and high-level semantic features. The proposed WFCNN model uses a variable-weight fusion strategy that adjusts features using a weight adjustment layer to ensure their stability. This strategy is more conducive to extracting effective features than those adopted by other models.

Our method requires training images to be marked pixel-by-pixel, creating a large workload. In subsequent research, we intend to introduce a semi-supervised training method to reduce this requirement and make the model more applicable to real-world situations.

Author Contributions: Conceptualization, C.Z. and X.Y.; methodology, C.Z.; software, Y.C.; validation, S.G., X.Y. and F.L.; formal analysis, C.Z. and L.S.; investigation, A.K. and D.Z.; resources, F.L.; data curation, A.K.; writing—original draft preparation, C.Z.; writing—review and editing, C.Z.; visualization, S.G. and D.Z.; supervision, C.Z.; project administration, C.Z. and S.G.; funding acquisition, C.Z. All authors have read and agreed to the published version of the manuscript.

Funding: This research was funded by the Science Foundation of Shandong, grant numbers ZR2017MD018; the Key Research and Development Program of Ningxia, Grant numbers 2019BEH03008; the National Key R and D Program of China, grant number 2017YFA0603004; the Open Research Project of the Key Laboratory for Meteorological Disaster Monitoring, Early Warning and Risk Management of Characteristic Agriculture in Arid Regions, Grant numbers CAMF-201701 and CAMF-201803; the arid meteorological science research fund project by the Key Open Laboratory of Arid Climate Change and Disaster Reduction of CMA, Grant numbers IAM201801. The APC was funded by ZR2017MD018.

Conflicts of Interest: The authors declare no conflict of interest.

References

1. Mhangara, P.; Odindi, J. Potential of texture-based classification in urban landscapes using multispectral aerial photos. *S. Afr. J. Sci.* **2013**, *109*, 1–8. [CrossRef]

2. Wang, F.; Kerekes, J.P.; Xu, Z.Y.; Wang, Y.D. Residential roof condition assessment system using deep learning. *J. Appl. Remote Sens.* **2018**, *12*, 016040. [CrossRef]

3. Jiang, T.; Liu, X.N.; Wu, L. Method for mapping rice fields in complex landscape areas based on pre-trained Convolutional Neural Network from HJ-1 A/B data. *ISPRS Int. J. Geo Inf.* **2018**, *7*, 418. [CrossRef]

4. Du, S.; Du, S.; Liu, B.; Zhang, X. Context-Enabled Extraction of Large-Scale Urban Functional Zones from Very-High-Resolution Images: A Multiscale Segmentation Approach. *Remote Sens.* **2019**, *11*, 1902. [CrossRef]

5. Kavzoglu, T.; Erdemir, M.Y.; Tonbul, H. Classification of semiurban landscapes from very high-resolution satellite images using a regionalized multiscale segmentation approach. *J. Appl. Remote Sens.* **2017**, *11*, 035016. [CrossRef]

6. Pan, Y.Z.; Li, L.; Zhang, J.S.; Liang, S.L.; Hou, D. Crop area estimation based on MODIS-EVI time series according to distinct characteristics of key phenology phases: A case study of winter wheat area estimation in small-scale area. *J. Remote Sens.* **2011**, *15*, 578–594.

7. Zhang, J.Y.; Liu, X.; Liang, Y.; Cao, Q.; Tian, Y.C.; Zhu, Y.; Cao, W.X.; Liu, X.J. Using a portable active sensor to monitor growth parameters and predict grain yield of winter wheat. *Sensors* **2019**, *19*, 1108. [CrossRef] [PubMed]

8. Ma, Y.; Fang, S.H.; Peng, Y.; Gong, Y.; Wang, D. Remote estimation of biomass in winter oilseed rape (*Brassica napus* L.) using canopy hyperspectral data at different growth stages. *Appl. Sci.* **2019**, *9*, 545. [CrossRef]

9. Padmanaban, R.; Bhowmik, A.K.; Cabral, P. Satellite image fusion to detect changing surface permeability and emerging urban heat islands in a fast-growing city. *PLoS ONE* **2019**, *14*, e0208949. [CrossRef]

10. Liu, L.Y.; Dong, Y.Y.; Huang, W.J.; Du, X.P.; Luo, J.H.; Shi, Y.; Ma, H.Q. Enhanced regional monitoring of wheat powdery mildew based on an instance-based transfer learning method. *Remote Sens.* **2019**, *11*, 298. [CrossRef]

11. Wang, L.; Chang, Q.; Yang, J.; Zhang, X.H.; Li, F. Estimation of paddy rice leaf area index using machine learning methods based on hyperspectral data from multi-year experiments. *PLoS ONE* **2018**, *13*, e0207624. [CrossRef] [PubMed]

12. Xu, B.; Liang, C.; Chai, D.; Shi, W.; Sun, G. Inversion of natural grassland productivity from remote sensor imagery in Zulihe River Basin. *Arid Zone Res.* **2014**, *31*, 1147–1152. (In Chinese) [CrossRef]

13. Wang, X.T.; Chen, D.D. Interannual variability of GNDVI and its relationship with altitudinal in the Three-River Headwater Region. *Ecol. Environ. Sci.* **2018**, *27*, 1411–1416. (In Chinese) [CrossRef]

14. Zhang, L.; Gong, Z.N.; Wang, Q.W.; Jin, D.D.; Wang, X. Wetland mapping of Yellow River Delta wetlands based on multi-feature optimization of Sentinel-2 images. *J. Remote Sens.* **2019**, *23*, 313–326. (In Chinese) [CrossRef]

15. Rao, P.; Wang, J.L.; Wang, Y. Extraction of information on construction land based on multi-feature decision tree classification. *Trans. Chin. Soc. Agric. Eng.* **2014**, *30*, 233–240. (In Chinese) [CrossRef]

16. Liu, Z.C.; Xu, H.Q.; Li, L.; Tang, F.; Lin, Z.L. Ecological change in the Hangzhou area using the remote sensing based ecological index. *J. Basic Sci. Eng.* **2015**, *23*, 728–739. (In Chinese) [CrossRef]

17. Wang, W.; Zhang, X.; Zhao, Y.; Wang, S. Cotton extraction method of integrated multi-features based on multi-temporal Landsat 8 images. *J. Remote Sens.* **2017**, *21*, 115–124. (In Chinese) [CrossRef]

18. Sun, C.L.; Bian, Y.; Zhou, T.; Pan, J.J. Using of multi-source and multi-temporal remote sensing data improves crop-type mapping in the subtropical agriculture region. *Sensors* **2019**, *19*, 2401. [CrossRef]

19. Moya, L.; Zakeri, H.; Yamazaki, F.; Liu, W.; Mas, E.; Koshimura, S. 3D gray level co-occurrence matrix and its application to identifying collapsed buildings. *ISPRS J. Photogramm. Remote Sens.* **2019**, *149*, 14–28. [CrossRef]

20. Chen, J.; Deng, M.; Xiao, P.; Yang, M.; Mei, X. Rough set theory based object-oriented classification of high resolution remotely sensed imagery. *J. Remote Sens.* **2010**, *14*, 1139–1155.

21. Zhao, Y.D.; Zhang, L.P.; Li, P.X. Universal Markov random fields and its application in multispectral textured image classification. *J. Remote Sens.* **2006**, *10*, 123–129.

22. Reis, S.; Tasdemir, K. Identification of hazelnut fields using spectral and Gabor textural features. *ISPRS J. Photogramm. Remote Sens.* **2011**, *66*, 652–661. [CrossRef]

23. Wu, J.; Zhao, Z.M. Scale co-occurrence matrix for texture analysis using wavelet transform. *J. Remote Sens.* **2001**, *5*, 100–103. (In Chinese)

24. Mao, L.; Zhang, G.M. Complex cue visual attention model for harbor detection in high-resolution remote sensing images. *J. Remote Sens.* **2017**, *21*, 300–309. (In Chinese) [CrossRef]

25. Liu, P.H.; Liu, X.P.; Liu, M.X.; Shi, Q.; Yang, J.X.; Xu, X.C.; Zhang, Y.Y. Building footprint extraction from high-resolution images via Spatial Residual Inception Convolutional Neural Network. *Remote Sens.* **2019**, *11*, 830. [CrossRef]

26. Ball, J.E.; Anderson, D.T.; Chan, C.S. Comprehensive survey of deep learning in remote sensing: Theories, tools and challenges for the community. *J. Appl. Remote Sens.* **2017**, *11*, 042609. [CrossRef]

27. Jia, K.; Liang, S.; Gu, X.; Baret, F.; Wei, X.; Wang, X.; Yao, Y.; Yang, L.; Li, Y. Fractional vegetation cover estimation algorithm for Chinese GF-1 wide field view data. *Remote Sens. Environ.* **2016**, *177*, 184–191. [CrossRef]

28. Zhang, F.; Ni, J.; Yin, Q.; Li, W.; Li, Z.; Liu, Y.F.; Hong, W. Nearest-regularized subspace classification for PolSAR imagery using polarimetric feature vector and spatial information. *Remote Sens.* **2017**, *9*, 1114. [CrossRef]

29. Zhang, J.S.; He, C.Y.; Pan, Y.Z.; Li, J. The high spatial resolution RS image classification based on SVM method with the multi-source data. *J. Remote Sens.* **2006**, *10*, 49–57. (In Chinese) [CrossRef]

30. Al-Obeidat, F.; Al-Taani, A.T.; Belacel, N.; Feltrin, L.; Banerjee, N. A fuzzy decision tree for processing satellite images and Landsat data. *Procedia Comput. Sci.* **2015**, *52*, 1192–1197. [CrossRef]

31. Chen, J.Y.; Tian, Q.J. Vegetation classification based on high-resolution satellite image. *J. Remote Sens.* **2007**, *11*, 221–227. (In Chinese)

32. Pereira, L.F.S.; Barbon, S.; Valous, N.A.; Barbin, D.F. Predicting the ripening of papaya fruit with digital imaging and random forests. *Comput. Electron. Agric.* **2018**, *145*, 76–82. [CrossRef]

33. Wang, N.; Li, Q.Z.; Du, X.; Zhang, Y.; Zhao, L.C.; Wang, H.Y. Identification of main crops based on the univariate feature selection in Subei. *J. Remote Sens.* **2017**, *21*, 519–530. (In Chinese) [CrossRef]

34. Krizhevsky, A.; Sutskever, I.; Hinton, G.E. ImageNet classification with deep convolutional neural networks. *Commun. ACM* **2017**, *60*, 84–90. [CrossRef]

35. Szegedy, C.; Liu, W.; Jia, Y.; Sermanet, P.; Reed, S.; Anguelov, D.; Erhan, D.; Vanhoucke, V.; Rabinovich, A. Going deeper with convolutions. In Proceedings of the 2015 IEEE Conference on Computer Vision and Pattern Recognition (CVPR), Boston, MA, USA, 7–12 June 2015. [CrossRef]
36. Simonyan, K.; Zisserman, A. Very deep convolutional networks for large-scale image recognition. *arXiv* **2014**, arXiv:1409.1556.
37. He, K.; Zhang, X.; Ren, S.; Sun, J. Deep residual learning for image recognition. In Proceedings of the IEEE Conference on Computer Vision and Pattern Recognition, Las Vegas, NV, USA, 26 June–1 July 2016; pp. 770–778. [CrossRef]
38. Long, J.; Shelhamer, E.; Darrell, T.; Berkeley, U.C. Fully Convolutional Networks for Semantic Segmentation. *arXiv* **2015**, arXiv:1411.4038v2.
39. Badrinarayanan, V.; Kendall, A.; Cipolla, R. SegNet: A deep convolutional encoder-decoder architecture for image segmentation. *arXiv* **2015**, arXiv:1505.07293. [CrossRef]
40. Chen, L.; Papandreou, G.; Kokkinos, I.; Murphy, K.; Yuille, A.L. DeepLab: Semantic image segmentation with deep convolutional nets, Atrous convolution, and fully connected CRFs. *IEEE Trans. Pattern Anal. Mach. Intell.* **2018**, *40*, 834–848. [CrossRef]
41. Lin, G.; Milan, A.; Shen, C.; Reid, I. RefineNet: Multi-path refinement networks for high-resolution semantic segmentation. *arXiv* **2016**, arXiv:1611.06612v3.
42. Ronneberger, O.; Fischer, P.; Brox, T. U-Net: Convolutional networks for biomedical image segmentation. *arXiv* **2015**, arXiv:1505.04597.
43. Shustanov, A.; Yakimov, P. CNN design for real-time traffic sign recognition. *Procedia Eng.* **2017**, *201*, 718–725. [CrossRef]
44. Dai, X.B.; Duan, Y.X.; Hu, J.P.; Liu, S.C.; Hu, C.Q.; He, Y.Z.; Chen, D.P.; Luo, C.L.; Meng, J.Q. Near infrared nighttime road pedestrians recognition based on convolutional neural network. *Infrared Phys. Technol.* **2018**, *97*, 25–32. [CrossRef]
45. Wang, D.D.; He, D.J. Recognition of apple targets before fruits thinning by robot based on R-FCN deep convolution neural network. *Trans. Chin. Soc. Agric. Eng.* **2019**, *35*, 156–163. (In Chinese) [CrossRef]
46. Ferentinos, K.P. Deep learning models for plant disease detection and diagnosis. *Comput. Electron. Agric.* **2018**, *145*, 311–318. [CrossRef]
47. Cheng, X.; Zhang, Y.; Chen, Y.; Wu, Y.Z.; Yue, Y. Pest identification via deep residual learning in complex background. *Comput. Electron. Agric.* **2017**, *141*, 351–356. [CrossRef]
48. Krylov, V.A.; Kenny, E.; Dahyot, R. Automatic discovery and geotagging of objects from street view imagery. *Remote Sens.* **2018**, *10*, 661. [CrossRef]
49. Ahmad, K.; Conci, N. How Deep Features Have Improved Event Recognition in Multimedia: A Survey. *ACM Trans. Multimed. Comput. Commun. Appl.* **2019**, *15*, 39. [CrossRef]
50. Gaetano, R.; Ienco, D.; Ose, K.; Cresson, R. A two-branch CNN architecture for land cover classification of PAN and MS imagery. *Remote Sens.* **2018**, *10*, 1746. [CrossRef]
51. Liu, Y.; Ren, Q.R.; Geng, J.H.; Ding, M.; Li, J.Y. Efficient patch-wise semantic segmentation for large-scale remote sensing images. *Sensors* **2018**, *18*, 3232. [CrossRef]
52. Alonzo, M.; Andersen, H.E.; Morton, D.C.; Cook, B.D. Quantifying boreal forest structure and composition using UAV structure from motion. *Forests* **2018**, *9*, 119. [CrossRef]
53. Fu, G.; Liu, C.; Zhou, R.; Sun, T.; Zhang, Q. Classification for high resolution remote sensing imagery using a fully convolutional network. *Remote Sens.* **2017**, *9*, 498. [CrossRef]
54. Fu, K.; Lu, W.; Diao, W.; Yan, M.; Sun, H.; Zhang, Y.; Sun, X. WSF-NET: Weakly supervised feature-fusion network for binary segmentation in remote sensing image. *Remote Sens.* **2018**, *10*, 1970. [CrossRef]
55. Sharma, A.; Liu, X.; Yang, X.; Shi, D. A patch-based convolutional neural network for remote sensing image classification. *Neural Netw.* **2017**, *95*, 19–28. [CrossRef] [PubMed]
56. Zhang, C.; Pan, X.; Li, H.; Gardiner, A.; Sargent, I.; Hare, J.; Atkinson, P.M. A hybrid MLP-CNN classifier for very fine resolution remotely sensed image classification. *ISPRS J. Photogramm. Remote Sens.* **2018**, *140*, 133–144. [CrossRef]

57. Helber, P.; Bischke, B.; Dengel, A.; Borth, D. Eurosat: A novel dataset and deep learning benchmark for land use and land cover classification. *IEEE J. Sel. Top. Appl. Earth Obs. Remote Sens.* **2019**, *12*, 2217–2226. [CrossRef]
58. Maggiori, E.; Tarabalka, Y.; Charpiat, G.; Alliez, P. Can Semantic Labeling Methods Generalize to Any City? The Inria Aerial Image Labeling Benchmark. In Proceedings of the IEEE International Symposium on Geoscience and Remote Sensing (IGARSS), Fort Worth, TX, USA, 23–28 July 2017; Volume 6.

MDPI

St. Alban-Anlage 66

4052 Basel

Switzerland

Tel. +41 61 683 77 34

Fax +41 61 302 89 18

www.mdpi.com

Remote Sensing Editorial Office

E-mail: remotesensing@mdpi.com

www.mdpi.com/journal/remotesensing